Heating Systems, Plant and Control

Heating Systems, Plant and Control

A.R. Day
M.S. Ratcliffe
K.J. Shepherd

Blackwell
Science

Editorial Offices:
9600 Garsington Road, Oxford OX4 2DQ, UK
Tel: +44 (0)1865 776868
Blackwell Publishing, Inc., 350 Main Street, Malden, MA 02148-5018, USA
Tel: +1 781 388 8250
Iowa State Press, a Blackwell Publishing Company, 2121 State Avenue, Ames, Iowa 50014-8300, USA
Tel: +1 515 292 0140
Blackwell Publishing Asia Pty, 550 Swanston Street, Carlton South, Victoria 3053, Australia
Tel: +61 (0)3 9347 0300
Blackwell Wissenschafts Verlag, Kurfürstendamm 57, 10707 Berlin, Germany
Tel: +49 (0)30 32 79 060

First published 2003 by Blackwell Science Ltd

Library of Congress
Cataloging-in-Publication Data
Day, Anthony R.
Heating systems : plant and control/
Anthony R. Day, Martin S. Ratcliffe & Keith J. Shepherd.
p. cm.
ISBN 0-632-05937-0 (hardcover)
1. Heating. I. Ratcliffe, Martin S. II. Shepherd, Keith. III. Title.

TH7223 .D38 2003
697–dc21

2002151907

ISBN 0-632-05937-0

A catalogue record for this title is available from the British Library

Set in 10 on 11pt Sabon
by SNP Best-set Typesetter Ltd., Hong Kong
Printed and bound in Great Britain by
MPG Books Ltd, Bodmin, Cornwall

For further information on
Blackwell Publishing, visit our website:
www.blackwellpublishing.com

Contents

1 Introduction

1.1 Heating: the fundamental building service

Climates vary significantly around the world. Some regions experience sharp swings from summer to winter, whereas others are more temperate and have less severe variations. In many regions of the world buildings will require space heating systems to make them habitable. This is true even of buildings in which the predominant need, year-round, is for space cooling. A heating system is therefore *the* fundamental element of any building services strategy for providing thermal comfort. Whatever other facilities are required for a building to be occupied, none of these are much use if the heating system fails at low ambient temperature. Furthermore, beyond space heating itself, few of us would nowadays be willing to accept working in an environment in which hot water was not more-or-less readily available for sanitary and domestic use. This water must be heated, by one means or another, whether this be the same that is used to heat the spaces within the building or by independent means.

1.2 Low-pressure hot water

While a variety of means may be employed to provide the space and water heating that buildings typically need, none have so far proved quite as flexible and effective as systems based on the use of low-pressure hot water (LPHW), generated by a heating plant comprising one or more boilers. This is not, of course, to suggest that all LPHW heating systems *are* uniquely flexible and effective in every application. In such an arrangement the LPHW is used as a heat transfer medium, and is typically generated by the combustion of natural gas or fuel oil in the boilers. While medium-pressure hot water (MPHW), high-pressure hot water (HPHW) and steam have also been employed as heat transfer media, the cost, complexity and health and safety issues associated with their use will typically mean that they will only be considered in the rare commercial heating applications in which the sheer size of the heating loads make it either economically advantageous or technically essential. They will be more commonly found in industrial process heating applications where high temperatures must be attained and the health and safety issues may be more effectively accommodated.

Therefore, if heating is the fundamental element of building services HVAC systems, boiler-based LPHW systems represent the most fundamental form of heating employed in a wide range of commercial, public, institutional, educational, healthcare, retail and residential buildings, hotels and leisure facilities.

The fundamental element of any heating system is a means of generating heat, and in the case of LPHW heating systems this invariably, though not *exclusively*, takes the form of a boiler plant comprising one or more individual boilers. While very many buildings can operate quite satisfactorily without refrigeration or air handling plant, in practice few can work effectively without a boiler plant of some kind.

Once the heat has been generated, it must then be distributed around the building to the heat emitters. The heat emitters are the interface between the heating system and the occupants and determine the thermal comfort of the latter. 'Radiators' are, of course, the most common form of emitter in use for many building types (other than industrial), having a good reputation for thermal comfort, and their massive production has resulted in low cost. However, changes in building construction standards and the increased rental value of floor space in commercial buildings may make other forms of emitter preferable.

Achieving thermal comfort also requires a controllable heat output rate. This involves the design of a control and distribution system which can respond quickly and effectively to changes in heating demand. The second part of this book is concerned with the analysis of heat emission, distribution and control. This book is aimed particularly at non-domestic heating and hot water systems. However, many of the principles discussed will relate to domestic systems. Simple domestic systems are often a good starting point to explain the fundamental operation of more complex systems.

1.3 The need for efficient heating systems

Creating the levels of thermal comfort that we have come to expect of our public and private built environments consumes a prodigious amount of energy, virtually all in the form of fossil fuels. Currently natural gas is the favoured boiler fuel for new installations, or LPG where this is not available, although fuel oil continues to be used as well. Apart from any consideration of the economic cost of all of this gas and oil, their use contributes massively to the emission of carbon dioxide into the atmosphere. Indeed, in the UK and Northern Europe, heating systems are the single largest contributor, higher than transport or industry.

Even where good practice is employed in the design of efficient and effective space and water heating systems, a modern office building may annually consume some 100 kWh of fossil fuel per m^2 of treated floor area[1]. As each kWh of natural gas burnt in a boiler produces typically 0.19 kg of carbon dioxide, (0.052 kg carbon equivalent), this results in 19 kg of carbon dioxide per m^2 being emitted into the atmosphere each year, (or 5.2 kg carbon equivalent). This level of emission will be increased by some 35–40% if the boiler is fired on fuel oil.

Add to this that quantities of oxides of nitrogen (NO_x) are also emitted for each kWh of boiler fuel burnt. These too are greenhouse gases, with nitrous oxide being some 310 times more powerful in its global warming effect than carbon dioxide. There is increasing concern that NO_x emissions are also detrimental to health, with possible links to childhood asthma. At even the low level of emission of 70 mg NO_x kWh^{-1} for natural gas burnt in a modern boiler, 100 kWh releases 7 g of NO_x, equivalent to nearly another 2.2 kg of carbon dioxide. (Some boilers may emit two to three times this level of NO_x.) Studies in the early-to-mid 1990s put the UK stock of private-sector office space alone at some 275 000 buildings totalling over 72.5 million m^2, with public-sector offices and industrial buildings increasing this substantially to c.90 million m^2, and retail space alone more than doubling this total[2,3]. This pales into insignificance, of course, with the floor area of some 20 million dwellings.

The building services industry is therefore responsible for a large environmental liability, and has a duty to ensure the lowest environmental impact from systems that it installs. However, perhaps the most sobering thought for readers of this book as building services engineers is that the figures for energy consumption considered above represent the *good practice cases*. It may (or may not always!) be reasonable to assume that these performance levels are now routinely achieved in new buildings. Yet the 'typical' cases that represent much of the existing building stock may contribute twice as much emission of carbon dioxide and NO_x per m^2 floor area.

Viewed in this light the importance of ensuring that new buildings do indeed possess efficient and effective heating systems is clear, and that wherever possible the standard of existing installations is brought up to current standards of good practice when significant modification or refurbishment works are carried out. As in many disciplines, one of the keys to achieving improved performance of heating systems is a thorough understanding of both the theoretical and the practical issues involved in their design, installation, operation and control. The aim of this book is to provide that understanding.

1.4 Scope of the book

This book examines the key components of building heating systems, with an emphasis on how to select plant and design distribution circuits and controls to provide good indoor comfort conditions while minimising energy consumption. There are a number of texts that deal with the design and sizing of heating systems. These often focus on the procedures for heat loss calculations and pipe sizing, an essential part of design, yet they rarely explore the bewildering range of equipment available today, nor the operational implications for various design decisions, nor do they look at the complex dynamics of systems in use.

This book attempts to fill this gap, and is aimed at both experienced practitioners and students of building services engineering alike. It assumes a basic knowledge of heating systems design – steady state heat loss, pipe and pump sizing, the principles of control – and builds on this to facilitate a deeper understanding of how systems work in practice.

Much of what we understand about heating systems is based on common sense and day-to-day familiarity with heated buildings; however, there are many elusive questions that standard theory and practices cannot answer: what is the best boiler for a particular application, what is the relationship between energy and boiler firing patterns, what is the best emitter control strategy? The answers are not straightforward, but with a bit of practical analysis it is possible to select plant and controls that will best suit a particular application, and minimise the risk of failure or sub-standard performance.

1.5 Content of the book: an overview

The book is divided into two parts. Part A deals with the heat generating components of the central plant – boilers and flues, as well as various alternatives; part B deals with system design and control – emitters, system and central plant arrangements, efficient operation and energy performance. These areas have not been brought together before so comprehensively, and in order to deal with them in depth some of the fundamentals have been left out of the book where they are adequately dealt with in existing publications.

The reader is encouraged to refer to these where necessary. A brief outline of the chapters in this book is given below.

Part A deals with the heat generating equipment. Chapter 2 looks at low-pressure hot water (LPHW) boilers and burners from a generic perspective. The chapter opens with a brief consideration of how a boiler may be defined, and then sets out principal functional elements of gas-fired and oil-fired boilers, with particular regard to function, characteristics and general design. The function of the *boiler block* is explained, with a discussion of how heat transfer is maximised. Issues such as temperature rise across a boiler and the implications for low water content boilers are also dealt with. Burner function, combustion processes and fuel types are introduced, and specific attention is given to atmospheric and forced-draught natural gas burners and atomizing fuel oil burners. The control of boilers is introduced here in preparation for more detailed coverage in Chapter 10.

Chapter 3 looks at the types of LPHW boiler that may be encountered in modern commercial building services space and water heating applications. Types of boiler, and the materials used in their construction, are reviewed, with explanations of why typical boiler types are used in specific applications. This is extended to the appropriate use of single, multiple and modular boilers. The chapter also examines the principles, benefits and requirements of condensing operation and the design of condensing boilers. Boiler efficiency is defined, in terms of both gross and net calorific value, and the concepts of carbon intensity and carbon performance rating as future performance parameters are introduced. All of the operational needs of the boiler installation are analysed with regard to water flow rate, temperature and pressure, water quality, supply of the common boiler fuels and ventilation requirements. The chapter concludes by discussing the range of standards and recommendations that are potentially applicable to the design of LPHW boiler installations.

Chapter 4 looks at alternatives to the gas and oil-fired boilers of Chapter 3, suitable for LPHW heating applications. Combined heat and power (CHP) is an increasingly popular choice, with its potential economic and environmental advantages over conventional boilers. The relatively high capital cost means that CHP is rarely sized to meet peak heating demand, being used instead in conjunction with conventional boilers. This leads to difficulties in selecting the most appropriate size of CHP and it is hoped that this chapter will shed some light on that, together with how it should be integrated within the heating system. Another increasingly popular system is the heat pump, again offering potential economic and environmental advantages with higher capital costs. The principles of the heat pump cycle are briefly explained, with how it may be used as a source of LPHW and a discussion on the environmental and economic advantages. However, CHP and heat pumps are both large subjects and cannot be dealt with fully here. Finally, alternatives to fossil fuel are discussed. Sustainability is a buzz-word in building services but is yet to make much impact on heating systems. Nevertheless, it is felt that a discussion of the various renewable fuels suitable for LPHW generation should be included in preparation for the changes that are bound to take place in the near future.

Chapter 5 deals with the flueing of boilers. Types of boiler flue are defined, and the nature of *flue draught* is explained. Each type of boiler flue or flue system is then reviewed in turn, including natural-draught and mechanical-draught flues, balanced flues and flue dilution systems. Sizing of a flue or chimney is considered, together with the typical features of its general design and construction. The specific needs of condensing boilers are reviewed, and the chapter closes by noting some acoustic implications for flue design.

Part B looks at the way LPHW systems are designed and operated in order to distribute the heat around the building effectively and efficiently. Chapters 6 to 8 cover emitters, system layout and domestic hot water systems. Chapter 6 deals with room heat emitters including natural and fan convectors, radiators and radiant panels and heated floors. It begins with a review of the types of heat emitter and their construction, giving typical heat output rates. Standard empirical relationships between mean water temperature and heat output are presented, allowing for corrections to manufacturers' data for non-standard conditions. The ratio of radiant to convective heat produced by the emitter has an effect on both thermal comfort and running costs, depending on the thermal properties of the room. This is discussed in detail with examples provided for two very different buildings: a modern office and a factory. The control of emitters is then discussed, looking at variable water flow rate and flow temperature compensation and how they can be usefully combined. The chapter concludes with a discussion of the special case of heated floors which differs substantially from other heat emitters due to the low surface temperature and high thermal mass. Much of the specialist theory used to carry out the analysis, where this is not readily available elsewhere, has been placed in appendices at the end of the chapter to aid fluency of the discussions. Readers may wish to skip some or all of these on a first reading.

Chapter 7 deals with heating circuit design. It begins with a discussion of the importance in choosing appropriate flow and return water temperatures and how this affects many aspects of design. A method of determining the economic thickness of pipework insulation is developed, demonstrating that there may be advantages in exceeding currently recommended thicknesses for environmental reasons at little or no extra cost. Pipework circuit layouts are discussed including variable flow and variable temperature. This leads on to a discussion of pumping methods, particularly using variable speed drives for improved control and reduced running costs. Control valve selection and their performance is covered. Means of allowing for pipework expansion is then examined, together with methods of support and anchoring the forces acting on the pipe. Appendices at the end of the chapter contain the theory used in the analysis of insulation thicknesses, control valves and pumping.

Chapter 8 is dedicated to hot water services. The merits of instantaneous and storage systems are discussed, with methods of determining hot water consumption, instantaneous loads, storage volume and boiler power. The key design aspects of storage systems and hot water generators are discussed. Solar panels, slowly increasing in popularity, are described, together with how the efficiency and output varies with design and weather conditions. (A thermal analysis of a solar panel is provided in an appendix at the end of the chapter.) Finally, the legionella issue, as it applies to the design of heating systems, is discussed.

Chapters 9 to 11 are also analytical in approach and present some (relatively simple) mathematical models to assist in analysing and understanding design decisions. Chapter 9 deals with issues surrounding the installed output capacity of the central boiler plant. It shows the relationship between boiler size and energy consumption in intermittently heated buildings, and uses this to develop a rational method for sizing the central plant. The emphasis is on the designer making decisions based on fundamental principles rather than relying on rules of thumb or tables for which the source of the original data is often uncertain. The principles of intermittent operation are used to explain the operation of optimum start control for switching on the plant, and the chapter concludes by examining the difference between ideal and practical optimum start algorithms.

Chapter 10 looks at the way central boiler plant should be configured and controlled to achieve stable and energy efficient operation. In order to do this it looks at how the configuration of the system can impact on the central plant, in particular how the use of compensator circuits affects temperatures and flow rates around the system. This in turn has ramifications for the way central plant is configured, and the control of boilers will often be dictated by this mix of factors. The importance of temperature sensor location is explained, and boiler modelling is used to show the dynamic effects of different control settings. Guidance is given on how to set up the controls, and mention is made of add-on control devices designed to save energy.

Chapter 11 concentrates on the overall energy performance of a heating system. It presents a modified degree-day approach to energy estimation, and shows how degree-days may be used to monitor a building's energy consumption. The chapter also discusses benchmarking and normalisation of energy use, and discusses ways in which, operationally, a building can be maintained to give maximum performance from the heating system.

Taken as a whole the book should give the reader a deeper understanding of a building services system often taken for granted. Heating installations are so ubiquitous in temperate climates that it is easy to design systems on the basis of what has gone before. This book should explode any myth (if one truly exists) that there is such a thing as a standard heating system, and should strengthen the need for sound design practices. Given that so much of the carbon dioxide emission in Northern Europe comes from heating systems, it is of paramount importance that they are designed and installed to the highest standards that can deliver good indoor environmental quality with the greatest energy efficiency.

References

1. Building Research Energy Conservation Support Unit (BRECSU) (2000) *Energy Use in Offices*. Energy Consumption Guide 19. BRECSU, Garston.
2. Samuelsson-Brown, G. & Whittome, S. (1992) *The Office Sector – A Review of Floor Space, Building Type, Market Potential and Construction Activity in Great Britain*. Report no. MS 7/91, Building Services Research and Information Association, Bracknell.
3. Rickaby, P.A, Bruhns, H.R., Mortimer, N.D., Clark, E. & Steadman, J.P. (1995) *Mechanical Ventilation and Air Conditioning Systems in UK Buildings*, vol. 1–3. Project Report of the Department of the Environment Scoping Study for HVAC Systems. Extracts in *Building Services Journal*, 18, June, 12–13.

PART A
HEAT GENERATION

2 Boilers and Burners

2.1 Definition of a boiler

What is a boiler? Despite the name, the one thing that the boilers serving the vast majority of building heating, ventilation and air conditioning (HVAC) systems are rarely called upon to do these days is to *boil water*. Indeed this is typically the one thing that they must not be allowed to do. The tenth edition of *The Concise Oxford Dictionary* defines a boiler as 'a fuel-burning apparatus for heating water, especially a device providing a domestic hot water supply or serving a central heating system; a tank for generating steam in a steam engine'. This is a basic definition but one that is nonetheless accurate and comprehensive, and the latter part naturally displays the origins of boilers as components of the early steam engines. The definition of a boiler given by the American Society of Heating, Refrigeration and Air Conditioning Engineers (ASHRAE)[1] is, as one would expect, more detailed and technically specific: 'a cast-iron, steel or copper pressure vessel heat exchanger, designed with and for fuel-burning devices and other equipment to (1) burn fossil fuels (or use electric current) and (2) transfer the released heat to water (in water boilers) or to water and steam (in steam boilers)'.

What then are the essential distinguishing features that enable a 'fuel-burning apparatus for heating water' to function as a practical heating boiler, and to be the kind of heating plant that most HVAC engineers would see as implicit within the ASHRAE definition? First and foremost, it should be capable of heating a continuous supply of water. Second, it should be capable of burning fuel cleanly and efficiently, and in a stable and controlled manner that automatically maintains a specified water temperature within close limits under all operating conditions. This requires a burner and an automatic control system. Finally, the products of the combustion process must be continuously removed and safely dispersed into the general atmosphere, irrespective of the vagaries of wind and weather. This requires the boiler to be associated with a flue or chimney. These defining characteristics of a boiler are illustrated diagrammatically in Figure 2.1. In the following sections of this chapter we will consider boilers in terms of the functional elements that they need to possess in order to achieve the first two of these defining characteristics. The flue or chimney that is necessary to achieve the third is the subject of Chapter 5.

2.2 Principal functional elements of a boiler

2.2.1 Gas-fired boilers

A typical conventional (non-condensing) gas-fired low-pressure hot water (LPHW) heating boiler comprises five principal functional elements:

(1) the boiler 'block'
(2) the burner
(3) the burner gas line
(4) the control system
(5) the boiler casing.

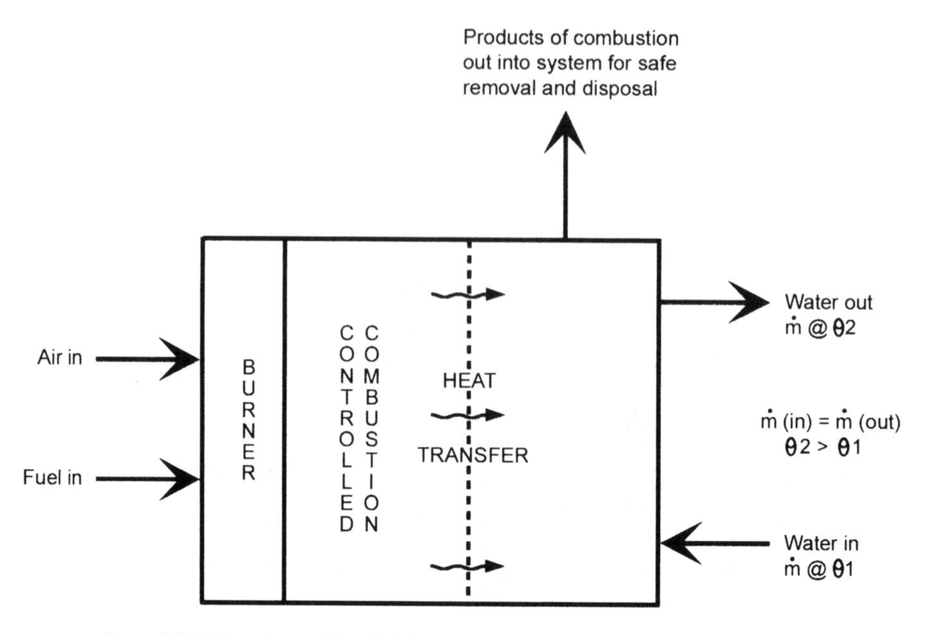

Figure 2.1 Defining characteristics of a boiler.

These are illustrated schematically in Figure 2.2. Each is an assembly of individual components that is more or less complex according to its function. The functions of each of these components, and of the functional elements themselves, are closely linked, and their design and operation must be carefully and thoroughly matched to ensure the safe and efficient performance of the boiler as a whole. Each of the functional elements of Figure 2.2 is considered in more detail in the following sections of this chapter. The point of reference for much of this will inevitably be the conventional *cast-iron sectional boiler* that remains numerically predominant in typical commercial building services heating applications. However, the characteristics of other types of boiler will be considered wherever these differ in principle from this point of reference.

It will make little difference whether the gas on which the boiler in Figure 2.2 is fired is natural gas, manufactured gas (town gas) or liquified petroleum gas (LPG), although within the UK as a whole the choice is effectively between the first and last of these. A complete installation for boilers fired on LPG will differ significantly from one intended for natural gas in its provision of on-site fuel storage.

2.2.2 Oil-fired boilers

In terms of its functional elements a typical oil-fired heating boiler differs little from its gas-fired counterpart. Principally it is more appropriate to consider the burner oil line as part of the fuel supply installation. A complete installation for an oil-fired boiler will again differ significantly from one intended for natural gas in its provision of on-site fuel storage.

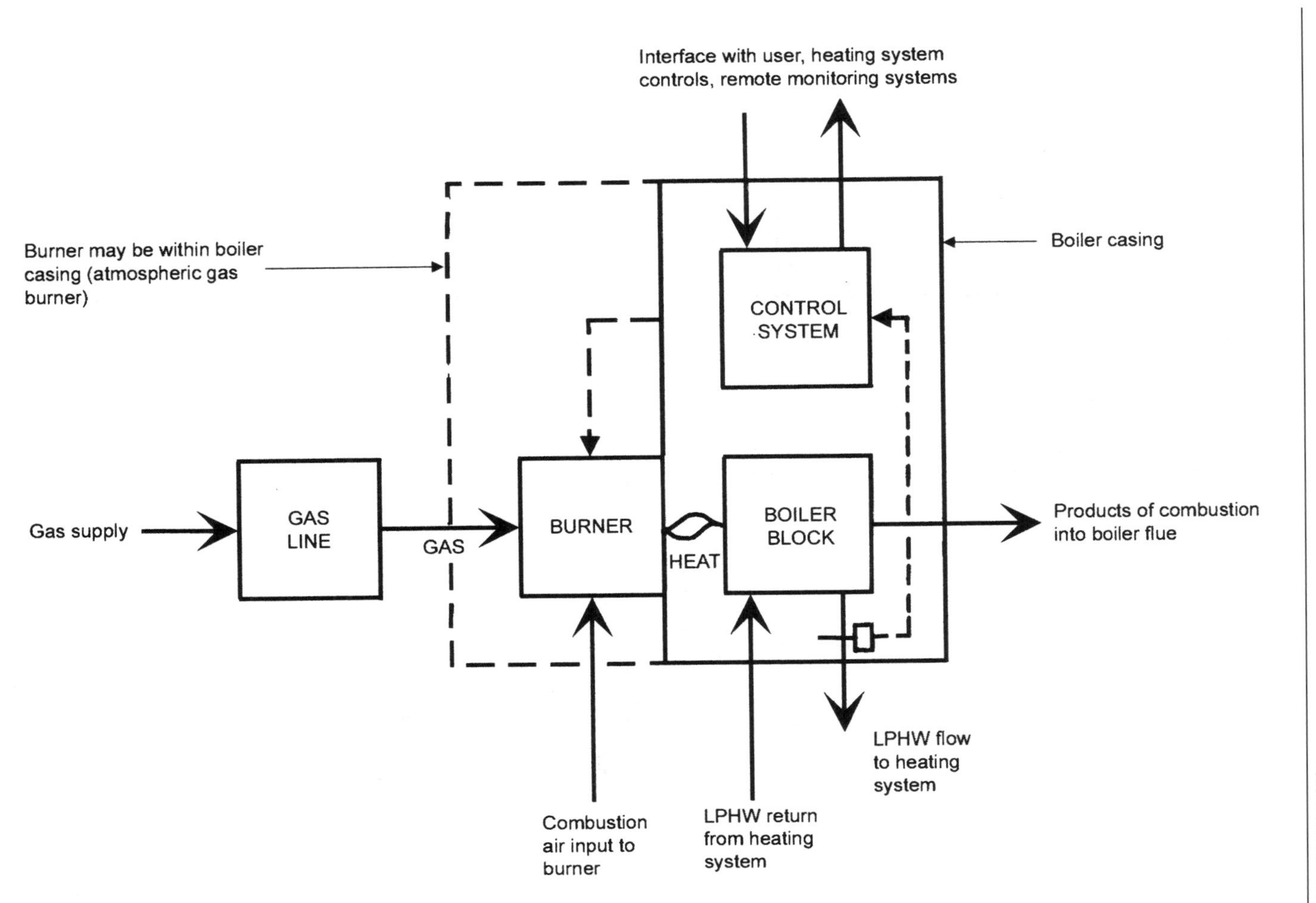

Figure 2.2 Principal functional elements of a gas-fired boiler.

2.2.3 Solid fuel boilers

Within the UK the use of boilers fired on the 'fossil' solid fuels, in particular coal and its related fuel 'family', is now restricted to the large-scale generation of electrical power or to niche domestic applications where alternative fuel sources are unavailable, or simply where it is the *visual attributes* of a fire that are desired. Furthermore, the simple logistics of storing and handling solid fuels make it unlikely that, even should (or when) existing supplies of gas and oil finally run out, there would be a return to widespread coal-firing for commercial or industrial building services applications. Hence for today's HVAC engineers, coal-fired boilers are effectively extinct, and their specific functional requirements and operational and performance characteristics will not be considered in this book.

There is another class of solid fuels for boilers, on which environmental considerations have refocused positive, albeit limited, attention. These are the *renewable* solid fuels, both conventional and new. The former category includes wood derived from forestry operations, the waste products of furniture manufacturing and other woodworking industries, and disposable pallets or other wooden packaging materials. The latter category includes the so-called 'energy crops', of which the most promising appear to be short-rotation willow coppice and miscanthus.

While a building services engineer practising in the UK today may indeed encounter wood-burning in rather specific commercial and industrial heating applications, its scale and technology typically owe more to the stove tradition than to modern boilers. Current thinking regarding energy crops (since there is, as yet, no widespread practice) is directed towards electrical power generation, rather than towards space, air or domestic hot water heating in buildings. Hence boilers or other heating plant capable of firing on the renewable solid fuels will not be considered in this book.

2.3 The boiler block

2.3.1 Function of the boiler block

The boiler 'block' is the primary component of any boiler. It is here that combustion of fuel takes place, and where the high-temperature products of this combustion are brought into sufficient proximity with water from the heating system to allow the transfer of heat from the combustion gases to the water in a controlled manner.

2.3.2 Configuration and design

In order to carry out its function the boiler block contains an area where the combustion process itself may take place. In a cast-iron sectional boiler this typically takes the form of a cylindrical chamber, the *combustion chamber*, which is located either along a horizontal axis in the lower part of the block (Figure 2.3) or along its central axis. In order to transfer the heat generated by the combustion process to water, the boiler block must also contain a network of passages that will allow the hot combustion gases and cooler boiler water to be brought into close proximity, separated only by a relatively thin division formed by the metal of the block itself. These gas and water passages extend throughout the length and depth of the block beyond the

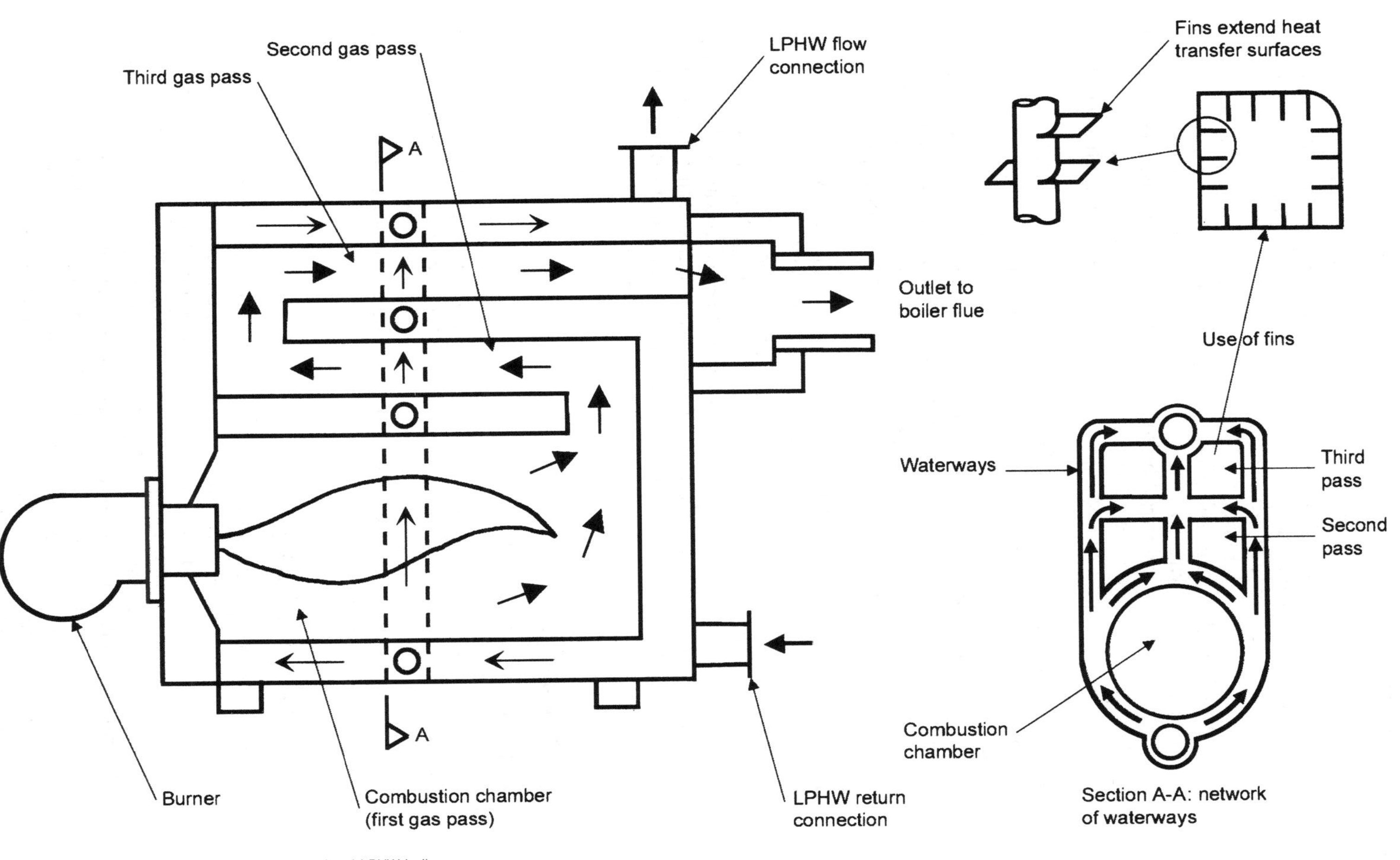

Figure 2.3 The boiler block in a multi-pass sectional LPHW boiler.

combustion chamber. Those passages that carry boiler water are commonly referred to as the *waterways* of the boiler, while those that carry the combustion gases are sometimes referred to as *flueways*. An arrangement of the kind that may be found in many cast-iron sectional boilers is illustrated in Figure 2.3, although the detailed configuration of a boiler block will naturally tend to reflect the specific requirements and characteristics of a particular boiler design. The ways in which types of boiler other than cast-iron sectional differ from the layout shown in Figure 2.3 are outlined in Chapter 3.

The arrangement of vertical waterways shown in Figure 2.3 provides a number of parallel paths for water to flow through the boiler block. It is important therefore that the design of the block ensures that the flow rate of water through each of these parallel paths is matched to the rate of heat input in that part of the block, so that excessive local water or metal temperatures are avoided. The existence of local hot spots, which may result if there are any areas of low water flow rate in the block, can result in local boiling of the water and *thermal overstress* of the metal of the block. Such localised boiling of water in the block is sometimes referred to as 'kettling', since its occurrence is accompanied by a noise similar to that made by a domestic kettle at the onset of boiling. Achievement of the correct water distribution throughout the boiler, and the avoidance of short-circuiting close to the boiler flow and return connections, relies on natural balance through attention to the resistance to flow of the various elements that make up the network of waterways. Conversely it is equally important that there should be no areas in the boiler block that are subject to temperatures low enough to cause condensation of the water vapour in the combustion gases. The risk here is that any such 'cold spots' will be the site of heavy corrosion that will ultimately lead to failure of the boiler.

The nature of the heat transfer process occurring within the boiler differs between the combustion chamber and the gas passages or flueways. In the former the high temperature of the flame, in which the chemical reactions of the combustion process itself are occurring, generates both radiant and convective heat transfer into the combustion chamber wall that is the primary heating surface of the boiler. The reacting gases themselves generate a low-intensity radiation that is discernible by a faint blue colour. Where particles of unburnt carbon are present suspended in the flame, either as an intermediate stage in the combustion process or as a result of incomplete combustion, they radiate as tiny 'black bodies', and their incandescence gives the flame a yellow colour. A flame that is rich in suspended particles of carbon is *luminous*, and radiates intensely. A flame that is relatively poor in suspended carbon particles is *non-luminous*, although it may be appreciably hotter than a luminous one. Natural gas typically burns with a non-luminous 'bluish' flame, while fuel oil burns with a luminous flame. In the combustion gas passages or flueways there is no flame, and heat transfer from the combustion gases into the waterways is predominantly convective. The nature of this process involves the operation of complex convective effects on both the gas and water sides of the metal of the boiler block. On the gas side, common practice with heat exchanger design generally employs extended surface areas outside the combustion chamber, and sometimes even within it. In modern boilers these extended heat transfer surfaces are typically achieved by the use of *fins* (inset to Figure 2.3), although thimble-shaped projections or 'pips' have also been used. These extended surface areas may also be used to optimise levels of turbulence within the flow of combustion gases, for improved heat transfer performance. On the water side of the boiler metal the transfer of heat depends on the velocity of water flow over the heat transfer surface, which is (as far as the water is concerned) the circumference or perimeter of the waterways. For a given temperature rise of the boiler water higher heat transfer rates typically require higher water velocities. In all modern commercial building services applications water will flow though the boiler as a result of a pumped circulation. However, the velocity of water flow over the heat transfer surfaces of the boiler block will result not only from the effect of the external pumped circulation but also from local convective and 'thermosyphonic' effects within the waterways themselves. Setting up these local convective effects is naturally favoured by a more generous dimensioning of the waterways.

The *water content* of the boiler, i.e. the volume of water that is contained within the boiler block at any instant of time, is naturally a function of the total volume of the waterways within the block. Water content is *independent of the volume flow rate* at which water enters and leaves the boiler via the flow and return connections to the heating system. A low water content produces a more rapid response to changes in the rate of heat input to the boiler, including the important case when firing of the burner commences and heat input changes from zero to the thermal output of the burner. Where the size and/or number of the waterways are greater, the water content of the boiler will be increased. The overall size and cost of the boiler are also likely to be greater. The trend in modern boiler design is towards the achievement of high heat transfer rates in compact designs with low water content, which can provide both a high overall efficiency of conversion between the heat input to the boiler (as fuel) and the heat output from it to water, and rapid response to changes in the rate of heat input. The implications of this for boiler control are considered in Chapter 10.

In order to fulfil its function the boiler block must include two other provisions. First, it must include flow and return pipe connections, so that a continuous supply of water may be received from the building heating system, heated and returned to it. Second, it must collect the 'spent' combustion gases and provide an outlet for them to be discharged into a boiler *flue*, which will ultimately disperse them safely to atmosphere.

2.3.3 The multi-pass principle

Water circulated around the circumference of the combustion chamber provides the first, and primary, stage of heat transfer within the boiler. It is clear that a major aim of boiler design will be to achieve the maximum possible transfer of heat from combustion gases to heating system water for any given rate of combustion of fuel. Since the combustion gases must ultimately be discharged from the boiler and rejected safely to atmosphere, any heat that is not transferred will be wastefully rejected with them. It is also clear that, with plant room space typically at a premium in modern buildings, it is desirable to achieve the maximum heat transfer within a boiler of compact physical dimensions. Modern boilers do not, therefore, rely solely on heat transfer from the combustion chamber. Secondary heat transfer is introduced outside the combustion chamber through the technique of *multiple passing* of the boiler (Figure 2.3). On reaching the end of the combustion chamber, in their first pass along the length of the boiler, the hot gases are typically directed back along the length of the boiler in a *second pass* that takes them through passages that either surround, or are surrounded by, water from the heating system, to achieve additional heat transfer. In the quest for high heat transfer efficiencies, modern boilers have typically introduced further sets of passages that pass the combustion gases back and forth along the length of the boiler a third and even fourth time before discharging them to atmosphere via the boiler flue. Figure 2.3 illustrates a *three-pass* boiler. The configuration of the boiler 'passes', and their relationship to the waterways that distribute water from the heating system through the boiler, is a characteristic of the type and particular design of boiler. At the end of their final pass of the boiler the combustion gases are collected in a manifold or *smoke box* (*smoke hood*), from which they are discharged into a boiler flue. The use of multiple passes produces a complex relationship between the flow of combustion gases and water within the boiler.

The boiler block is thermally insulated to reduce heat loss to its surroundings by convection and radiation from its outer surface. Since the shape of the outer surface of the boiler block may be complex, thermal insulation is typically applied in the form of a flexible blanket. In order to minimise what is, after all, a wasteful loss of some of the energy released by the combustion process, up to 125 mm thickness of thermal insulation may be applied in this way.

2.3.4 Water content and temperature differential

In the UK hydronic ('wet') systems for space and air heating typically employ Low-Pressure Hot Water (LPHW). Boilers are therefore designed with this in mind. The design temperature differential (or rise) of a boiler is the temperature difference through which water from the heating system is raised while passing through the boiler block, with the burner firing at its maximum thermal output. This is a function of both boiler and heating system design considerations. The system design considerations, which relate to the choice of heating system flow and return temperatures, will be examined in Part B of this book. For present purposes it will be sufficient to note that UK practice is typically based on a 'standard' design temperature differential of 11 K across the boiler block, and that this is traditionally assumed to be between nominal flow and return temperatures of 82°C and 71°C respectively. These flow and return temperatures are simple metric conversions from earlier imperial values of 180°F and 160°F respectively (giving a design temperature differential of 20°F). These had been in traditional use in the UK for many years, and there seems little reason to consider that they represent any kind of optimum operating parameters in the context of *either* modern boiler *or* heating system design.

In any event the boiler design temperature differential defines the design flow rate of water from the heating system through the boiler block. For a given heat output to water from the boiler block the water flow rate and temperature differential are, naturally, inversely proportional – halve the former, double the latter. For a conventional (non-condensing) boiler this may be shown by the equation:

$$\dot{Q} = \dot{m}_w c_p(\theta_F - \theta_R) \tag{2.1}$$

where $\dot{Q}$ = heat output rate to water (kW); $\dot{m}_w$ = water mass flow rate ($kg\,s^{-1}$); c_p = specific heat of water ($kJ\,kg^{-1}\,K^{-1}$); θ_F = flow water temperature from the boiler (°C); θ_R = return water temperature to the boiler (°C).

With regard to the temperature range encountered in building services LPHW space and water heating systems and their boilers, it is typically acceptable to regard the density of water as 'fixed' at $1000\,kg\,m^{-3}$ ($1\,kg\,l^{-1}$), so that mass flow rate and volume flow rate are numerically the same.

The minimum water flow rate that will transfer away from the boiler block, under firing conditions, sufficient heat to avoid thermally overstressing it, is typically defined as that resulting in a temperature differential of 20 K, although this may be reduced to as low as 14–15 K for some designs of high-efficiency modular boilers. In the former case the minimum flow rate of water to be maintained through the boiler block under firing conditions would be 55% of that at the design temperature differential of 11 K. In the latter case it would be 70–75%. We have already noted that the volume (and therefore mass) of water contained within the boiler at any instant of time is independent of the rate at which it enters and leaves (and therefore flows through) the boiler. Hence a given mass flow rate in equation 2.1 may be associated with either a high or low water content of the boiler block. Since it is the flow of water through the boiler that carries away or dissipates heat from the boiler block, for a given boiler design temperature differential a high heat output to water inevitably requires a high volume flow rate of water through the block.

Within the boiler, however, we may for simplicity represent the overall heat transfer process by an equation in the form

$$\dot{Q} = UA\Delta\theta \qquad (2.2)$$

where $\dot{Q}$ = total heat transfer rate to water (kW); U = an overall heat transfer coefficient for the design of the boiler ($kW\,m^{-2}\,K^{-1}$); A = surface area available for heat transfer (m^2); $\Delta\theta$ = a characteristic temperature difference (K).

Under steady-state conditions, the heat transfer rate in equation 2.2 and the heat output rate to water in equation 2.1 are the same. The *characteristic temperature difference* in equation 2.2 will typically not be the simple difference between boiler flow and return temperatures in equation 2.1, but is more likely to be a *log mean temperature difference* (*LMTD*) based on water and gas-side temperatures. However, neither this nor the origin of the overall heat transfer coefficient need concern us here. What *is* important is to note two things. First, the water content of the boiler block is a function of the value of term A in equation 2.2. This depends on the number and size (typically diameter) of the waterways. The water content of the boiler depends on the *volume* of these same waterways. Second, for a given heat transfer rate and characteristic temperature difference the value of A in equation 2.2 is inversely proportional to that of U. Therefore, with a high value of U the value of A may be reduced for a given heat output to water, and with it the water content of the boiler. When we talk of achieving a high heat transfer rate in a boiler of compact design we are implying that the rate of heat transfer per unit area of heat exchange surface (U) is high, and it is this trend in boiler design that has lead to the typically low water content of many modern boilers. While this is only one aspect of boiler development, it nevertheless has significant implications for the design and control of boiler installations.

2.3.5 Wet-base and dry-base types

In boiler designs where downward heat flow from the combustion chamber is not transferred into waterways, an insulating base will be required to protect the builder's work base or other non-combustible surface on which the boiler is mounted. Designs in which the combustion chamber is surrounded by waterways are defined as 'wet-base' types, those in which the combustion chamber is below the waterways are 'dry-base' types, and those in which waterways wrap around its top and sides (but not below it) are 'wet-leg' types. All three types are illustrated in Figure 2.4.

2.4 The burner

2.4.1 Function of the burner

At the burner the boiler fuel is mixed with combustion air and ignited to produce the flame in which the combustion process takes place. In its simplest form the function of the burner is to safely initiate, maintain and finally terminate the combustion process in response to the requirements of the boiler control system. In the quest for higher seasonal boiler efficiency through improved load matching, the burner may be required not only to maintain the combustion process, but to vary it in a controlled manner according to the demand of the boiler control system.

2.4.2 Boiler fuels and the combustion process

As far as modern boilers are concerned, the common boiler fuels are either gaseous or liquid hydrocarbons. As far as almost all of mainland UK is concerned, the former category typically comprises natural gas or liquified petroleum gas (LPG), while the latter comprises various grades of fuel oil. The 'lightest' grades of fuel oil, 'C' (kerosene) and 'D' ('gas oil'), are those most likely to be employed in modern commercial building services applications. Chemically natural gas and LPG have relatively simple molecular structures. The natural gas supplied for public use in the UK is principally (in excess of 90%) *methane* (CH_4), while the two main types of LPG that are commercially available are *propane* (C_3H_8) and *butane* (C_4H_{10}). The chemical structure of the fuel oils is more complex and may contain various levels of impurities, sulphur typically being that of particular concern in building services applications. In any event complete combustion of the fuel input to the burner is necessary both to obtain the maximum release of energy and to prevent the production of carbon monoxide, which is the poisonous product of the partial combustion of hydrocarbon fuels.

The choice of fuel will be made on the basis of availability, the cost of the fuel itself, the impact of its adoption on the capital cost of the boiler installation and its environmental impact (emission of carbon and pollutants such as oxides of nitrogen and sulphur dioxide). The cost per kWh of heat input to a boiler is broadly comparable for natural gas and the light grades of fuel oil, although absolute costs are typically influenced by annual consumption (consumer purchasing power). On a similar basis the cost of LPG is typically two to three times that of natural gas, which will make it broadly comparable with the cost of 'off-peak' electricity. While for identical heat outputs the capital cost of oil-fired commercial space and water heating boilers is

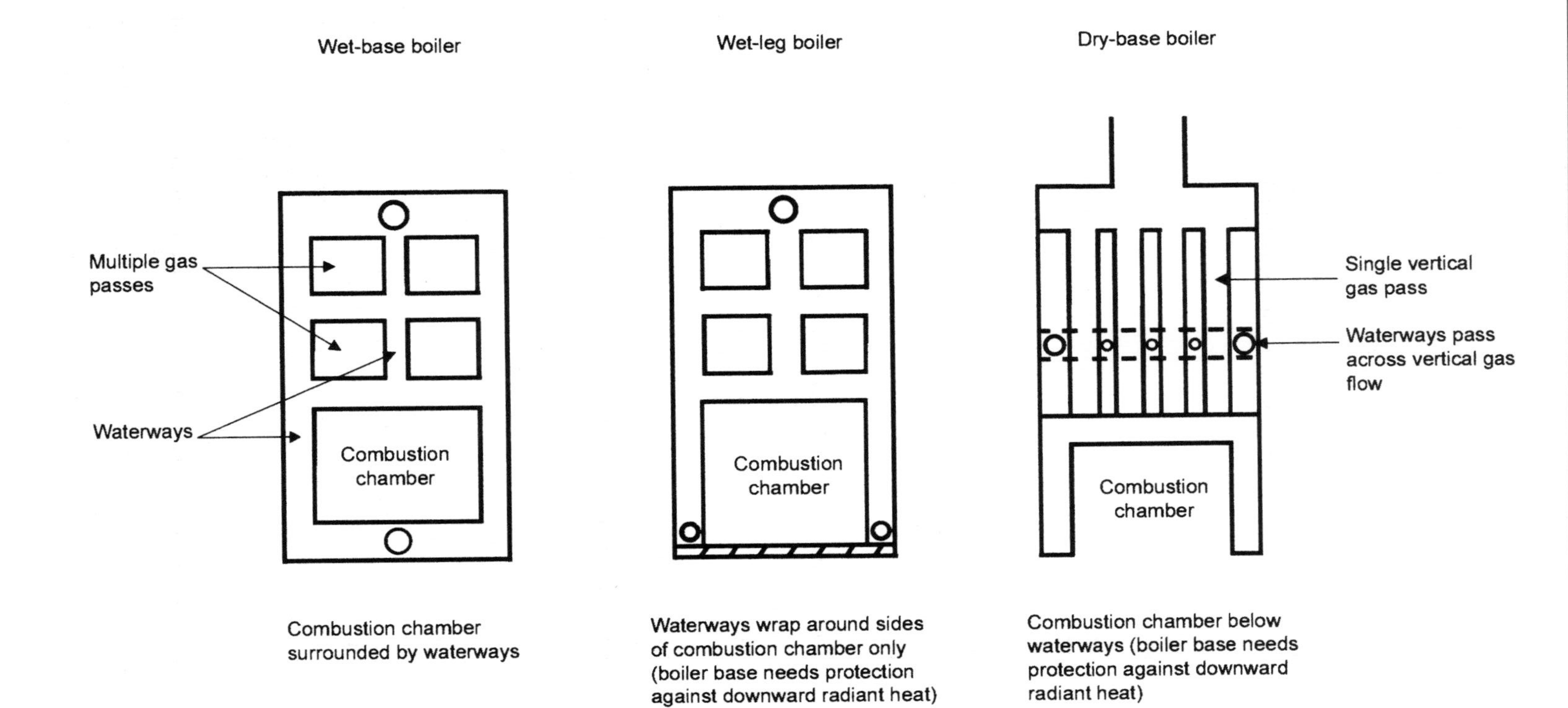

Figure 2.4 Wet-base, dry-base and wet-leg boilers.

typically lower than that of natural gas boilers, oil firing incurs the additional cost of a fuel storage installation. The use of LPG also incurs the cost of fuel storage. Carbon emission per kWh of delivered fuel is significantly lower for natural gas and LPG than for fuel oil (c.0.052 kgC per delivered kWh for natural gas, which is approximately 73% of the value for fuel oil, c.0.071 kgC per delivered kWh). Furthermore, the gaseous boiler fuels have a negligible sulphur content and are therefore effectively 'free' from emissions of sulphur dioxide. As far as the production of the other significant pollutants, oxides of nitrogen, are concerned, there is likely to be a greater variation in performance between different boiler designs for a given fuel than between fuels. However, natural gas is typically considered to be the lesser polluter or 'cleaner' fuel of the two in this respect. Where a supply is available, natural gas will typically be the preferred choice for the fuel input to the burner. The use of alternative boiler fuels is considered in Chapter 4. However, one additional fuel that may ultimately come into the reckoning for use in boilers is the so-called 'bio-diesel', which is a liquid (diesel) fuel derived from certain energy crops (*rapeseed* being an example).

At this point it will be worth briefly considering what our 'wish-list' would be for the ideal combustion process to occur within an LPHW boiler. Most fundamentally we should ideally like the fuel input to a boiler to take the form of a pure hydrocarbon, comprising only carbon and hydrogen without any undesirable impurities, so that its complete combustion results only in the formation of carbon dioxide (CO_2) and water vapour (H_2O). In this respect it is interesting to note that we typically think of the use of 'fossil' fuels as being based on the combustion of carbon itself. However, for a pure hydrocarbon a substantial proportion of the energy released derives from the combustion of hydrogen to form water vapour. The energy released from this process is only approximately 60% of that from the combustion of carbon to form carbon dioxide, or approximately 75% if all of the water vapour formed may be condensed. However, all of the common boiler fuels have more hydrogen atoms than carbon atoms in each molecule. In the case of methane, which we have seen is the principal constituent of natural gas, the ratio of hydrogen atoms to carbon atoms is at its greatest (4), and over 50% of the energy released derives from the combustion of the hydrogen content of the fuel. The combustion of hydrogen to water vapour has the nice point that, unlike that of carbon, it cannot be incomplete (you either get water vapour or you do not get anything). Our ideal fuel would also be a *single* pure hydrocarbon, since this makes the qualitative and quantitative analysis of the chemical reactions of its combustion simpler. Moving on one stage, with pure hydrocarbon fuel in the burner we should ideally like to mix it (perfectly, of course) with *pure oxygen* and then to burn the thoroughly homogeneous mixture. The energy released by the combustion process arises only as a result of the combination of the carbon atoms and hydrogen molecules with oxygen, and the presence of anything else is, at least in simple combustion terms, superfluous. Therefore when a fuel is completely burnt in pure oxygen the combustion gases will achieve the highest possible temperature.

The natural gas supplied for public use in the UK makes a reasonably good approximation to a pure hydrocarbon. Despite the predominance of methane as a constituent, its typical *combustible* content is, strictly-speaking, a mixture of hydrocarbons (including a few percent by volume of ethane and traces of propane, butane and benzene). These are typically supplemented by a few percent of *non-combustibles*, principally nitrogen with a trace of carbon dioxide, to make up the overall mixture that is referred to as 'natural gas'. So in practice we have a pretty good boiler fuel. However, it is certainly not practical to supply the burner with only pure oxygen. We have to make do using air as a source of the oxygen required. Since only approximately 21% of air by volume is oxygen, with more-or-less the whole of the remainder being nitrogen, we naturally have to supply quite a lot of air to provide the oxygen necessary for combustion – in fact approximately 4.76 m^3 of air for every m^3 of oxygen required. The *gravimetric* analysis of air gives it as approximately 23.3% oxygen *by mass*, so that approximately 4.29 kg of air are required to provide 1 kg of oxygen. Apart from the undesirable formation of small quantities of some of its oxides, the 3.76 m^3 of nitrogen supplied with every m^3 of oxygen necessary for 'just-complete' combustion of the fuel is merely present throughout the combustion process. While not contributing usefully to the energy released by this, it is heated by it. Since this heat would otherwise have gone into raising the temperature of the products of combustion (the CO_2 and H_2O) alone, a lower final temperature is attained now that the combustion gases comprise a mixture of these with the 'spare' nitrogen. Heat transfer rate from the combustion gases to the boiler water is reduced in consequence. For a given temperature at outlet to the boiler flue, the combustion gases also carry more heat wastefully away from the boiler, so the boiler's efficiency is reduced.

The amount of air that must be supplied to provide sufficient oxygen to permit the complete combustion of a given quantity of boiler fuel is expressed in terms of an *air-to-fuel ratio*. In principle this may be the ratio of kg of air per kg of fuel or m^3 of air per m^3 of fuel. For natural gas and LPG, the air-to-fuel ratio is expressed in terms of m^3 of air per m^3 of gas (with LPG being in the gaseous state, ready for the burner). However, in the case of fuel oil the ratio may be expressed in terms of m^3 per kg or per litre of fuel, in which case the units of each quantity must be defined. For methane, which is the principal constituent of natural gas, the approximate theoretical or *stoichiometric* air-to-fuel ratios are 9.5 (in volumetric terms) or 17.2 (in gravimetric terms, i.e. kg of air per kg of methane). Since natural gas contains minor quantities of other more complex hydrocarbons, the stoichiometric air-to-fuel ratio is actually slightly in excess of the value for pure methane (c.9.7 in volumetric terms, a value of 10 being adequate for most approximate calculations). In practice it is not enough simply to supply air for combustion according to the stoichiometric air-to-fuel ratio, since in a practical boiler and burner combination inaccuracies in the metering of both the fuel and air supplies and imperfect mixing of the fuel and air (i.e. combustion of a non-homogeneous fuel/air mixture) will result in incomplete combustion and the formation of poisonous *carbon monoxide* (CO) in addition to the carbon dioxide and water vapour. Hence an air-to-fuel ratio is actually used that is higher than the stoichiometric value. This provides a sufficient excess of air to ensure that, effectively, complete combustion takes place. How much excess air is required will naturally depend on the design of the burner and the efficiency with which it mixes the fuel and combustion air, but it may be of the order of 15–20%. In this respect modern boiler and burner combinations exhibit improved performance over earlier designs. The extent of the excess air that is supplied to a burner is reflected by the volumetric content of CO_2 in the combustion gases leaving the boiler, with values of the order of 9–10% typically being sought for a natural gas boiler, and somewhat higher than this for oil-fired boilers (c.12–13% for grade 'D' gas oil).

Excess air does not contribute usefully to the energy released by the combustion process, but is again heated by it. The final temperature of the combustion gases is therefore further reduced, again resulting in a reduction in heat transfer rate to the boiler water. For a given temperature at outlet to the boiler flue, even more heat is carried wastefully away from the boiler by the combustion gases. The greater the air-to-fuel ratio is above the stoichiometric value, the greater will be the temperature dilution due to the presence of excess air in the combustion gases.

We can represent the general combustion of a *pure hydrocarbon* boiler fuel in excess air by the chemical formula:

$$C_xH_y + E(x + y/4)O_2 + 3.29E(x + y/4)N_2$$
$$= xCO_2 + (y/2)H_2O + (E - 1)(x + y/4)O_2 + 3.29E(x + y/4)N_2 \qquad (2.3)$$

where x = the number of carbon atoms in a molecule of the fuel (e.g. 1 for methane, CH_4); y = the number of hydrogen atoms in a molecule of the fuel (e.g. 4 for methane); E = actual air-to-fuel ratio ÷ stoichiometric air-to-fuel ratio, so that E – 1 is the excess air for combustion expressed as a decimal fraction of the stoichiometric air quantity (or % excess air ÷ 100).

The factor 3.29 in equation 2.3 is the gravimetric or mass ratio of nitrogen to oxygen in the combustion air, i.e. the number of kg of nitrogen that accompany each kg of oxygen in the combustion air. Equation 2.3 may be used to analyse the combustion process in terms of the masses of the fuel and air input to the burner, and of the contents of the combustion gases. It may also be used to consider volumes of the substances if it is remembered that the mass of a gaseous substance that is equal to its molecular weight in kg (1 *kmol*) occupies 22.41 m^3 at 1.01325 bar and 0°C (273 K), and appropriate *Charles' Law* corrections are made for the temperatures of the substances. For a fuller description of the combustion process the interested reader is referred to standard thermodynamics texts, such as that by Rogers and Mayhew[2].

One further aspect of the combustion process affected by the burner must be considered, and this is the production of *pollutants*. These are unintentional and undesired by-products of the combustion process that typically arise either from the use of air as the source of oxygen for combustion or from the combustion of impurities present in the fuel. In some instances the chemical reactions that generate pollutants may be net 'consumers' of energy. As far as the use of air as a source of oxygen for combustion is concerned, we have already seen that the largest constituent of the air supplied to the burner is nitrogen. As far as the air that we breathe (and burn things in) is concerned, its nitrogen content is typically considered to be chemically 'inert'. However, at high temperatures nitrogen combines with the oxygen in air to form oxides of nitrogen. These are atmospheric pollutants and greenhouse gases (nitrous oxide has a *global warming potential* (*GWP*) of over 300 times that of carbon dioxide, taken over a timescale of 100 years), and the temperatures necessary for their production are encountered in the combustion zone produced by the burner in a boiler. While chemically there are several different oxides of nitrogen, it is typically the practice concerning boiler emissions to cover them by the 'umbrella' term NO_x. Burner design is crucial in minimising the level of NO_x emissions from boilers. With regard to emissions resulting from *impurities in boiler fuels*, of principal concern is the production of sulphur dioxide (SO_2) from sulphur present in fuel oils.

While the popularity of natural gas as the 'first-choice' boiler fuel has led to a reduction in the significance of SO_2 production from commercial space and water heating boilers, at least as far as new boiler installations are concerned, modern burner design places a considerable emphasis on the reduction of NO_x emissions.

2.4.3 Burner design

The aim of burner design is to achieve an intimate and completely homogeneous mixing of the fuel with the combustion air, and its complete combustion to release the maximum energy with the minimum production of unwanted emissions.

Burner design is dependent on the type of fuel, whether natural gas, LPG, fuel oil or 'dual-fuel'. Dual-fuel burners are able to burn either natural gas or fuel oil. Wherever proximity to a mains supply of adequate capacity permits, or the establishment of such a supply is economically viable, natural gas will inevitably be preferred over LPG. The economic basis for this has already been noted. Hence the discussion of gas burners that follows will be based primarily on burners designed to fire on natural gas.

Burners designed to fire on natural gas may be *atmospheric, fan-assisted, forced-draught or premix type*. Burners designed to fire on fuel oil may be *vaporising, mechanical atomizing, rotary or twin-fluid atomizing type*. In vaporising oil burners, as the name suggests, the fuel oil is heated to produce a vapour. Use of this type of oil burner, together with its variants and derivatives, has typically been confined to domestic boilers, and it will not be considered further here.

The position of the burner in relation to the boiler block depends on the type of burner. Fan-assisted and forced-draught natural gas and oil burners typically fire horizontally ('inshot', Figure 2.5a). A cutout is provided in the front of the boiler to accommodate a penetration necessary for the burner. This may be carried out at the boiler manufacturer's works or on site. The burner itself is fitted to the front of the boiler via a mounting plate or flange. It is vital that a gas-tight seal is established between the burner mounting plate or flange and the boiler. If this cannot be achieved with the basic cast or machined finishes of the components, an additional 'soft' or gasket-type seal will be required. This part of the boiler block is typically also hinged to provide an access door for cleaning of the combustion chamber, flue gas passes and waterways. Hence the positioning of boilers on site must allow adequate clearance for the opening swing of this access section with the burner in place. On large boilers, where the burner assembly is too bulky for convenient manual handling, traversing arms or mounting trolleys may be provided to allow mechanical retraction and reseating of the assembly. Atmospheric gas burners typically fire upwards ('upshot', Figure 2.5b), and are therefore mounted at low level in a boiler block or beneath a gas-to-water heat exchanger. Access to the burner is typically from the front of the boiler.

The boiler manufacturer matches the performance and physical characteristics of the burner to those of the boiler. A vital part of this process, and of the installation and commissioning of the burner, is to ensure that firing of the burner takes place along the central axis of the combustion chamber, and that the combustion flame is contained (or 'burns out') within its length. It is essential to ensure that impingement of the burner flame on the metal surfaces of the combustion chamber does not occur. The overheating of the

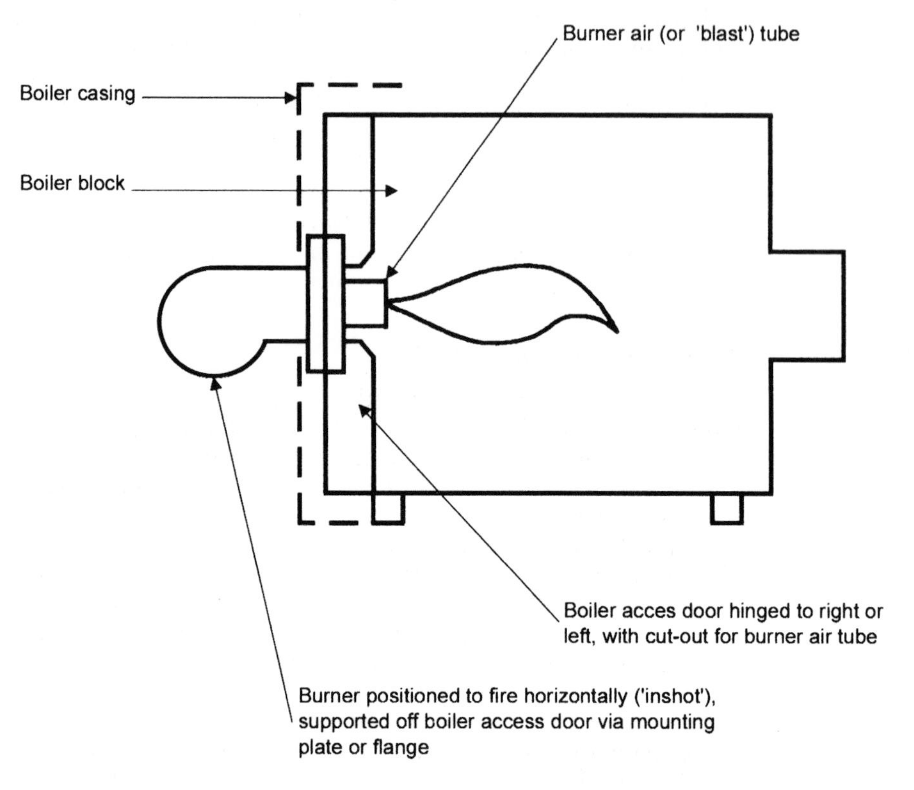

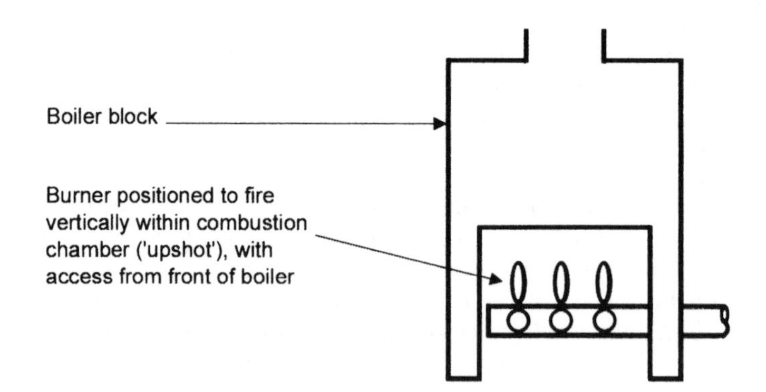

Figure 2.5 Position of the burner.

metal surfaces that flame impingement causes leads to thermal overstressing and may result in crack development and (ultimately) premature boiler failure. In cast-iron boilers prolonged overheating may even produce local modification of the structure of the metal that impairs its physical properties and strength. The availability of a viewport that allows the characteristics and appearance of the burner flame to be observed may prove helpful not only in ensuring correct installation, alignment and commissioning of the burner, but also in investigating any subsequent firing problems or (in extreme cases) instances of boiler failure. Deviations from the idealised shape and colouration of the burner flame may be symptomatic of incomplete combustion or flame impingement.

2.4.4 Atmospheric natural gas burners

The 'atmospheric' type of natural gas burner was the first to be developed following the introduction of natural gas as a fuel for space and water heating boilers. It utilises the static pressure of the natural gas itself to deliver gas from the supply connection (gas meter) to the boiler, and thence to the burner heads (or 'ports'), where it entrains and mixes with combustion air and is finally burnt. Air for combustion is typically drawn from the boiler plant room into the combustion chamber via openings positioned at low level around the base of the boiler. The burner relies on the *natural draught of the boiler flue* to generate a slightly negative pressure in the combustion chamber, which is required to achieve the necessary induction of combustion air.

Atmospheric gas burners are typically manufactured from stainless steel and generally comprise a number of horizontal burner 'bars' that are held together in a burner assembly or 'basket' that is fed with natural gas by a common injection manifold (Figure 2.6). Since atmospheric gas burners are typically positioned at low level in the boiler block, the base on which the boiler is mounted may need to be protected against downward radiant heat from the combustion zone and burner assembly. Where the combustion chamber is not surrounded by waterways, the provision of plate heat shields or ceramic insulation may be required. An alternative approach places the burner bars above a system of baffles that reflect downward radiant heat from the combustion zone and burner bars back up into the combustion chamber. The result is a reduction in the temperature that would otherwise exist below the boiler, and a reduction in heat loss from the boiler. Such baffles will typically also preheat the combustion air that is drawn through them. This helps the combustion process and will typically result in some increase in boiler efficiency.

In order to perform its function of initiating combustion when required to do so by the boiler control system, the burner must be provided with a means of igniting the gas fed to it. This may be achieved by a 'pilot' burner. A small supply of natural gas is continuously fed to this in order to maintain a permanent flame that will ignite gas supplied to the main burner. The pilot burner itself is initially lit by an ignition spark. A safety device is essential to ensure that the gas supply to the pilot burner is immediately shut off in the event of failure of the pilot flame. A *permanent-pilot* ignition system of this kind is relatively simple but continues to consume gas, albeit at a low rate, even when the boiler is not firing. *Spark ignition systems* ignite gas supplied to the burner directly by means of a spark generated by passing a high voltage between

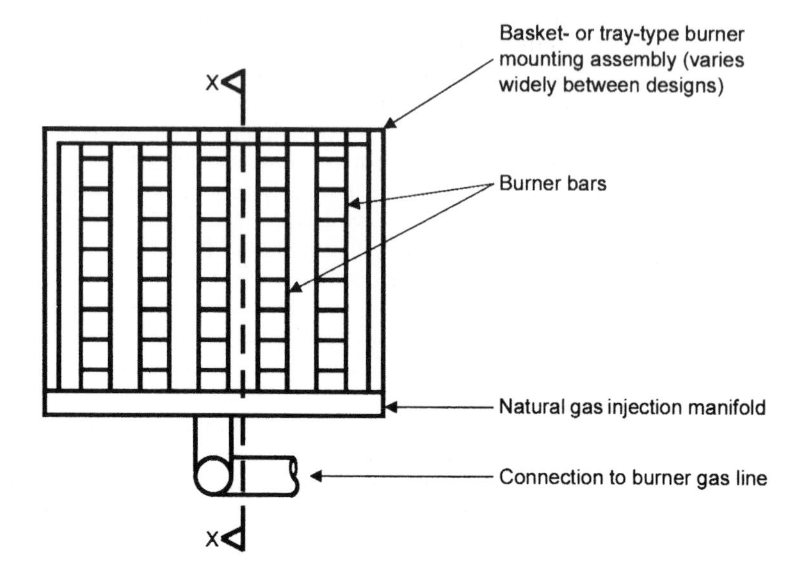

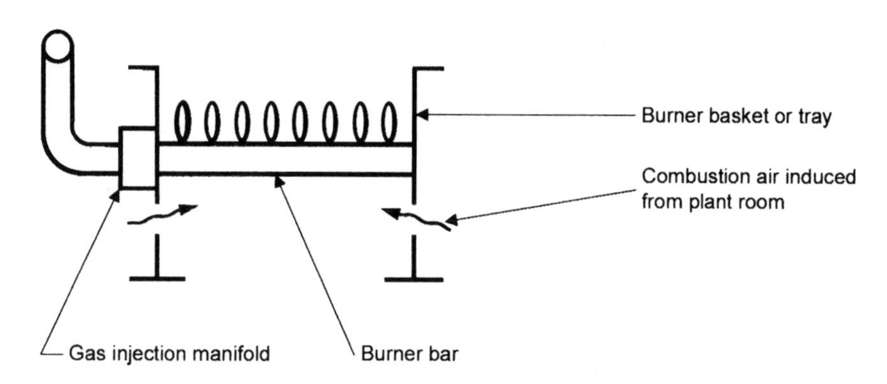

Figure 2.6 Configuration of a typical atmospheric natural gas burner.

suitable electrodes. This voltage (of the order of 10 kV or higher) is generated by an internal transformer that is fed from the mains electrical supply to the boiler. Since the need is for brief repetitive operation, rather than continuous-load service, the transformer is of the intermittently-rated type. As an alternative to spark ignition systems, the gas supplied to the burner may be ignited directly by means of an element that is heated electrically to a sufficiently high temperature. This technique is claimed both to have lower maintenance requirements than spark ignition systems and to remove the risk to control systems of radio frequency interference and electrical 'spikes' that may be associated with the former. Whichever burner ignition technique is employed, in the event that combustion is not established within a critical safety period the demand for the burner to fire must be overridden and the gas supply to the burner cut off. The techniques used to achieve this safety function are discussed later in this chapter.

It is the static pressure of the natural gas that generates flow through the gas line (or gas 'train') to the burner and through it, and atmospheric gas burners therefore need to operate within a fairly narrow range of gas pressure. Under conditions of maximum burner output a static pressure of the order of 20 mbar (2 kPa) gauge will typically be required at the upstream connection to the gas line, in order to give a static pressure in the range 9–13 mbar (0.9–1.3 kPa) gauge at the burner itself, depending on the particular design. Atmospheric gas burners should generally not be exposed to static pressures significantly greater than 25 mbar (2.5 kPa) gauge at the upstream connection to the gas line. Therefore care should be taken during testing of gas pipework to ensure that the burner and its gas line are adequately isolated from pressure that might damage them.

For a given draught the boiler flue, burner assembly and gas line operate at constant volume flow rates, and not at constant mass flow rates. If the density of the gas supplied reduces, as for example in installations at altitude, its thermal content, in $MJ\,m^{-3}$ of gas supplied, will be reduced. For a given heat output to water, a greater volume flow rate of gas will therefore need to be supplied to the burner. For a given pipe or valve diameter the associated pressure losses will also be increased. Hence, unless either the design of the burner and gas line can take this into consideration or specific design modifications are possible (both of which are unlikely in equipment from a standard product range), the entire assembly ought to be subject to a derating factor when operated in locations that are significantly above sea level. ASHRAE[3] quotes a recommended derating factor of 4% per 300 m of altitude.

In the UK, low-pressure supplies of natural gas typically provide a static pressure of 21 mbar (2.1 kPa) gauge at the outlet from the supply utility's meter. If a pressure of 20 mbar (2 kPa) gauge is required at the connection to the gas line, 1 mbar (100 Pa) is available for all pressure losses due to the flow of gas in the pipework between the meter outlet and the gas line of the furthest boiler. If the boiler plant can only be located physically remote from the supply utility's meter, such that the available gas pressure remains inadequate, even with generous sizing of the pipework, the supply pressure may be increased or 'boosted' by means of a gas booster. Gas boosters are considered in Chapter 3. The common practice of locating boiler plant in a roof-level plant room may effectively place it as far away from the gas meter as possible, and also introduces an 'altitude effect'. The nature of this effect is different according to the gas used. In the case of natural gas, which is less dense than air, altitude results in an increase in the gauge pressure of the gas. In the case of LPG, which is denser than air, altitude results in a reduction in the gauge pressure that is equivalent to an additional pressure loss in the pipework.

Atmospheric gas burners are, relatively-speaking, both mechanically simple and quiet in operation. However, they are sensitive to flue performance and need adequate and stable natural draught for satisfactory combustion performance. Any reduction of the draught, or the existence of *down-draught conditions* within the flue, will lead to incomplete combustion and the formation of poisonous carbon monoxide gas (CO). Flue gases may be trapped

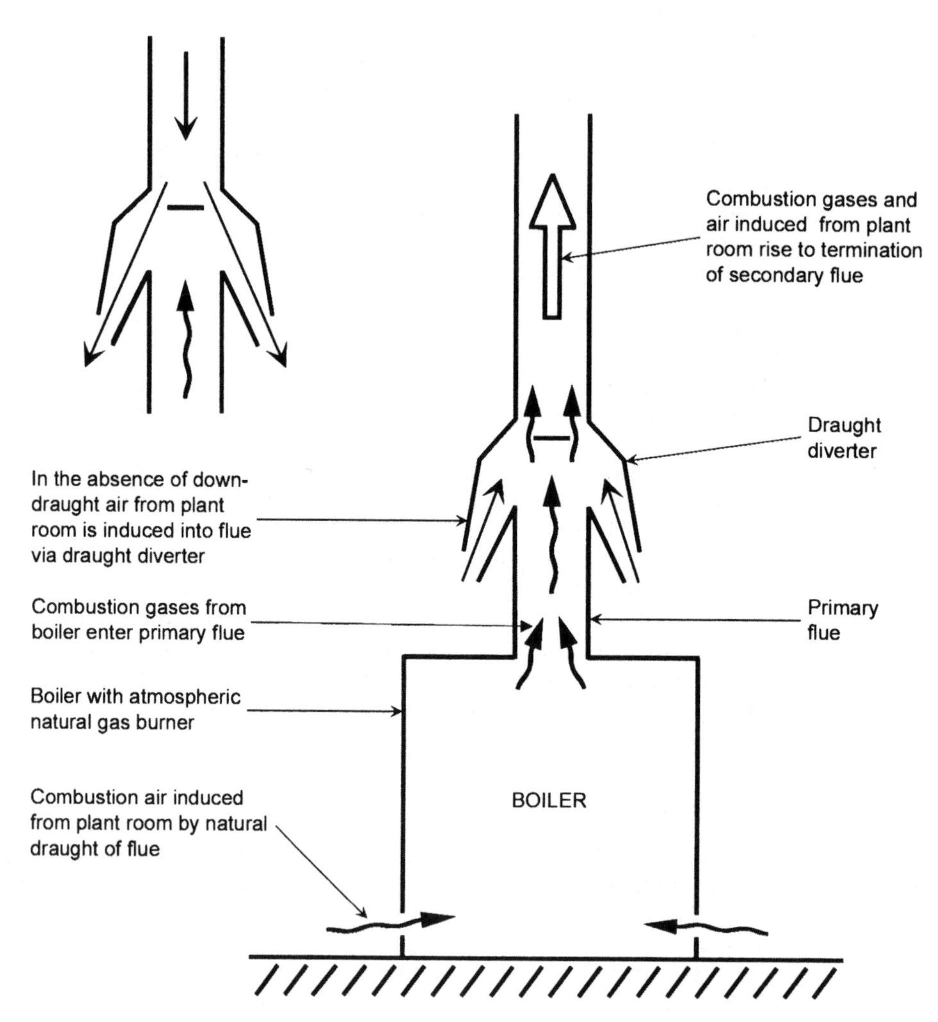

Figure 2.7 Principle of operation of a draught diverter.

within the boiler, and boiler performance will naturally suffer. Since it is so essential both to prevent down-draughts within the flue from reaching the boiler, and to regularise the natural draught, atmospheric gas boilers are fitted with *draught diverters*. While the detailed design of these devices varies from boiler to boiler, the principle of their operation is illustrated in Figure 2.7. This shows an arrangement in which the boiler is supplied complete with a short *primary flue* that terminates in the draught diverter itself. The shape of this is designed to spill any down-draught outside the primary flue before it can adversely influence the boiler. The draught diverter also provides a low-resistance path for air from the boiler room to bypass the combustion air path through the boiler and enter directly into the flue. Hence any tendency for draught to increase will result in an induction of boiler room air via the draught diverter in a corrective manner that will tend to stabilise draught conditions. However, this induction of boiler room air into the flue, and its resultant mixing with the combustion gases from the boiler, will inevitably lower the flue gas temperature and may result in condensation within the flue. Furthermore the air path into the flue via the draught diverter is permanently open, so that some induction of boiler room air will also occur under normal operating conditions. Mounting the draught diverter in a primary flue naturally adds to the height of the installation. Hence some atmospheric gas boilers provide the draught diverter function within the boiler *flue hood*. The latter is located above the boiler block to collect the spent combustion gases and channel them into the boiler flue. Protection against down-draught conditions may also be provided by a *down-draught thermostat* mounted within the flue. This is arranged to shut down the boiler on sensing a suitable low flue gas temperature that would give a clear indication of the existence of down-draught conditions. One manufacturer recommends a fixed set-point of 70°C.

Under correct firing conditions atmospheric gas burners typically exhibit a blue flame without yellowing of the flame tip. This normal flame presents a well-defined inner core and there is no 'lifting' of the flame away from the burner bars. Yellowing of the flame tip or body typically suggests a shortage of primary combustion air and incomplete combustion. This may be a result of blockages in the combustion air path or improper installation of the burner. Flames may lift away from the burner bars as a result of overfiring of the burner, due to excess gas pressure. Yellowing and lifting of the burner flames are both signs of improper burner operation that should be corrected without delay.

2.4.5 Fan-assisted and forced-draught natural gas burners

In *fan-assisted (or 'blown')* natural gas burners all the air required for combustion is supplied by a fan driven by an electric motor. The fan is sized to overcome the air-side pressure drop associated with the burner. Its use allows the designer of the burner to define both the manner in which combustion air is supplied (number, position and type of supply point) and the rate at which it is supplied. This allows a more thorough, uniform and closely controlled mixing of the gas and combustion air across the burner head than is possible in an atmospheric gas burner, and combustion performance is improved. Stable fan characteristics are essential for good combustion.

The general layout of a typical fan-assisted natural gas burner is shown in Figure 2.8. The motor of the combustion air fan may be located either within the combustion air stream itself or outside it. In the latter case a motor cooling fan will be required. Fan-assisted gas burners are noisier than atmospheric types, and the combustion air fan and its motor cooling fan (where applicable) are obvious sources of noise that contribute to this. An air intake silencer may be fitted to the burner, and in low-noise applications the burner assembly itself may be contained within an acoustic shroud.

Where the heat output or *firing rate* of the burner is staged (high/low/off) or modulated, the flow rates of both gas and combustion air to the burner are also staged or modulated according to the demands of the boiler control

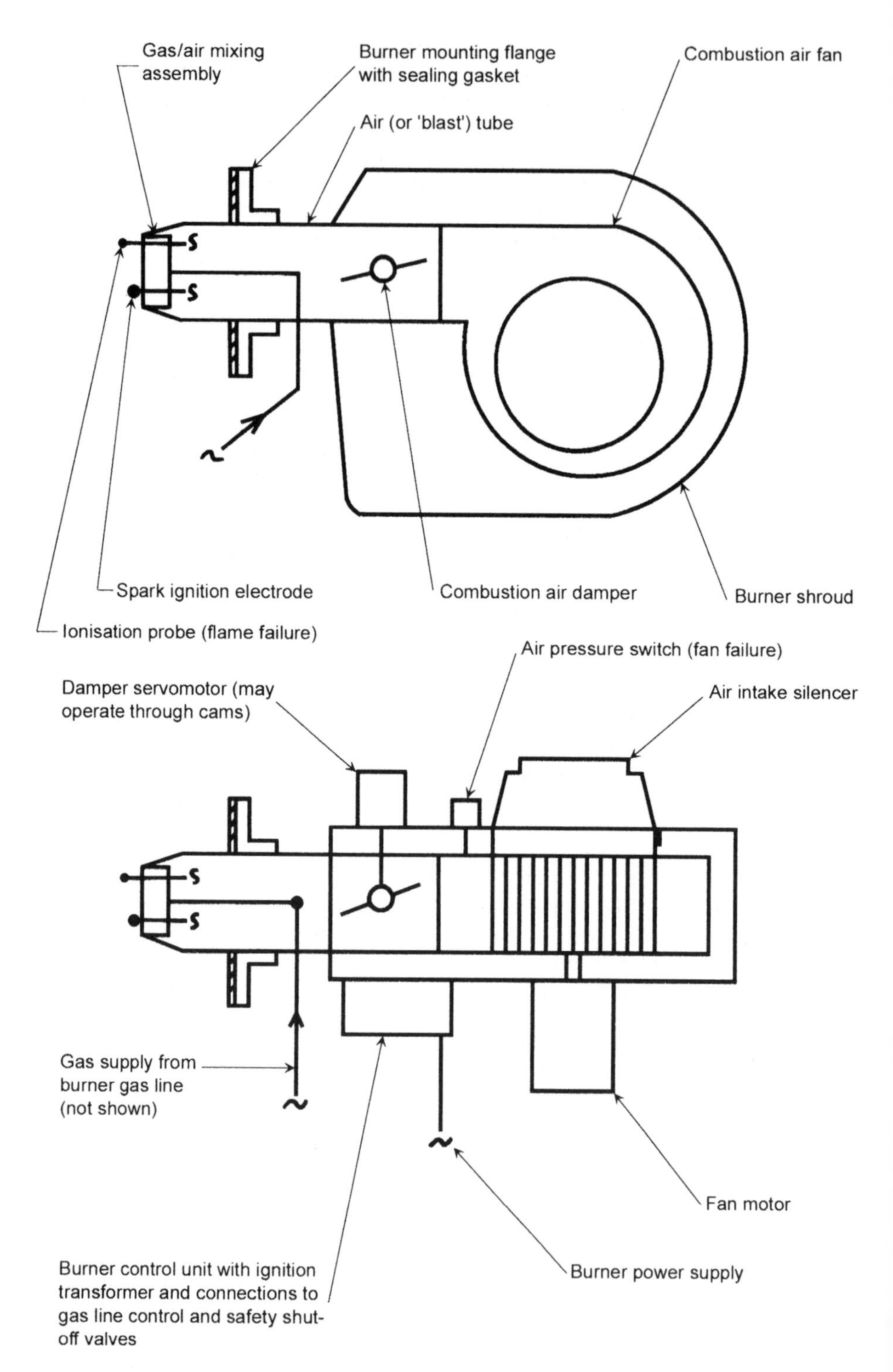

Figure 2.8 Configuration of a typical fan-assisted / forced-draught natural gas burner.

system for heat output. Burner ignition is by high-voltage spark generated by an internal transformer, which is rated for intermittent operation and fed either from the mains electrical supply to the boiler (typically single-phase) or from the burner's own dedicated power supply, as the case may be. As is the case for atmospheric gas burners, if combustion is not established within a critical safety period, or should the process fail for any reason once established, the gas supply to the burner must be overridden to full shut-off. Additionally, since the burner is completely dependent on the fan for its supply of combustion air, fan failure must also result in a complete shut-off of the gas supply to the burner.

The fan-assisted gas burner still depends on the natural draught developed by the boiler flue to draw the products of combustion from the combustion chamber through the boiler. In order to enable a proper combustion process to occur, the fan-assisted type of burner will typically expect to work with a pressure at the boiler flue connection that is between atmospheric (zero gauge) and a slight suction (i.e. a slightly negative gauge pressure). The minimum value of this suction is typically of the order of –0.3 mbar (–30 Pa) gauge, although the exact value may vary from boiler to boiler, and the manufacturer's recommendations for a particular boiler/burner combination should always be observed. Where the flue can generate excessive suction pressures at the boiler connection, or stable draught conditions cannot be reasonably assured, a draught stabilizer may be fitted to the flue. This acts to stabilize draught conditions by the induction of boiler room air into the flue, in a manner that is analogous to the operation of the draught diverter in an atmospheric gas boiler. Where used with fan-assisted or forced-draught gas burners, a draught stabiliser is, however, a component part of the flue, rather than of the boiler. Again any induction of cooler boiler room air into the flue will lower the flue gas temperature and may result in condensation occurring within the flue. The role of draught stabilisers is considered further in Chapter 5.

The burner provides the only path for air into the boiler. When the boiler control system no longer demands a heat output, the gas supply to the burner is shut off and the boiler block begins to cool down. Even though the combustion air fan is also shut down, there will still be a tendency for air from the boiler room to be drawn through the boiler by residual draught in the flue. This air readily picks up residual heat stored in the boiler block and carries it into the flue. The heated air develops a natural draught within the flue in the same way that the hot combustion gases do when the burner is firing, although naturally to a lesser extent. The air is thus drawn through the flue and dispersed to atmosphere, the heat it carries being lost from the boiler. This process will continue as long as heat can be picked up from the boiler block or sufficient draught is available from other boilers firing into a common flue system. While the relatively high resistance to air flow through the burner will typically mean that the flow generated by the available natural draught in the flue will be much less than the combustion air supplied by the fan, nevertheless heat lost to this air is potentially a major component of the heat loss from a non-firing boiler – the so-called 'standing' (heat) losses. If the fan-assisted burner is equipped with an air damper or valve that can be driven to full closure (shut-off) when the burner is not firing, air will not be drawn through the boiler, and this source of standing heat losses will be removed. A similar function can be performed by a *flue damper*, and this type of device is considered further in Chapter 5. For our present purposes it will

be useful to bear in mind that the heat losses resulting from allowing air to be drawn unchecked through a non-firing boiler may represent several percent of its annual fuel consumption. Preventing this air flow on the burner side of the boiler, rather than on the flue side, puts the closure device in a more benign operating environment. This potential source of boiler standing losses is naturally also active in the case of atmospheric gas boilers, and here it is evident that shut-off on the flue side of the boiler is the only option.

Fan assistance gives burner designers some degree of control over the physical size and shape of the combustion flame, and may allow boiler manufacturers to adopt more compact combustion chambers. In *forced-draught* natural gas burners the combustion air fan is sized to overcome not only the air-side pressure drop associated with the burner itself, but also the pressure drop associated with the flow of combustion gases through the boiler block. This means that the combustion chamber operates at a positive pressure, and the pressure at the boiler outlet connection to the flue may be 'balanced' (i.e. zero draught) or positive. In the latter case part of the flue will also be under positive pressure, and the implications of this are again considered in Chapter 5. Any positive pressure available at the boiler outlet connection may be used in the design of the flue. With standard burner types positive pressures between 25 and approximately 90 Pa may be achieved at the boiler outlet connection to the flue, depending on boiler output.

The design of modern forced-draught natural gas burners is increasingly concerned with minimising the production of NO_x. In this respect a typical aim is to achieve as thorough a mixing as possible of the gas and excess combustion air at the burner head, in order to avoid zones where high local combustion temperatures may exist which favour the production of NO_x. However, other techniques have also been proposed to achieve the desired low-NO_x operation. One such technique is the recirculation of some of the combustion gases through the burner. This may be achieved by *external recirculation*, employing a duct that links the flue gas manifold or collector at the rear of the boiler to the burner. This flue gas return duct is located outside the boiler block but within the casing, and a mechanical recirculation of flue gas is achieved using a fan that is mounted above the burner at the front of the boiler. The pressure drop associated with the flow of combustion gases through the 'passes' of the boiler block is increased by this form of recirculation. The recirculation of some of the combustion gases to the burner may also be achieved *internally within the combustion chamber*, via openings in the burner 'blast tube'. Both methods of combustion gas recirculation are illustrated schematically in Figure 2.9. Low-NO_x operation with forced-draught burners generally may be associated with an increase in combustion noise, and values of 8–12 dBA are sometimes quoted for the level of this increase. Burner acoustic shrouds and flue gas silencers may, of course, be applied in appropriate circumstances.

2.4.6 Premix natural gas burners

In a *premix* natural gas burner gas and excess air are mixed in the correct proportions *before* being delivered to the burner for combustion. 'Premixing' ensures that the excess air required for combustion is evenly distributed throughout the fuel/air mixture, and a more even combustion temperature is achieved. This will typically result in a low emission of NO_x, ($<70\,mg\,kWh^{-1}$ of natural gas delivered to site), and may achieve a very low or ultra-low

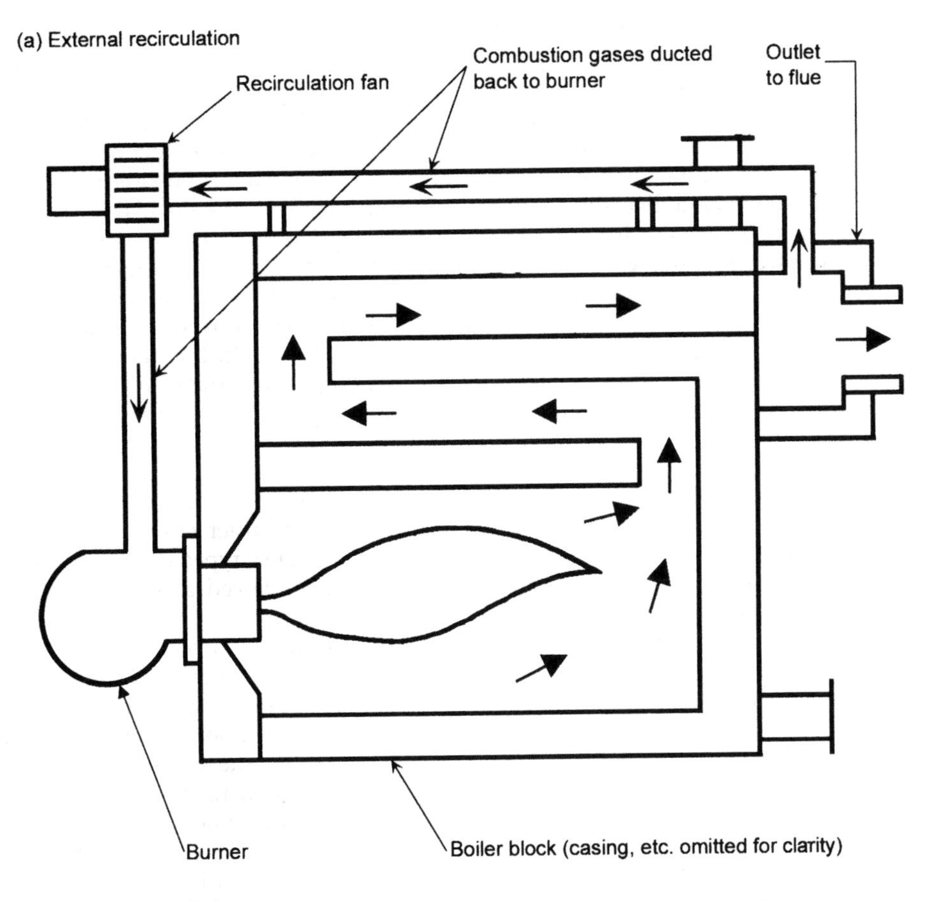

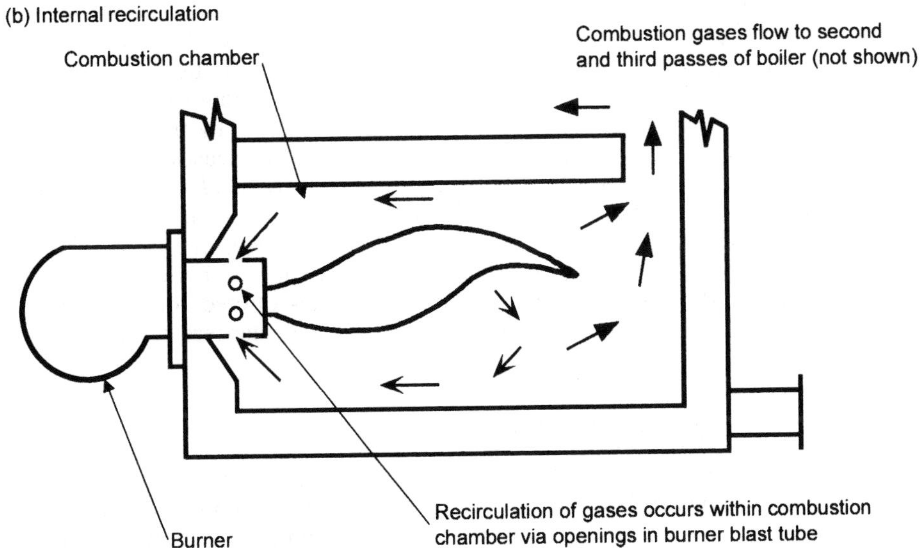

Figure 2.9 Recirculation of combustion gases to a forced-draught natural gas burner.

emission (down to <20 mg kWh^{-1}). Noise level is also reduced. Premix burners may be of fan-assisted or forced-draught type.

Premix burners take a variety of forms. In the so-called 'sectional' approach to boiler design a common combustion air fan supplies separate burners in each individual section of the boiler, with the gas supply to each being injected into the combustion air upstream of the burner. In the so-called 'modular' approach to boiler design premixing of the gas and excess air is achieved by injecting gas into the inlet of a fan (the *premix fan*) serving the premix burner in an individual boiler heat exchange module. This is illustrated in Figure 2.10, which shows the cylindrical heat exchanger of an individual module of a *modular boiler*. In this arrangement the premix burner typically takes the form of a perforated stainless steel tube or cylinder that is positioned concentrically within the heat exchanger. The gas/air mixture flows radially outwards through the perforations of the burner tube, where it is ignited. The gas/air mixture burns around the whole circumference of the burner tube, and the hot combustion gases flow radially across the extended surface of the cylindrical heat exchanger into the insulated boiler casing. Some form of distribution device is typically required within the burner tube to ensure an even distribution of the gas/air mixture along its length. Gas is injected into the inlet of the premix fan via a suitable distribution plate or orifice arrangement. The gas supply to the burner must be immediately and completely shut off in the event of failure of the premix fan. In some designs this may be achieved by feedback of the fan differential pressure to a suitable diaphragm-type gas valve. Such an arrangement can also allow the gas supply to be adjusted to maintain a constant air-to-fuel ratio in the face of variations in flue draught that might otherwise affect burner performance. In a similar way, sensing of a pressure drop associated with the volume flow of the gas/air mixture through the burner can allow shut-off of the gas supply in the event of a blockage occurring either in the flue or in the burner itself. This type of burner typically employs spark ignition, and flame failure (or failure to establish) is sensed by an ionisation probe. Prior to firing, the combustion chamber is typically 'purged' with air (only) from the premix fan, to remove any traces of gas/air mixture from the previous firing cycle that may linger within the chamber.

2.4.7 Other natural gas burners

In some countries *pulse combustion* technology is available for some commercial heating equipment. 'Pulse' combustion systems do not burn natural gas in a steady flame, but rather a gas/air mixture is ignited at a frequency of typically 60–80 Hz. It is claimed that improved heat transfer results from the oscillating flow, together with a reduction in NO_x emissions.

2.4.8 Burners for other gases

In principle any of the types of gas burners described above can be designed or configured to fire on gaseous fuels other than natural gas. In the UK the alternatives are effectively limited to Liquified Petroleum Gas (LPG) or manufactured gas (town gas). In general, however, use of either of these would only be considered where it is the only gas-firing option.

Since calorific value of the fuel will differ from that of natural gas, the rate of supply to the burner will vary for a given heat output to water from the

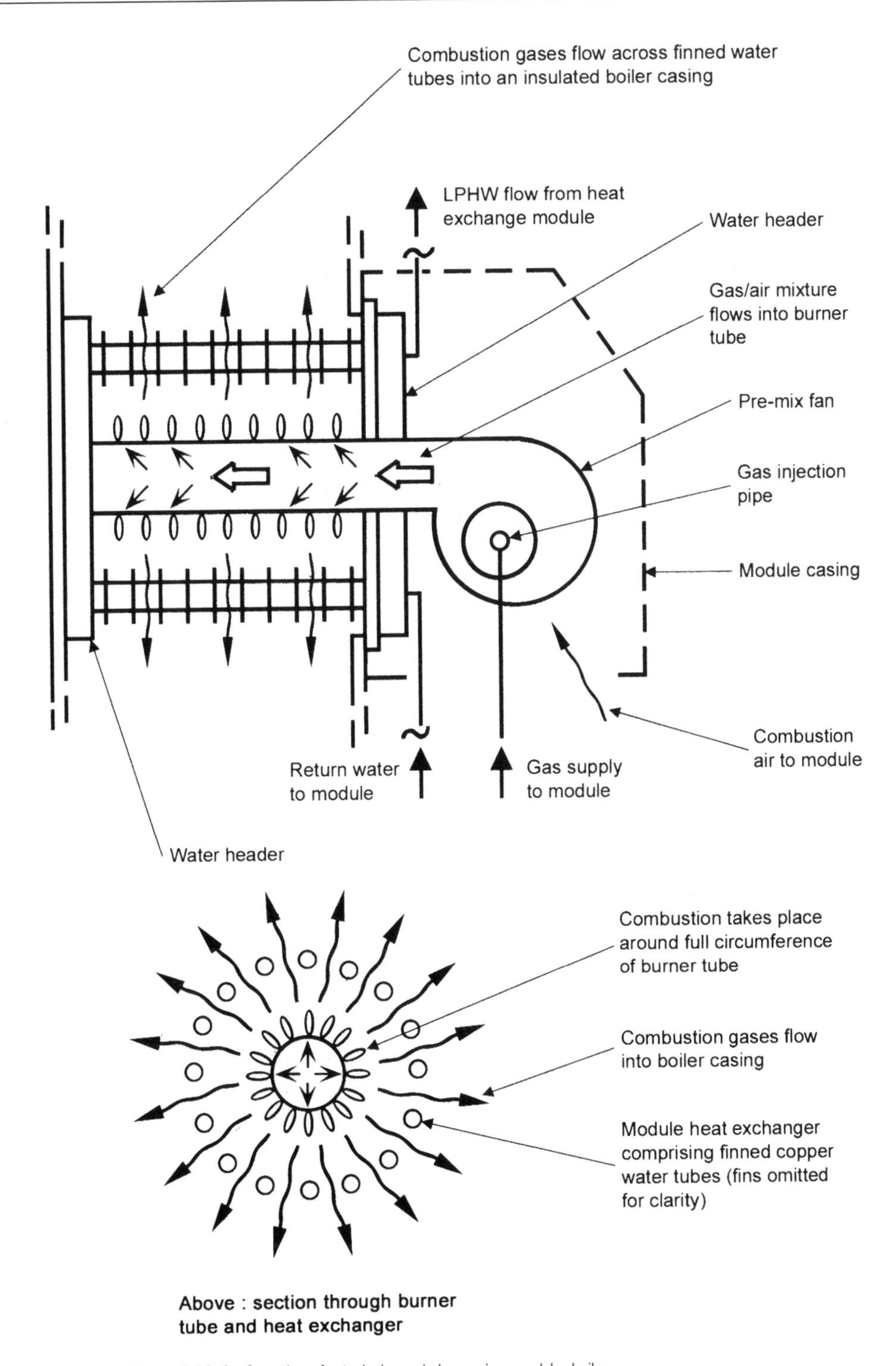

Figure 2.10 Configuration of a typical premix burner in a modular boiler.

boiler. Alternatively, boiler ratings may be effected for the same rate of gas supply to the burner. LPG has a substantially higher calorific value than natural gas. For propane and butane, the two principal types of commercial LPG, the respective values are approximately 2.5 and 3.17 times that for natural gas. Hence, for boilers of identical rated heat output to water and efficiency, the required flow rates of these gases will be approximately 40% and 32%, respectively, of that for natural gas. Similarly combustion air requirements will differ from those for natural gas. There are thus implications for the sizing of 'pilot' and main burner gas orifices, the available pressure of the gas supply, the sizing of combustion air fans in fan-assisted or forced-draught burners, and the availability of adequate flue draught with atmospheric gas burners. While permanent conversion from natural gas to LPG may be possible (and vice versa), mere *substitution* of one gas for another will *never* be acceptable. In any event, whenever it is proposed to fire a boiler on other than natural gas, the advice and recommendations of the boiler and burner manufacturers should always be sought.

2.4.9 Pressure-jet oil burners

The burners employed in commercial and industrial oil-fired boilers typically operate on the principle of atomization, to which there are two elements:

(1) A supply of fuel oil is delivered into the combustion chamber as a conical spray or 'mist' of finely-divided ('atomized') droplets
(2) A swirling motion is imparted to a suitable supply of combustion air, which is then directed into the cone of finely-divided oil droplets to achieve an intimate mixing of the oil and air prior to ignition of the mixture.

Pressure-jet burners are probably the type of atomizing oil burner most likely to be encountered in modern boiler plant serving building HVAC applications. They operate on the principle of mechanical atomization and are the simplest of the atomizing burners. The configuration and principal elements of a typical pressure-jet oil burner are shown in Figure 2.11. The burner essentially consists of an air tube or air director (also referred to as a 'blast tube') with a specially-designed nozzle mounted coaxially within it at the combustion chamber end. Oil is pumped through this nozzle at a relatively high pressure in the range 0.45–1.5 MPa (gauge). The nozzle is so designed that the oil forced through it is given a high rotational velocity. The shear forces associated with this imparted motion break the oil up, so that it leaves the nozzle as a conical spray or mist of finely-divided oil droplets. The apex of this conical spray is at the nozzle or 'atomizer'. Multiple nozzles may be employed and are claimed to improve atomization of the oil.

Combustion air is blown along the burner air tube by a low-pressure centrifugal fan and exits through ports that surround the oil nozzle. This achieves an intimate (turbulent) mixing with the atomized fuel leaving the nozzle, and the fuel/air mixture then enters the combustion chamber. As in the case of fan-assisted and forced-draught gas burners, it is essential for good combustion that the combustion air fan exhibits stable performance characteristics. A common electric motor typically drives both combustion air fan and oil pump, the latter via a suitable gearing arrangement. Ignition of the fuel/air mixture is achieved by a high-voltage spark generated by a pair of electrodes mounted adjacent to the nozzle. Once established, combustion is

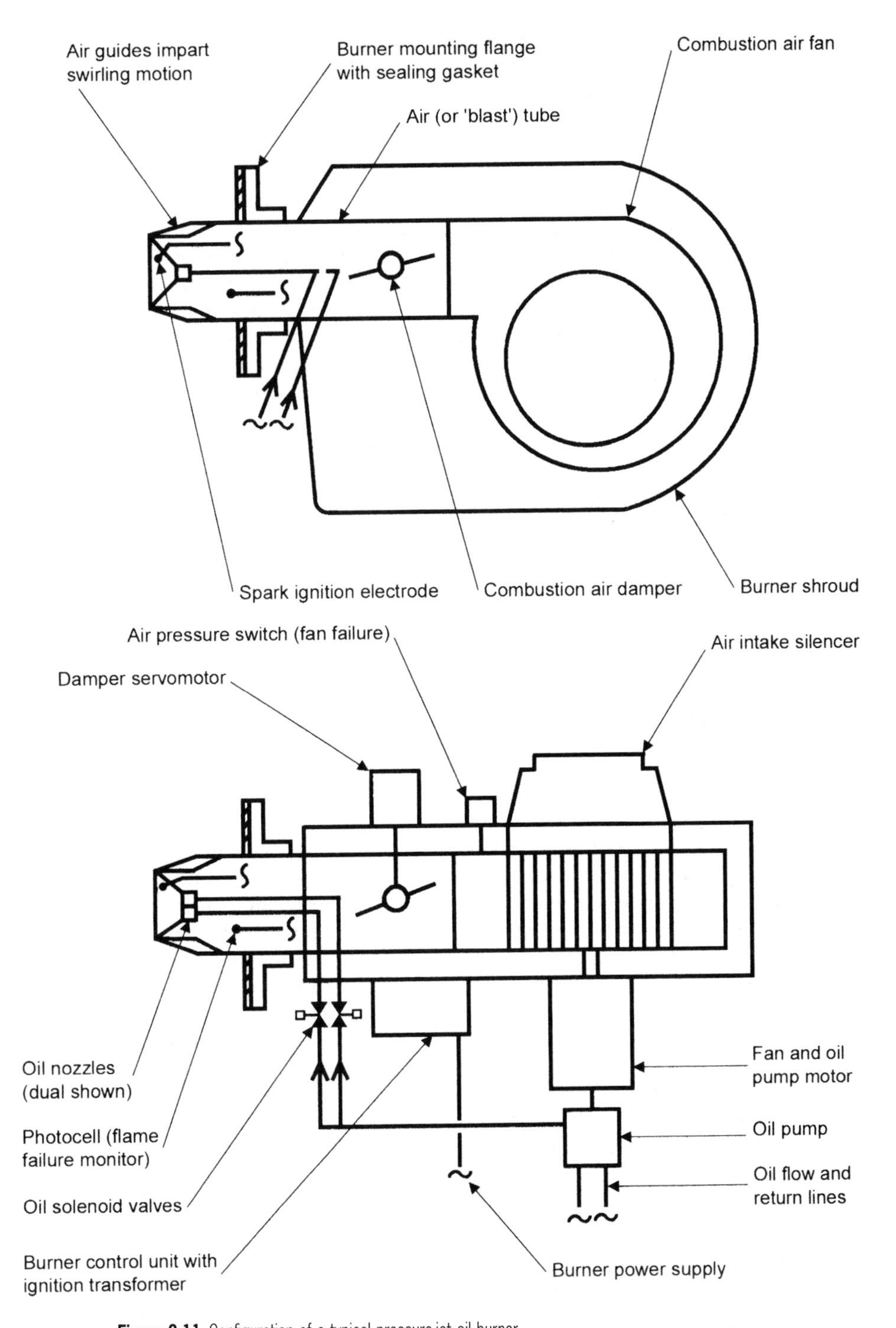

Figure 2.11 Configuration of a typical pressure-jet oil burner.

self-supporting and the electrical ignition system may be switched off. In the event that the burner fails to ignite, the flow of oil to the burner must of course be shut down. The design of the air tube and combustion air ports is vital in ensuring complete mixing of combustion air with the atomized oil. Due to the conical shape of the oil spray from the nozzle, it is also vital in providing the desired flame shape within the combustion chamber. Designs are available to suit various configurations of combustion chamber and heat transfer surface.

Pressure-jet burners are typically available in almost any size, burning between a few litres per hour, for nominal heat outputs to water of a few *tens of kilowatts*, to hundreds of litres per hour and *multi-megawatt* outputs. Burner heat output may be varied by changing the pressure (and hence flow rate) of the oil at the nozzle. The volume flow rate of combustion air is simultaneously varied to suit this by means of an air damper. The variation in burner output is limited by the possible variation in pressure available at the nozzle. For a simple pressure-jet burner a maximum turndown ratio in the range 1.6–2:1 is typically available over the range of nozzle pressures between a minimum of 450 kPa and a maximum of 1500–2000 kPa.

Where fuel oils are used that exhibit high viscosity at normal storage temperatures, the oil must be heated to attain a temperature at which it is practical to achieve atomization. Pressure jet burners designed to fire on these grades of 'medium' and 'heavy' fuel oil may be fitted with an oil pre-heater and/or a heated nozzle head, with burner operation enabled thermostatically on achieving the correct oil temperature. Evenly-heated oil assists the establishment of combustion. Overheating is to be avoided, since this may lead to carbonisation of the oil.

As in the case of fan-assisted and forced-draught natural gas burners, full closure of the combustion air damper will again remove a potentially significant source of boiler 'standing' heat losses. Optimum burner and combustion performance again requires a relatively constant draught to be maintained at the boiler connection to the flue. This may be a balanced condition or slight suction in natural-draught applications, or a positive pressure where the burner is designed for forced-draught application. Since oil has a potential for fouling and clogging that is largely absent from the gaseous fuels, optimum performance will typically only be achieved if all components of the burner are maintained in a clean condition.

Atomization of the oil and the turbulent oil/air mixing process are an obvious source of burner noise, as is the combustion air fan itself. While this may be reduced by careful attention to the aerodynamic design of the components, and by full static and dynamic balancing of all rotating components, most burner manufacturers offer models that either incorporate acoustic absorbers/materials in critical parts of the burner or are fitted with acoustically-treated 'shrouds' for noise-sensitive applications.

2.4.10 Other atomizing oil burners

Other types of atomizing oil burner do exist. The list, which does not claim to be exhaustive, includes the following:

- Rotary or rotary cup burners
- Twin-fluid oil burners
- Return-flow pressure jet or pressure atomizing burners
- Mechanical atomizing oil burners
- Return-flow mechanical atomizing oil burners.

Since all of these burner types are more likely to be encountered in industrial or marine applications, or in 'heavy' electrical power generation, than in the typical building services applications with which this book is concerned, none of them will be considered further here.

2.4.11 Dual-fuel burners

If the same burner may fire alternatively on either natural gas or fuel oil, it is termed a *dual-fuel burner*. Dual-fuel burners are of fan-assisted or forced-draught type and must incorporate the features of both natural gas and oil burners necessary for the safe and efficient combustion of either fuel. The burner mechanisms and their controls must be able to achieve the correct fuel/air ratios for firing with either fuel and at any firing rate that the burner is capable of. This naturally increases both the complexity and cost of dual-fuel burners in comparison with single-fuel types. Changeover between fuels may be effected manually, or automatically on failure of the fuel supply that the burner is firing on at the time. As an alternative to a dual-fuel burner, dual-fuel *operation* may also be achieved by a physical change of burner.

Dual-fuel burners are only likely to be preferred to a single-fuel (typically natural gas) burner where:

- The available supply of natural gas is unreliable
- Continuous availability of gas-fired boiler plant is critical
- There is an economic case for taking a supply of natural gas on an 'interruptible' tariff.

Under present conditions of fuel availability and pricing, the dual-fuel burner option will typically be seen as an alternative to a natural gas burner, rather than to an oil burner. Interruptible natural gas tariffs are only likely to be relevant to users with long-term large steady demands at individual sites. Oil storage in dual-fuel installations may be sized on the basis of the likely frequency and duration of interruptions in the natural gas supply, rather than on sole-fuel patterns of consumption. In doing so there are clear advantages in minimising the capital cost and space requirements of such an installation, and of the cost tied up in the stored 'standby' oil itself. As far as building services applications in the UK's climate are concerned, the continuous availability of boiler plant may be critical in facilities caring for the very young, the elderly and the sick or physically frail, or where the population is 'tied' to location, as in the case of various custodial or hostel-type facilities. However, since neither gas nor oil firing will work without an *electrical power supply*, the ability to utilise alternative fuel supplies will not, in itself, provide a guarantee of security of heat generation. This would require, in addition, that the electrical power supplies to the boiler and burner are supported by a standby generator.

2.5 Burner operation and control

2.5.1 Functions of burner control

Modern gas and oil burners are varied and complex assemblies of mechanical, electrical and electronic components, and it is beyond the scope of the present work to consider the detailed mechanisms by which these complex assemblies are controlled. However, the functions of burner control may be split into two distinct categories:

- Control of burner heat output or 'firing rate'
- Control of burner safety.

Control of burner output or firing rate is concerned with ensuring that the boiler can respond to, and meet, demands made on it for heat output to the water of the heating system to which it is connected. The search for improved energy efficiency and reduced operating costs, coupled latterly to a realisation of the potential consequences of global warming, increasingly mean that it is no longer sufficient simply to meet the demands for heat output from the boiler, but to meet them *with optimum efficiency*. Control of burner safety is primarily concerned with preventing the continued supply of fuel in the absence, for whatever reason, of a sustained combustion process.

The operation and control of modern commercial gas and oil burners is typically automatic, although manual intervention is commonly required following the activation or 'tripping' of some safety control functions. Atmospheric gas burners are typically supplied as part of an integrated boiler/burner package that includes controls. Fan-assisted and forced-draught gas and atomizing oil burners typically have their own dedicated control systems which interface with the control systems of the boilers to which they are fitted. In Chapter 10 it will be seen that the latter are themselves interfaced with the heating system controls.

2.5.2 Modes of control for burner output

Three main modes are typically available for the control of burner heat output in fan-assisted and forced-draught natural gas burners. These are:

- on/off
- high/low/off
- modulating.

The same modes of control are typically available for atomizing oil burners. In the case of atmospheric gas burners control is typically on/off, although boiler/burner packages that offer high/low/off, high/medium/low/off and modulating control are available from some manufacturers. The provision of 'staged' or modulating control of atmospheric gas burners typically requires some measure of control to be exercised over the level of induction of combustion air under reduced-rate firing. In modular boilers control of premix natural gas burners is typically on/off. In other configurations it may be high/low/off or modulating.

2.5.3 On/off control of burner output

The on/off mode of control is the simplest form of burner control. The burner is either firing at its maximum heat output or it is not firing at all. On/off control of natural gas burners is achieved via the solenoid valves of the main gas line (Figure 2.12a). These are two-position devices that are normally held in the closed (gas shut-off) position for safety reasons. Powering them open releases the gas supply to the main burner for ignition. Valve operation is 'fail-safe', since failure or interruption of the power supply causes the valve to return to the gas shut-off position. Modern practice is to provide dual control and safety shut-off solenoid valves, one valve opening slowly and closing rapidly while the other both opens and closes rapidly. Starting of the

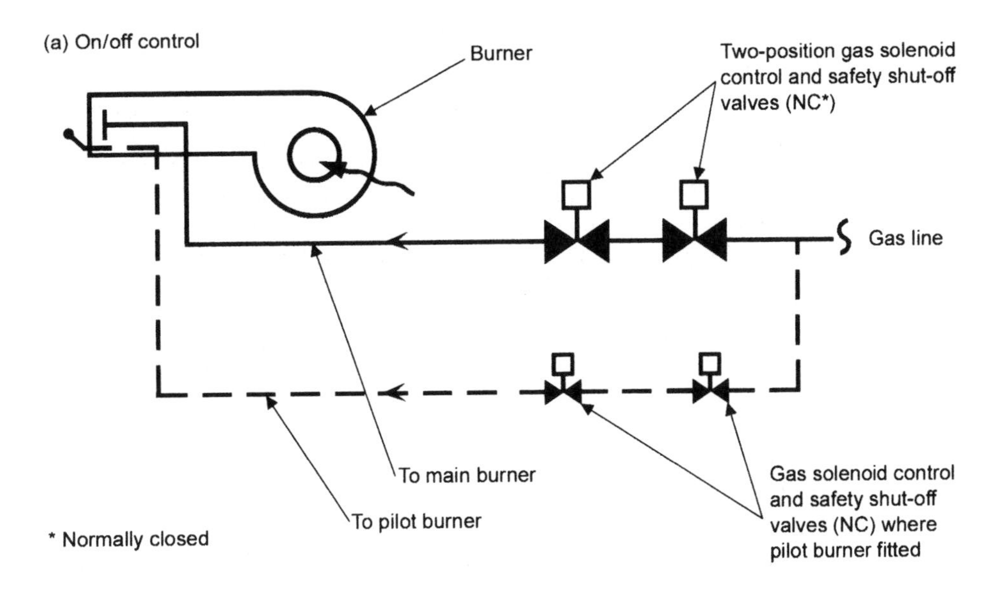

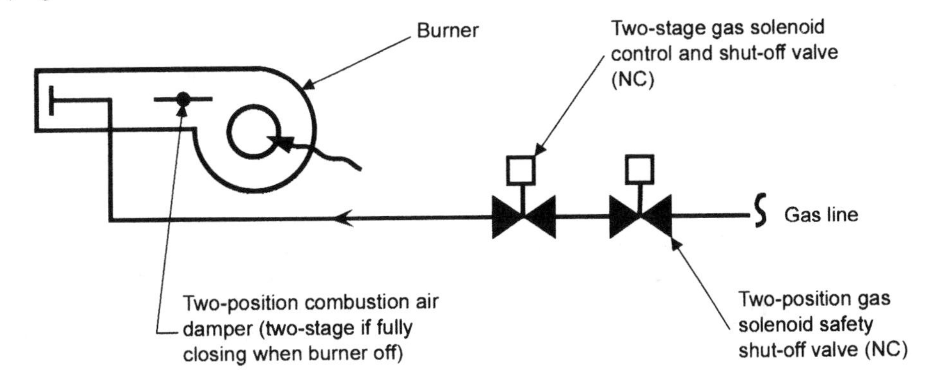

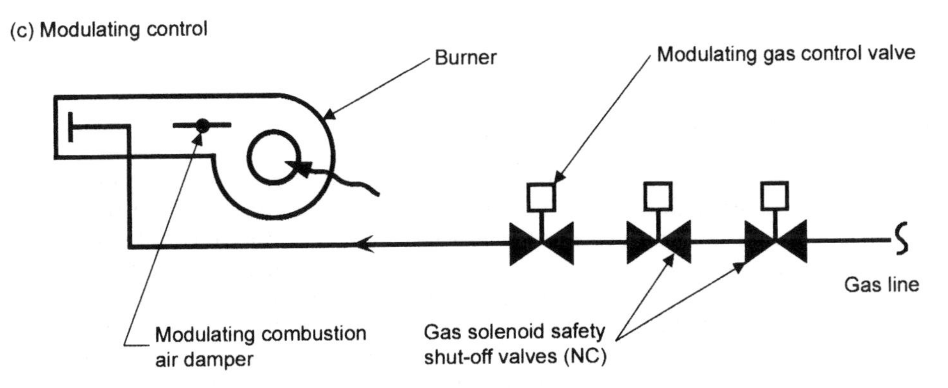

Figure 2.12 'On/off', 'high/low/off' and modulating control of burner output : natural gas burners.

combustion air fan will similarly form part of the control sequence for fan-assisted and forced-draught natural gas burners and atomizing oil burners. Flue dampers, where employed, must be opened prior to firing the burner.

On/off control of atomizing oil burners is again achieved via a solenoid valve, but in this case the device is an integral part of the burner assembly itself, and not part of the oil line (Figure 2.13a).

2.5.4 High/low/off control of burner output

In the high/low/off mode of control the burner has two firing rates, 'high' and 'low'. The high firing rate naturally provides the maximum heat output from the burner, while the low firing rate provides a heat output that is intermediate between the burner maximum and zero. The high/low/off mode of control is also termed *two-stage control*. Since a complete burner firing cycle would progress from 'off' through low firing to high firing, and back again through low firing to 'off', the mode of control should perhaps, strictly-speaking, be referred to as low/high/low/off. It is frequently assumed that the burner low firing rate will result in a rated heat output to water from the boiler that is approximately 50% of the maximum (design) rating with the burner at its high firing rate. However, this is not necessarily so, although values do typically lie in the range 40–60% of maximum, depending both on the particular type of boiler and the burner design.

For a single-boiler installation within the UK generally a low firing rate in the range 50–60% of full output is likely to satisfy the demands of the heating system to which it is connected for a large part of the heating season annually. In the presence of casual gains contributing 2–3 K of heating, and with an inevitable degree of boiler oversize, a 'low-fire' rating of 40–50% will typically be acceptable. In this respect it should be borne in mind that 2–3 K of casual heat gain will typically represent c.8–15% of the design heat load on most sedentary spaces in the UK, while 40% of a boiler capacity that is oversized to 120% of the design heating load is equivalent to *48% of that load*. For multiple-boiler installations the same consideration naturally does not apply, it being rather a question of ensuring that an adequate number of steps in boiler output are available to match the demand of the heating system as closely as possible for as much of the annual heating season as is practical. In this respect a burner 'low-fire' rating that delivers approximately 50% of the maximum design heat output probably represents the optimum.

At least one burner manufacturer offers a variant of high/low/off control that provides a 'sliding', rather than stepped, change between the high and low burner firing rates. In principle this is in fact intermediate between high/low/off and modulating control, the change in firing rate being slower-acting than the former and faster-acting than the latter. Rapid changes in the rate of fuel supply (whether gas or oil) are avoided. High/medium/low/off or *three-stage control* is not common for gas or oil burners. The one example of this form of control of which the authors are aware provides intermediate or part-load stages of burner output at 20% and 40% of the maximum firing rate.

High/low/off control of natural gas burners may be achieved by employing a two-stage solenoid valve to provide control and safety shut-off of the natural gas supply to the main burner (Figure 2.12b). From its normally

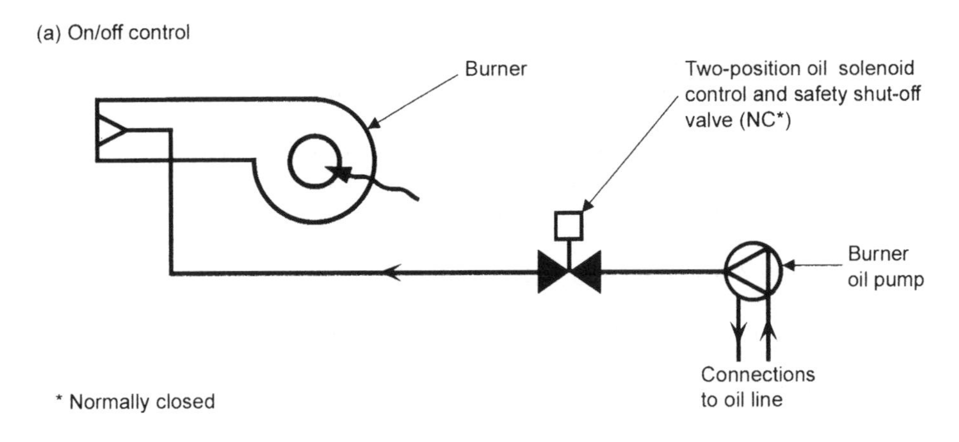

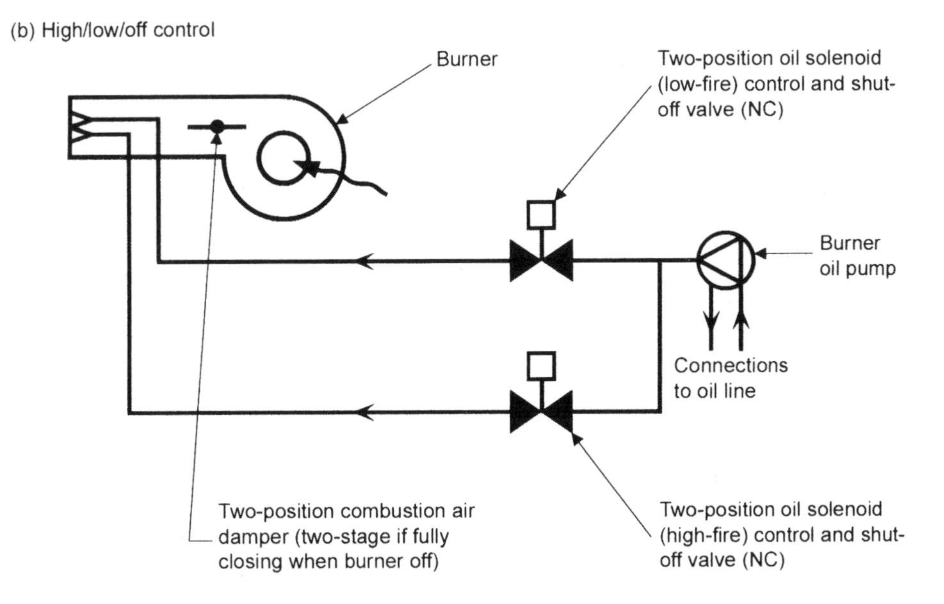

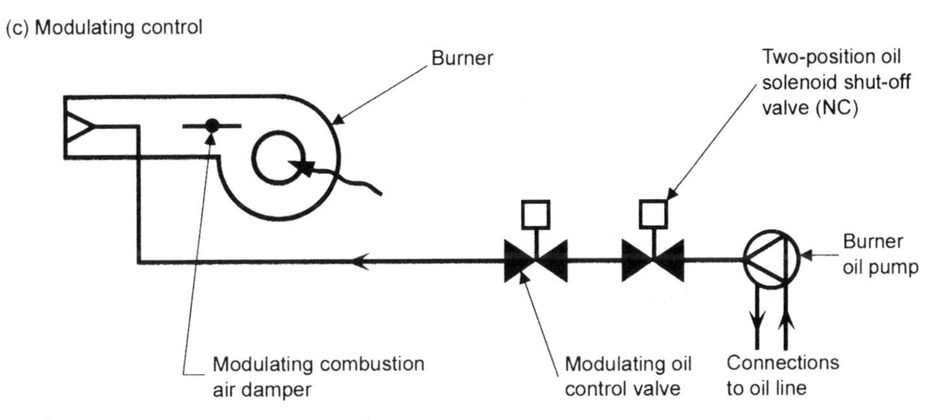

Figure 2.13 'On/off', 'high/low/off' and modulating control of burner output: oil burners.

closed gas shut-off position the first stage of valve operation releases the gas supply rate necessary for firing of the burner at its 'low' rating. The second stage of operation takes the valve to the fully-open position, releasing the gas supply rate necessary for firing of the burner at its maximum rating. Operation of the valve is again fail-safe to the closed position. An alternative approach divorces the gas control and safety shut-off functions. The former is handled by a gas control valve driven by a slow-running servomotor, while the latter is dealt with by dual solenoid safety shut-off valves.

The two stages of gas valve operation must be accompanied by a corresponding change in the supply rate of combustion air, so that the correct fuel/air ratio for stable and efficient combustion is maintained. In fan-assisted and forced-draught natural gas burners the change is achieved through repositioning of the combustion air damper (or air valve) that is part of the burner assembly itself. If the damper is not fully closed when the burner is not firing, a two-position type will suffice to provide the minimum-open and fully-open positions for the burner's low and high firing rates respectively. Where full closure of the combustion air damper is required when the burner is not firing (and we have seen earlier that this is desirable to reduce boiler standing heat losses), two-stage damper operation will be necessary. Two-stage natural gas solenoid control valves and combustion air dampers must be capable of accurate repetitive positioning, so that the correct fuel/air ratio will always be achieved during operation of the burner at its low firing rate. Nevertheless suitable interlocks should be provided with the two-stage gas valve to prevent the occurrence of excessively rich gas/air ratios. If a servomotor-driven gas control valve is employed in lieu of a two-stage solenoid control and shut-off type, the same servomotor may be used to reposition the combustion air damper or air valve, typically via a rotating cam arrangement. Where atmospheric natural gas burners offer high/low/off firing, a (two-position) damper arrangement may be applied to the combustion air path that typically enters at the base of the combustion chamber below the burner bars. This will inherently require a damper that is physically much larger than typically found in fan-assisted or forced-draught gas burners, and one that is located in a harsh operating environment. However, the damper may perform additional functions in reducing the flow of downward radiant heat from the combustion zone and burner bars, and in allowing pre-heating of the combustion air. Positioning of the two-stage solenoid gas control valve and combustion air damper must be accurately linked. An alternative approach employs a two-position flue damper in lieu of a low-level combustion air damper. (Two-stage operation will again be required if the flue damper is fully closing when the burner is not firing.)

Under high/low/off control, firing of the burner is typically initiated at its low firing rate. In multiple-boiler installations it is usual for the boilers to be controlled in sequence according to the load on the heating system. The ways in which this sequencing can be achieved, and how the measure of the load on the heating system may be derived, are considered in Chapter 10. For the purposes of the present discussion it is sufficient to acknowledge that sequence control is the norm for multiple boilers, and that the typical progression of such a sequence is for the burner in each individual boiler to be brought through its low firing rate to sustained operation at its full output before enabling the next burner in the sequence to fire at its low firing rate, should a continued demand from the heating system require it. However, there appears to be no reason why, if a particular boiler/burner combination offers an efficiency advantage at its low firing rate, the firing sequence may not bring all burners on at their low firing rate before bringing any unit up to full output. Such an approach is, it must be said, atypical.

In the case of atomizing oil burners a two-stage oil solenoid control and shut-off valve may be employed for high/low/off control. Alternatively a two-stage burner head with dual nozzles may be employed (Figure 2.13b). Each nozzle is fed with oil by its own solenoid valve. For operation at low firing rate a single nozzle is used, while for operation at full output both are used. The combustion air damper must again be repositioned at the change in firing rate.

2.5.5 Modulating control of burner output

Modulating control provides a burner heat output that is continuously and steplessly adjustable between its maximum (design) rating and some lower cut-off value. The latter represents the minimum limit of burner modulation, and the minimum firing rate of the burner. Any further reduction in boiler heat output to water can thereafter only be achieved by cycling the burner off/on. For fan-assisted and forced-draught natural gas and atomizing oil burners the lower limit of burner modulation may be as high as 50% of the design firing rate, although it typically lies in the range from one-quarter to one-third, with modern electronic controls tending to improve on (i.e. reduce) this.

Modulating control of fan-assisted and forced-draught natural gas burners is achieved by employing a fully-metering gas valve that may be continuously repositioned between the minimum-firing-rate position and fully open to suit the demand for heat output (Figure 2.12c). In order that the correct gas/air ratio for stable and efficient combustion is maintained, the change in the supply rate of natural gas must be matched precisely to the flow rate ~ opening characteristic of the combustion air damper or air valve, which must also be fully modulating. It will be important to minimise hysteresis in the positioning mechanisms of both the gas control valve and the combustion air damper.

Modulation of the gas control valve may be under *feedback* or *positioning (parallel) control*. The former employs a gas/air ratio controller that receives feedback of air pressure (or air pressure drop) due to the combustion air flow at some point in the burner air tube. The latter employs a predetermined positioning relationship between the gas control valve and the combustion air damper. The burner control system is typically interlocked to ensure that firing can only be initiated at the minimum firing rate (typically 25–33% of the design firing rate). The way in which modulating control of burners may be integrated into the sequence control of multiple boiler installations is considered in Chapter 10.

Where, exceptionally, modulating control of an atmospheric natural gas burner is provided, similar considerations will apply regarding positioning of the combustion air damper and its environment as noted for high/low/off control. The damper must now of course be fully modulating. The gas valve and combustion air damper will typically be modulated in parallel by means of a mechanical linkage. In some designs the combustion air is split between a primary and a secondary supply. The former is always available to meet the needs of the burner at its minimum firing rate, while the latter is adjusted to suit its needs in the modulating range of firing.

Modulating control of pressure jet oil burners is subject to similar considerations as those applicable to fan-assisted and forced-draught natural gas burners. Naturally in this case variation in the flow rate of oil to the burner nozzle is achieved via a modulating oil control valve (Figure 2.13c).

2.5.6 Control of burner safety

The principal control functions that are intended to ensure burner safety are summarised as follows:

- Fuel shut-off (gas or oil) on failure to detect a burner flame
- Gas shut-off on low supply pressure
- Gas shut-off on high supply pressure
- Oil shut-off on low pressure in the burner
- Fuel shut-off on failure of the combustion air fan
- Fuel shut-off on failure of burner electrical supply
- Proving of solenoid gas shut-off valves
- Interlocks to ensure that burner firing is initiated only at the low-fire or minimum firing rate for high/low/off or modulating burner control respectively.

Not all of these functions will be appropriate to all types of burner, and manufacturers may offer additional functions or interlocks as options to the specifier. For specific and specialist types of burner additional safety functions may be appropriate.

Fuel shut-off on failure to detect a burner flame is the most fundamental burner safety control function, and is universally provided. In natural gas burners this is typically achieved by monitoring the burner flame with an *ionisation probe*, although some simple atmospheric gas burners may still employ a *thermocouple* for flame monitoring. Ionisation probes utilise the local ionisation produced by the high temperature of the flame, which provides a conductive path for an electric current to flow between suitable electrodes. A thermocouple relies on the effect that an electric current is generated when two junctions of two dissimilar metals are held at different temperatures. In the case of a device for flame monitoring in a burner, one of the junctions, the *hot junction*, is subject to the heating effect of the combustion flame. Natural gas flames (which are bluish, rather than yellow) do not generate sufficient luminosity for *photocells* to be employed for flame detection and monitoring. Where atmospheric gas burners employ a permanent pilot flame for ignition of the main burner, both the pilot and main burners must be covered by flame monitoring, whether this is achieved by a common detection and monitoring device or independently for each burner. In the case of oil burners the burner flame is typically monitored by an optical sensor of photocell type. All of the techniques for flame monitoring share the common attribute of rapid response to a flame failure condition. Furthermore, they all generate an electric current *in the presence of a flame*, and must initiate shut-off of the burner fuel supply in its absence. Since such a condition inevitably exists at the start of the burner firing sequence, activation of the flame monitoring system must be inhibited for a period that is sufficient to allow ignition of the fuel/air mixture to take place and a burner flame to be established. Otherwise nuisance 'lock-outs' of boiler operation will occur.

Gas shut-off on low or high pressure may be achieved via pressure switches installed in the burner gas line. Of these two functions shut-off on low pressure is that most commonly provided by the burner manufacturer, with shut-off on high pressure a specifier option. *Proving* of solenoid gas shut-off valves as part of the burner firing sequence is a recommendation of some European Standards. This may be achieved by connecting the solenoid valves to a small diaphragm pump and pressure detector that are installed in the gas line (Figure 2.14). If the proving indicates leakage from either of the solenoid valves, the burner firing sequence is inhibited. The diaphragm pump must have its own solenoid valve, located upstream of the pump, to isolate it from the gas line on completion of valve proving.

In the case of fan-assisted/forced-draught natural gas burners or atomizing oil burners, the combustion chamber and flueways of the boiler block are typically swept free or 'purged' of any traces of unburnt gas or atomized oil by the combustion air fan prior to the burner firing. A similar purging process is typically carried out for premix natural gas burners.

Whatever the nature of the mechanical, electrical or electronic components involved, implementation of these safety functions typically relies on relay techniques that 'make' or 'break' an electrical supply. (It will typically be a case of breaking the supply, since an unrelated external failure of electrical power that cuts off the supply in question will then have the same effect as an activation of the safety function, which is a fail-safe feature.) Once activated or 'tripped', burner safety control functions may require manual reset of the tripping mechanism to allow the automatic sequence of burner operation either to continue from the point at which it was halted by the 'trip' or to 'initialise' and recommence from its initial starting point. In such circumstances the burner is typically described as being 'locked out', since it has effectively been locked out of action. Failure to detect a flame during burner ignition is an example of a safety function that will almost certainly require the tripping mechanism to be reset manually.

2.6 The burner gas line

The burner *gas line or gas 'train'* is connected to a supply of natural gas that is piped from the supply utility's gas meter. The burner fed by the gas line may be the only load served by this supply, or it may be one of several burner and gas line assemblies connected in parallel. We have seen in the previous section that the gas line is intimately involved in the control of gas burners. It is here that various components are installed that, while external to the burner itself, are nevertheless critical in ensuring, firstly, that it receives the correct flow rate of gas at the correct pressure whenever it is required to fire and, secondly, that the supply of gas is safely shut off either when its heat output is no longer needed or its continued firing cannot be allowed for safety reasons. In carrying out these functions, the components of the gas line typically provide for the following:

- Manual isolation of the burner and gas line itself
- Filtration of the gas supply
- Regulation of gas pressure to the main and (where fitted) pilot burners
- Gas control and safety shut-off for the main and (where fitted) pilot burners
- Gas pressure indication (optional)
- Shut-off of the gas supply on low-pressure
- Shut-off of the gas supply on high-pressure (typically an option)
- Manual measurement of gas pressure.

The gas line for a typical atmospheric natural gas burner is shown schematically in Figure 2.14. Gas filtration may be incorporated within the gas

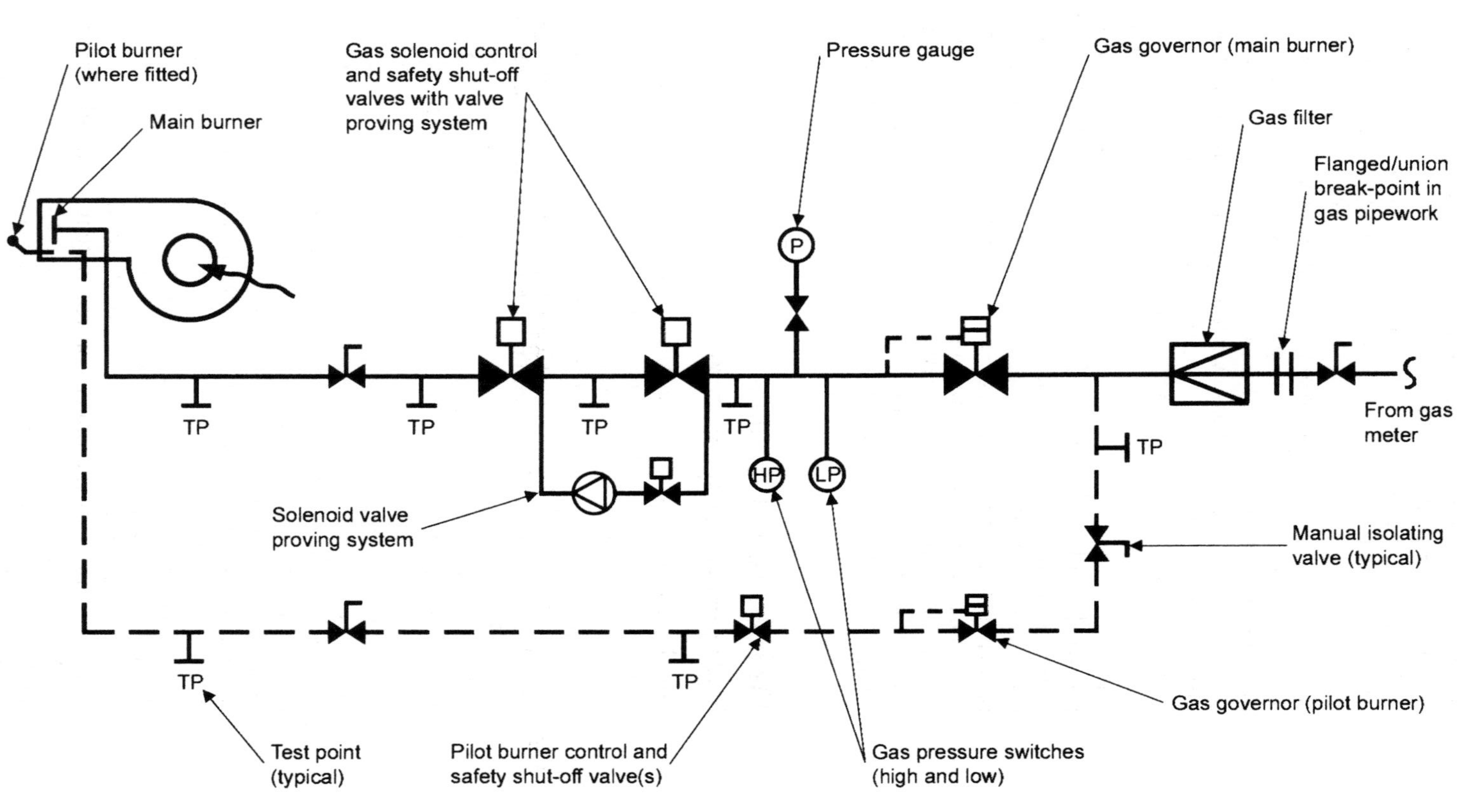

Figure 2.14 Schematic of the gas line for a typical natural gas burner.

solenoid valves themselves or may be provided as a separate dedicated filter within or upstream of the gas line itself.

As may be expected, there is some progression of complexity between the gas lines of atmospheric gas burners and those of fan-assisted or forced-draught types, and between burners in which the control of heat output is simply on/off and those in which staged or modulating control is employed. The UK boiler market is a competitive one that is served by a diverse range of products. The extent of the gas line components provided, and the form in which they are provided (whether pre-assembled or 'loose' for on-site installation, for example) may therefore differ between the packages provided by the various boiler and burner manufacturers.

Whatever its degree of complexity, suitable joints must be provided in the gas line that will allow both disconnection of the upstream gas supply pipework and withdrawal of the burner assembly. In particular, where fan-assisted or forced-draught burners with hinged access are employed, disconnection joints in the gas line must allow for the burner assembly to be swung clear of the boiler for access and inspection. Since the gas line contains sensitive control components, and the connections between these must not be overstressed (at the risk of subsequent leakage), careful attention should be paid to the support of the gas line, and to ensuring that all components remain fully supported even when the disconnection joints are broken.

2.7 The boiler control system

2.7.1 Boiler controls and system controls

For the purposes of this chapter we will draw a distinction between *boiler controls* and *heating system controls*. Boiler controls are an integral part of each boiler, while in commercial and industrial applications heating system controls are typically located in a control panel that is physically separate from the boiler. Heating system controls, and their relationship to the control of individual and multiple boilers, are considered in Chapter 10. Ultimately, however, it is one or more burners that are being controlled.

Typically the operation and control of modern boiler plant is fully automatic. The interface between operating personnel and the boiler controls is provided by a control and instrument panel that is mounted in or on the boiler casing. The extent and complexity of this panel varies according to the type, model and size of the boiler and the manufacturer's technical and commercial practice. While it may still make use of traditional 'knobs, gauges and lights', it is increasingly likely to employ a microprocessor-controlled, menu-driven liquid crystal display (LCD). Besides giving operator access to the boiler control functions, this panel typically also provides a visual display of key operating parameters. Using conventional gauges, the display of such parameters has typically been restricted to what can be presented on one or two dial-type gauges, namely boiler flow and return water temperatures and possibly operating pressure or 'head' ('altitude' gauge).

The functions of the boiler control system are divided between *control of the burner* and boiler *safety or limit control*. Control of the burner may be further sub-divided between single-boiler and multiple-boiler installations. Modern boiler control systems will typically also provide what might be termed *reporting functions* by the provision of suitable output signals to allow remote monitoring of various boiler operating status and fault conditions.

2.7.2 Control of the burner

The boiler achieves on/off control of the burner via a control thermostat located in the flow water connection from the boiler. In the simplest boilers with atmospheric gas burners and permanent-pilot ignition this thermostat may directly control the gas solenoid valve(s) in the gas line to the burner. In more sophisticated atmospheric gas boiler/burner packages, and where fan-assisted/forced-draught gas and atomizing oil burners are employed, the control thermostat initiates the burner's automatic firing sequence. The control thermostat may actually comprise a potentiometer in the boiler control panel linked to a remote sensor in the flow water connection.

For LPHW boilers employing fan-assisted or forced-draught natural gas burners the control thermostat will have an operating range of 0–120°C. The thermostat will typically be factory-preset by the boiler manufacturer to a nominal maximum value of less than 90°C. This setting will be adjusted during the on-site commissioning process to a value that suits the design flow temperature of the heating system. Where boilers are of conventional (non-condensing) design there will typically be a minimum required setting for the control thermostat in the region of 65°C. For boiler flow temperatures below this a conventional system design temperature differential of the order of 11K will generate a return water temperature to the boiler in the range (<55°C) that risks causing substantial and regular condensation of the moisture in the flue gases within the boiler, even under conditions of design heat load. Unless the boiler is designed to cope with routine condensation of this type, its life span will be shortened (perhaps very significantly) if it is permitted to occur. Since the return water connection has traditionally been at the rear of the boiler, protection of the boiler against low return water temperature is conventionally termed 'back-end protection'. This subject is considered further in Chapter 3.

In the UK a typical boiler flow temperature for conventional LPHW space and water heating applications is nominally 82°C, with an 11K heating system design temperature differential (more about this in Part B) generating a return water temperature to the boiler of 71°C under design heat load conditions (which, it must be said, are seldom encountered during the 'heating season'). Where alternative boiler flow temperatures are employed, these are typically matched with heating system design temperature differential to retain a return temperature to the boiler of c.70°C under design conditions. However, boiler flow temperatures higher than 82–85°C and system design temperature differentials greater than 11K have typically not been favoured by UK design practice. Theoretically the design temperature rise across the boiler will be the same as that across the pipe circuits of the heating system when a single boiler is piped in series with the load circuits, and when multiple boilers are piped in parallel with it. It will not, of course, be the same when more than one boiler is piped in series with the load circuits. The maintenance of minimum water flow rates through the boiler places an ultimate limit on the maximum system design temperature differential. While this limit varies according to the type of boiler, it is never greater than 22K for cast iron sectional boilers, and may be substantially less than this for some modular boilers. The choice of boiler flow water temperature will have implications for the design and control of the heating system, and these are considered in Chapters 6 and 7. For our present purposes it is sufficient to note that it is possible to 'step down' (reduce) LPHW flow temperature in the heating system, but not to 'step it up' (increase it).

Where boilers are designed for *condensing operation*, the setting of the boiler control thermostat will be substantially reduced to c.55°C or less, in order to ensure that a return water temperature is achieved that is low enough to maximise condensing operation. Water returning to the boiler may, therefore, be in the range 30–45°C for system design temperature differentials in the range of 10–20 K. Condensing operation of the boiler is considered further in Chapter 3.

It is an inherent characteristic of on/off control that the boiler thermostat must incorporate a *switching differential* that means that the flow water temperature from the boiler will fluctuate under part-load operation. The design of the boiler, in terms of the thermal capacity of the block or heat exchanger and its water content, plays an important part in determining the span of this switching differential. The implications of switching differential for the control of single boilers are considered further in Chapter 10.

Where firing of the burner is on a high/low/off basis (or, strictly-speaking, on a low/high/low/off cycle), the boiler control thermostat becomes an 'off/low' thermostat and a second, 'high/low' control thermostat is introduced in the flow connection from the boiler. Again both thermostats may comprise temperature sensors in the boiler flow connection linked to remote potentiometers in the boiler control panel. The off/low thermostat initiates and terminates operation of the burner at its low firing rate, while the high/low thermostat controls the change between this and full burner output. Effectively the two burner stages act like *two boilers in series*, each contributing to the overall temperature rise across the boiler in a cumulative way. If the heat output to water of the boiler at the burner low firing rate is 50% of that at its full output, then an overall boiler temperature rise of 11 K would be split equally between the firing stages, i.e. 5.5 K each. Thermostat settings and switching differentials for high/low/off burner control are again considered in Chapter 10.

Where the burner is under *modulating control* the boiler control thermostat is replaced by a boiler flow temperature sensor. The boiler incorporates a proportional controller that compares the temperature sensed by this with the desired set-point value and provides an output signal that continuously modulates the burner firing rate between its minimum level of modulation (maximum *turndown*) and full output. Below the maximum turndown in burner firing rate, the burner cycles on and off. In modern boilers the controller will typically offer *proportional-plus-integral* (PI) or *proportional-plus-integral-plus derivative* (PID) control. Modulation of the burner may alternatively be based either on sensing of return water temperature or the *temperature difference* between flow and return water, and in some cases an external control signal may be employed. The characteristics of modulating burner control are considered further in Chapter 10.

Irrespective of the type of burner control employed, many boiler plants for commercial space and water heating systems consist of a number of boilers connected in parallel. In such installations the burners of individual boilers are controlled in sequence to match the heating load as closely as possible throughout the heating season. Such sequence control is achieved through an external controller that is part of the heating system controls. Sequence control of boilers is considered in Chapter 10.

2.7.3 Boiler safety and limit controls

Two safety control functions are essential for any boiler. These are:

- Protection against high water flow temperature from the boiler
- Protection against zero or low water flow rate through the boiler.

Protection against (excessively) high flow water temperature is undoubtedly the most fundamental boiler safety control function. It is achieved by means of a boiler *high-limit thermostat or 'temperature limiter'* that is positioned in the system flow water connection from the boiler. Activation of this high-limit thermostat results in the immediate shut-down of the burner. Following activation, the high-limit thermostat can only be reset manually. For space and water heating applications employing LPHW, the high-limit thermostat will typically have an operating range of 90–110°C. Since activation of this thermostat effectively locks the boiler out of operation, 'nuisance' tripping is undesirable and is typically avoided by ensuring that an adequate differential exists between its setting and that of the boiler control thermostat. A minimum differential of 10K is typically recommended by boiler manufacturers. As in the case of the boiler control thermostat, the high-limit thermostat will typically be preset by the boiler manufacturer (a preset of 100°C is common), for final site adjustment to suit the design boiler flow temperature for the particular heating system. The high-limit thermostat is primarily intended to protect the heating system against delivery of water from the boiler at an excessive temperature. The prime scenario for this would be failure of the boiler control thermostat. Protection against high flow water temperature is a health and safety consideration in the UK.

The other truly fundamental boiler safety control function is protection against *zero or low water flow rate*. Serious, even irreparable, physical damage may be caused to the boiler if firing of the burner is initiated, or allowed to continue, in the absence of any water flow rate through the boiler block, or if this flow rate is excessively restricted. The trend in boiler design towards low water content has increased the need for protection against low-flow conditions, since the boiler depends on the correct throughput of water to remove the heat input by the burner. Without a substantial 'reservoir' of water, should this throughput of water be excessively restricted, the metal of the boiler block itself provides the only sink for burner heat output, with the real risk of thermal overstress.

In some cases individual boilers may be fitted with a *flow switch*, which is a mechanical device that is activated by the flow of water through the boiler. The flow switch will typically be located in either of the pipework connections to the boiler, return inlet or flow outlet. The flow switch is arranged to close (or 'make') an electrical circuit when a pre-determined threshold value of boiler flow rate is exceeded, and to open (or 'break') the circuit should the boiler flow rate drop below this threshold level for any reason. If the flow switch is incorporated into the circuit providing electrical power to the burner, firing of the burner will only be enabled if the minimum-flow condition is met, and will be terminated immediately should the flow rate through the boiler drop below the threshold level at any time during the firing cycle. Shut-down of the burner under zero/low-flow may be arranged to lock the burner out of operation. Where a flow switch is not fitted to the boiler itself, a pipeline-mounted type may, of course, be fitted remotely in the pipework to or from the boiler.

In modern commercial space and water heating systems the boiler inevitably depends on a circulating pump to initiate and maintain the required flow rate of water. Pump failure or operating problems are therefore a principal potential source of any zero-flow or low-flow conditions at the boiler. Indeed, on start-up of the heating system a zero-flow condition will inevitably exist at the boiler until the appropriate circulating pump is running. Where boilers are not individually fitted with flow switches, protection against a zero-flow condition caused by pump failure (or failure to start) is provided by start-up sequencing and interlock functions of the external control system. On start-up the pump that generates circulation through the boiler will be started, and its operation verified or *proved* prior to, and as a condition for, switching on the boiler. Satisfactory pump operation, at least as far as the boiler is concerned, is typically proved by means of a differential pressure switch that is installed across the pump. This differential pressure (or 'DP') switch is another type of mechanical switching device that can be arranged to close an electrical circuit providing power to the boiler and its burner when a pre-determined threshold value of differential pressure exists across the pump. A circulating pump generates a pressure rise (hence a differential pressure) between its inlet and its discharge connections, in the same way that a boiler generates a temperature rise between its return and flow connections. In the event of pump failure during burner firing, or if the differential pressure across the pump falls below the threshold value for its closure for any reason, the differential pressure switch will open and 'break' the electrical power supply to the boiler and its burner. However, a different situation exists on shut-down of the heating system. In this case the start-up sequence of pump, then boiler, must be reversed to dissipate residual heat from the boiler.

On start-up of the circulating pump, there may be a brief delay before water flow is fully developed through the boiler. The extent of this delay will depend on the total mass of water that must be set in motion by the pump, and hence on the hydraulic design of the installation. While for normal pump operation in any practical building services circumstance the delay will never be measured in terms of other than a relatively few seconds, it may still be necessary to allow a short interval between the starting of the pump and the switching on of the boiler, in order to allow for this, albeit brief, interregnum during which the inertia of the mass of water in the pumped circuit must be overcome. In particular, where individual boilers employ flow switches that lock the burner out of operation in the event of a low-flow condition, if this condition can be registered and acted upon faster than the threshold minimum flow rate through the boiler can be established by the circulating pump, repeated nuisance 'lockouts' of the boiler may occur on start-up of the heating system.

Other potential causes of excessively low boiler flow rate include the physical constriction or blockage of pipe bores or the boiler waterways themselves (whether caused by material failure or deterioration with age due to scale formation, corrosion or biological/bacterial contamination), the improper positioning of manual isolating or commissioning valves, or the malfunction of control valves. All of these potential causes may actually result in the existence of an increased differential pressure across the circulating pump. Boilers with individual flow switches will naturally respond to low-flow conditions resulting from these scenarios. Otherwise the boiler will be left to respond to them through the normal operation of its control and high-limit thermostats.

Where boilers are fitted with comprehensive microprocessor-based boiler controllers, suitable software-based control algorithms can analyse the monitored output of boiler flow and return water temperature sensors (not only in terms of simple absolute levels but also in terms of boiler temperature rise and rates of change) to provide both high-limit temperature and low-flow protection.

Thus far our consideration of boiler low-flow protection has confined itself to protection against specific identifiable events (start-up failures) and unforeseen occurrences (failures, blockages, error or misuse). However, a further type of protection against low flow rates through the boiler may be necessary. This constitutes protection against low flow rates through the boiler that can arise as a result of routine operation under part-load conditions. This is a function of the control of the heating system and the way in which the boiler is incorporated into its hydraulic design, and is considered in Chapter 10.

As far as many traditional boiler designs are concerned, protection against high water flow temperature and zero or low water flow rate will represent the extent of the safety controls provided. However, a number of modern boiler designs, and particularly those emanating from the European boiler market, incorporate additional safety control functions that may include some or all of the following:

- Protection against low water content in the boiler
- Protection against boiler down-draught
- Protection against high flue gas temperature.

As far as the first of these – protection against low water content – is concerned, serious physical damage may be caused to the boiler if firing of the burner is initiated, or allowed to continue, when the waterways and waterside volume of the boiler are not completely full. Protection against low water content may be provided either by a suitably-placed level sensor or by a temperature sensor positioned at a sensitive point of the boiler block.

As far as protection against down-draught is concerned, we saw earlier in the chapter (section 2.4.4) that atmospheric gas burners need to be protected against this condition. While physical protection against down-draught conditions is typically provided by fitting the boiler with a *draught diverter*, some boiler manufacturers recommend the use of a *down-draught thermostat* mounted within the flue itself, and which is arranged to shut down the boiler on sensing a low flue gas temperature.

Finally, the existence of a high flue gas temperature (FGT) implies either that combustion is taking place with a higher-than-design fuel/air ratio or that the gas-side heat exchange surfaces of the boiler are dirty. In either case the occurrence of a high flue gas temperature has implications for the continued trouble-free, safe, clean and efficient operation of the boiler, and its cause(s) should be resolved without delay. An FGT sensor may therefore be provided that is arranged to lock out boiler operation at a specified flue gas temperature.

2.7.4 Reporting functions: remote monitoring

Modern boilers typically provide volt-free contacts to allow signalling of various operating status and fault conditions to a remote *building manage-*

ment system (*BMS*). Basic operating status or fault conditions that potentially merit such monitoring typically include the following:

- Boiler 'running' (i.e. burner firing), or running at 'low-fire' and 'high-fire' where applicable to the type of burner
- Boiler operation 'locked out' by the high-limit thermostat ('locked out on high limit')
- Boiler operation 'locked out' (advice of a general condition, which may be triggered by more than one safety control function)
- Hours run (burner firing), or run at 'low-fire' and 'high-fire' where applicable to the type of burner.

The degree to which a particular boiler provides outputs for remote monitoring of these potential operating status or fault conditions depends on both the age of the boiler design itself and the manufacturer's own commercial and technical preferences. However, as with major items of HVAC plant generally, the trend appears to be towards increasing the range of status/fault conditions and operating parameters that may be remotely monitored.

2.8 The boiler casing

The boiler casing, which in modern boilers is typically manufactured from plastic-coated or 'stove-enamelled' mild steel sheet, performs several simple, but essential, functions. First, it provides physical protection to the boiler generally, and in particular to the vital thermal insulation covering the boiler block. The latter may be damaged relatively easily in the working environment of a plant room. Second, it provides a physical separation from potentially hot or physically hazardous items and electrically 'live' components for personnel requiring routine access to and around the boiler. Finally, it provides a suitable mounting for the boiler control and instrument panel.

Where atmospheric gas burners are employed, the boiler casing must take the form of a 'skirt' that terminates at a suitable height above the mounting surface to allow a supply of combustion air to reach the burner. The casing should not allow water to collect internally, where it might cause deterioration of the boiler thermal insulation or corrosion that could remain undetected. This applies particularly to the bottom of the casing and any potential low points. While the base on which the boiler stands must always be flat and level, care should nevertheless be taken to ensure that water from leakage or 'drain-downs' within the plant room cannot collect or 'puddle' around the bottom edge of the casing, causing corrosion. A common way of ensuring this is to place the boiler on a concrete plinth that is raised some 100 mm or so above the general level of the plant room floor.

References

1. American Society of Heating, Refrigeration and Air Conditioning Engineers (ASHRAE) (1996) Boilers, Chapter 27 in 1996 *Handbook: Systems and Equipment (SI)*, p. 27.1. ASHRAE, Atlanta.
2. Rogers, G.F.C. & Mayhew, Y.R. (1980) *Engineering Thermodynamics Work and Heat Transfer*, 3rd edn. Longman, London and New York.
3. American Society of Heating, Refrigeration and Air Conditioning Engineers (ASHRAE) (1996) Automatic Fuel-Burning Equipment, Chapter 26 in 1996 *Handbook: Systems and Equipment (SI)*, p. 26.3. ASHRAE, Atlanta.

3 Types of Boiler and Their Needs

3.1 Types of boiler

Boilers may be categorised according to a variety of criteria, principal among which are:

- Construction material – cast-iron, steel, copper or aluminium
- Method of construction – sectional or fabricated
- Physical relationship between the combustion gases and the water spaces – 'fire-tube' or 'water-tube'
- Fuel – principally natural gas, liquid petroleum gas, oil or dual-fuel
- Burner type – atmospheric gas, fan-assisted or forced-draught gas, premix gas, pressure jet oil
- Configuration – multiple or modular
- Application – heating, hot water service or a combination of the two
- Mode of operation – condensing or non-condensing.

3.2 Boiler construction materials

3.2.1 Range of materials

It is self-evident that a *range of materials* will be employed in the manufacture of any modern boiler, including both metals and plastics. Traditionally the *primary material* of a boiler has been regarded as the material from which the boiler 'block' or heat exchanger is formed, since with traditional cast-iron and steel construction this is where the bulk of the boiler mass resides. Cast-iron and steel remain overwhelmingly the most common materials for the construction of boilers, with cast-iron up to four times more common than steel. Copper may also be used as a material for boiler heat exchangers, and in this case does not fit the traditional concept of a boiler primary material.

Before considering the reasons for the continuing predominance of cast-iron as a boiler material, it will be prudent to note the distinction between 'iron' and 'steel', since this is a matter of common confusion. *Iron* is the predominant chemical element in both iron and steel. The output of a modern blast furnace is typically cast in the form of *pig iron*. However, the properties of this make it generally unacceptable for use in this form. It is therefore either remelted and modified in composition to produce an iron with acceptable properties for use in cast form, or is turned into steel. The term *cast iron* is applied to iron that contains c.4% carbon by weight, which is about the maximum proportion that the iron can hold. *Steel* is the term applied to iron with a low carbon content (typically less than 1% by weight), and nowadays implies *mild steel or carbon steel.* Steels that are alloyed with elements other than carbon are commonly referred to as *alloy steels.* The assumption is, therefore, that the composition of steel is more-or-less closely controlled.

3.2.2 Cast-iron

Cast-iron has been used in boiler manufacture since the late nineteenth century. As a boiler material it has the following characteristics:

- Heat transfer rate is high
- It may be cast and machined into complex forms with relative ease
- It has adequate physical strength and resistance to thermal and mechanical shock
- It has good resistance to corrosion.

Prior to the introduction of cast-iron, boiler manufacture had relied on the use of wrought iron, a material that was highly ductile and had good resistance to corrosion. However, while it offered enhanced physical properties, in comparison with traditional cast-iron, its production could not compete economically with that of steel (itself with improved physical properties over wrought iron) once the latter started to be produced cheaply and in large quantities from the middle of the nineteenth century. As a boiler material, therefore, wrought iron disappeared from the market, being replaced by cast-iron and steel.

The success of cast-iron as a boiler material has undoubtedly been due in large part to the manufacturing and economic advantages of the sectional method of boiler construction. This allowed a range of boiler outputs to be achieved by assembling a simplified range of standard system components. Since the sectional approach typically involves all but relatively small boilers being assembled on site, the components of the boiler can typically be manoeuvred into spaces which would be inaccessible for a boiler delivered to site as a more-or-less complete packaged unit. This may be an important factor where the replacement of an existing boiler plant is concerned.

The use of cast-iron as a boiler material is not, however, without its drawbacks. There is some risk of defects occurring in the manufacturing process, which may result in the development of 'pin-hole' leaks or cracks[1]. 'Pin-hole' leaks may develop as a result of three types of manufacturing defect. First, any impurities that become embedded in the metal of the casting and go undetected can result in a local thinning of the metal that can be crucial in some parts of the boiler. Second, in heavy sections the shrinkage in the volume of the metal as it sets may cause the development of a void or a series of small holes that produce a porous area of the section. Finally 'blow-holes' may develop if dissolved gases within the molten iron are unable to escape as it sets.

The other obvious defect that can occur in a cast-iron boiler is the cracking of a section. Cracks develop where metal failure occurs, either as a result of a fault in the metal itself or of thermal overstressing. Any restriction of water flow rate through the section, as may result if not all of the core moulding is removed following manufacture, will result in a heat transfer rate below the design value, and a consequent overheating that may produce thermal overstressing great enough to crack the section. Such defects may result in localised boiling (or 'kettling') of the water in a boiler. Changes in section thickness, bosses, projections and intersections in the material are a necessary feature of practical boiler design. However, their presence will influence the strength of the section in comparison with a simple shape of uniform thickness.

3.2.3 Steel

While cast-iron may have *taken over from* wrought iron in boiler manufacture, it may be said that steel was the more-or-less direct *replacement for* wrought iron. Steel is a more homogeneous material than cast-iron. Furthermore it is used in 'standard' forms, particularly sheet and plate, in which the parameters of thickness and density are controlled within specified limits. The use of steel as a boiler material is therefore not subject to some of the quality issues that may be associated with the casting of iron. The manufacture of steel boilers is typically based on welding techniques, and these of course have their own issues and requirements in respect of assuring the quality of the manufacturing process.

While cast-iron has great flexibility when associated with the sectional method of boiler construction, steel may be said to allow a greater flexibility in the overall design of boilers, since ultimately any configuration is possible if it can be fabricated. However, there are several 'downsides' to the use of steel as a boiler material. The first is an economic one, in that a range of boiler outputs may be achieved by using appropriately-sized components according to a common design. However, the potential for putting each size of boiler in a range together with varying numbers of components of a common size is much less for fabricated steel boilers than it is for the cast-iron sectional approach. For a given heat output to water, steel boilers are therefore typically more expensive than cast-iron boilers. The counter-argument is that the design flexibility of the steel boiler allows it to achieve a higher efficiency. The second (potential) 'downside' to the use of steel is a practical one. Steel boilers are fabricated off-site and delivered as more-or-less complete packaged units. While access to the designated boiler plant room can typically be ensured for 'new-build' applications, this may not be the case (or may only be possible with difficulty and at a cost premium) where the replacement of an existing boiler plant is concerned.

The final 'downside' to the use of steel as a boiler material is that it is both less resilient than cast-iron to the effects of scale formation and more susceptible to the effects of corrosion resulting from acidity in the boiler water or the presence in it of 'free oxygen'. These influences may of course be dealt with by appropriate chemical treatment of the water in the heating system. Steel is also less resilient than cast-iron where the formation of acidic deposits is likely as a result of the combustion of a sulphur-bearing fuel. Stainless steel is sometimes employed in the design of the heat exchangers of relatively small boilers.

3.2.4 Copper and aluminium

The great attraction of copper as a boiler material stems from its high thermal conductivity, which is approximately six times that of iron and 8.5 times that of mild steel. It is also a very ductile material and this combination of qualities therefore makes it well-suited to the construction of water-tube heat exchangers. As a boiler material it is not employed in the same way as cast-iron or steel, being used selectively within a finned-tube combustion gas-to-water heat exchanger, and with cast-iron or steel employed for the main body of the boiler. The type of boiler construction that employs copper will typically result in a boiler that is lighter, for a given heat output to water, than a cast-iron sectional boiler.

A similar situation exists for aluminium. This also has a higher thermal conductivity than either cast-iron or steel, although only approximately half that of copper. As a boiler material, its resistance to corrosion attack by the slightly acidic condensate that is formed when the combustion products of natural gas are condensed, makes it suitable for use in the water-tube secondary heat exchangers of condensing boilers.

3.3 Methods of construction

3.3.1 Cast-iron sectional boilers

In current application the sectional method of boiler construction is synonymous with the use of cast-iron. A conventional cast-iron sectional boiler that employs a forced-draught natural gas or pressure-jet oil burner is typically constructed in three parts. Perhaps not surprisingly, these comprise a front section, a middle and a rear section (Figure 3.1). The front section incorporates a hinged door which gives access to the combustion chamber and provides a mounting point for the burner. The boiler door incorporates a suitable opening for the burner 'blast tube', and the burner is located over this by means of a mounting plate. The rear section of the boiler incorporates the flue box or smoke box, from which a flue outlet terminates behind the boiler with a flange to allow connection of a boiler flue. The flue box may be removable and may incorporate an inspection door. The rear section of the boiler also incorporates the water flow and return connections to the boiler.

The middle section of the boiler comprises a number of standard sections that, when joined together, form the combustion chamber, flueways and waterways of the boiler. Each of these middle sections effectively consists of a 'slice' through the combustion chamber, the combustion gas passes and the main horizontal water distribution and collection passages, together with a pattern of vertical waterways that are integral to each section. The number of sections is varied to provide different sizes of boiler with a range of heat outputs to water. Typically from three to twenty such 'middle' sections may be combined in this way, although a single boiler model will normally offer fewer than this (from three to ten or from ten to twenty sections, for example, depending on their design and size and the intended range of heat output). Adding a section may allow an increment of from 20–50 kW in the maximum heat output to water available from the boiler, according to the design and size of the section. In the first instance the cast-iron sections of the boiler are joined mechanically by means of short mild steel pipe fittings called 'nipples', which fit into specially prepared openings formed in or around the main horizontal waterways of each section. These nipples may be tapered ('conical') in form to improve the 'positiveness' of the mechanical joint. Since on its own this mechanical joint cannot be relied upon to prevent leakage of combustion gases or water, the junction of each pair of sections is typically provided with machined surfaces or grooves that accept a non-metallic sealing material. At its simplest this takes the form of glass fibre cord that is packed into grooves in one of the 'mating' surfaces of each pair of sections. The whole assembly of cast-iron sections is typically anchored together by external 'tie rods' linking the front and rear sections of the boiler. The boiler as a whole is wrapped in a thermal insulation 'blanket' of high-density fibreglass (or other appropriate material), and is finally enclosed in a plastic-coated or stove-enamelled steel casing. The boiler naturally includes provisions to allow a full 'drain-down' of the water content of the sections. The 'front-to-back' disposition of a number of cast-iron sections will typically mean that the depth of this type of boiler may be substantially greater than its width, and this may need to be considered when planning the layout and overall space

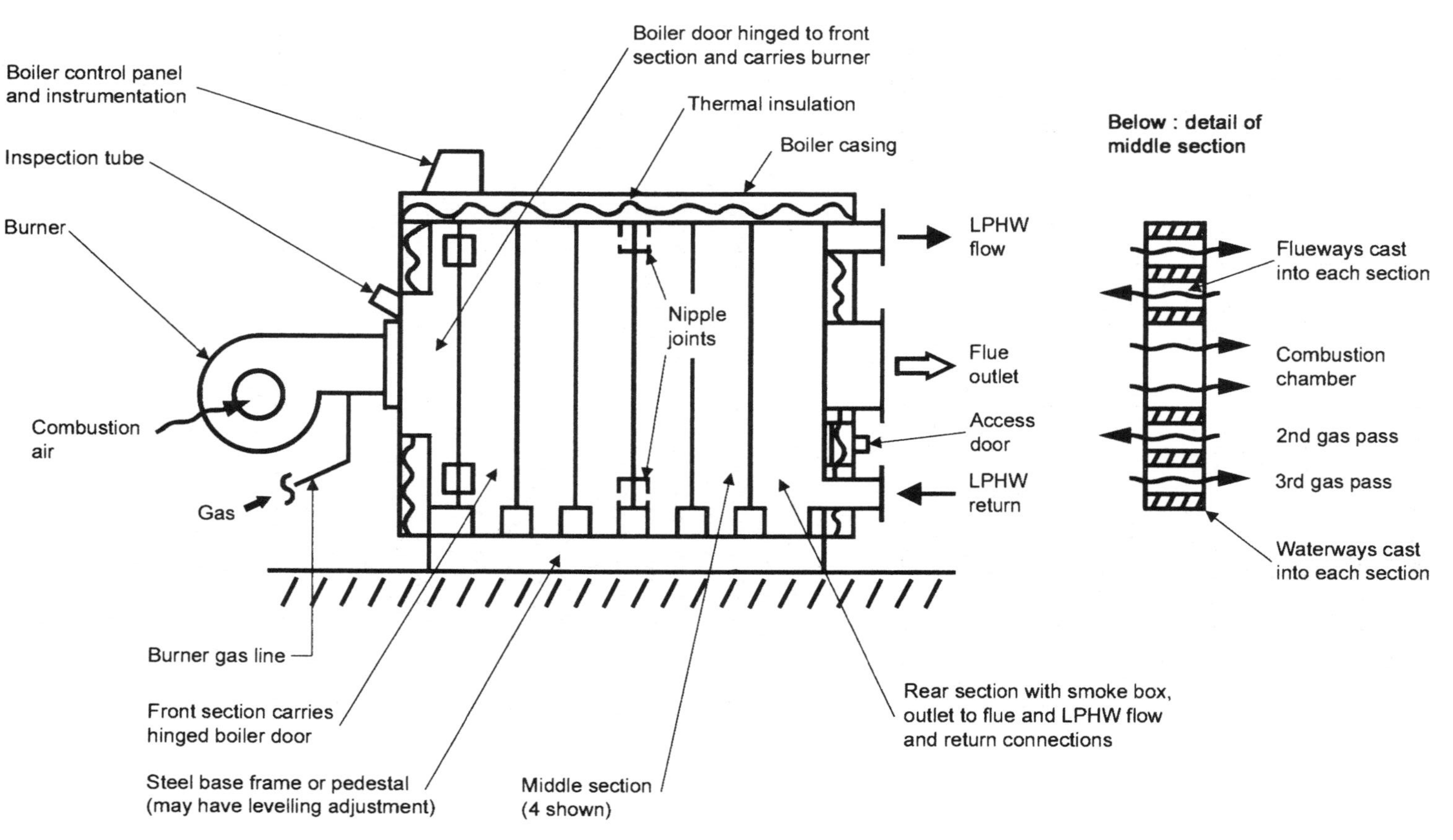

Figure 3.1 Layout of a typical cast-iron sectional boiler with forced-draught natural gas burner.

requirement for a boiler plant, and particularly when replacing boilers in an existing plant room.

A similar approach can be taken to the construction of cast-iron sectional boilers that employ atmospheric natural gas burners. In this case, however, the burner bars will be located at the base of the combustion chamber and the flow of combustion gases will take place in a single upwards vertical 'pass' through the boiler between the waterways of the cast-iron sections (Figure 3.2). In this case the flue box takes the form of a *flue hood* that is located above the boiler block, and which may be of fabricated steel construction rather than cast-iron. The top of the flue hood incorporates an outlet for connection to a boiler flue. The flue hood will typically also incorporate an integral *draught diverter*, the function of which has already been considered in Chapter 2. The design of such boilers must also provide openings at the base of the combustion chamber that will allow the induction of combustion air. The heat output of the boiler is again proportional to the number of sections. The atmospheric natural gas burner in such a boiler will typically comprise a number of parallel burner bars disposed along the sections of the boiler. Since access to the burner bars will be required, together with facilities for their withdrawal, it may again be necessary to consider the way in which a boiler plant comprising units of this type can be physically arranged within a plant room.

An alternative approach to the construction of cast-iron sectional boilers with atmospheric gas burners is possible which employs a *vertically-stacked arrangement* (Figure 3.3). In this the combustion gases from the burner bars flow vertically upwards to the flue hood through a number of horizontal sections. Each of these is a gas-to-water heat exchanger that directs the boiler water repeatedly back and forth in 'serpentine-fashion' across the vertical path of the combustion gases in its overall vertical movement from the lowest section to the uppermost. Except for the section that is most closely adjacent to the burner bars, these heat exchangers will typically provide extended heat transfer surface in the form of finning around the horizontal waterways. Sections are again joined by the use of 'nipples' and sealant material located in or against specially-machined surfaces, and anchored externally by tie rods. While fewer sections may typically be incorporated in this approach, it is well-suited to the production of a range of compact boilers which have both relatively low output and increments of output, but which can be used as the individual modules of a modular boiler system.

Since a wide variety of cast-iron sectional boilers are available to both the UK and the wider European boiler market, there are bound to be variations on the basic themes presented in this discussion, and some of these variations may be significant in their extent and the effect that they have on boiler construction and performance. In particular, the cast-iron sectional approach may be applied to *condensing boilers*, and the implications of this on boiler construction are considered later in this chapter. The cast-iron sectional approach may also be applied with premix natural gas burners. In the case of smaller boilers the cast-iron sections may be delivered to site pre-assembled.

3.3.2 Fabricated steel boilers

In building services space and water heating applications a fabricated steel boiler is typically constructed in a *shell-and-tube* configuration (Figure 3.4).

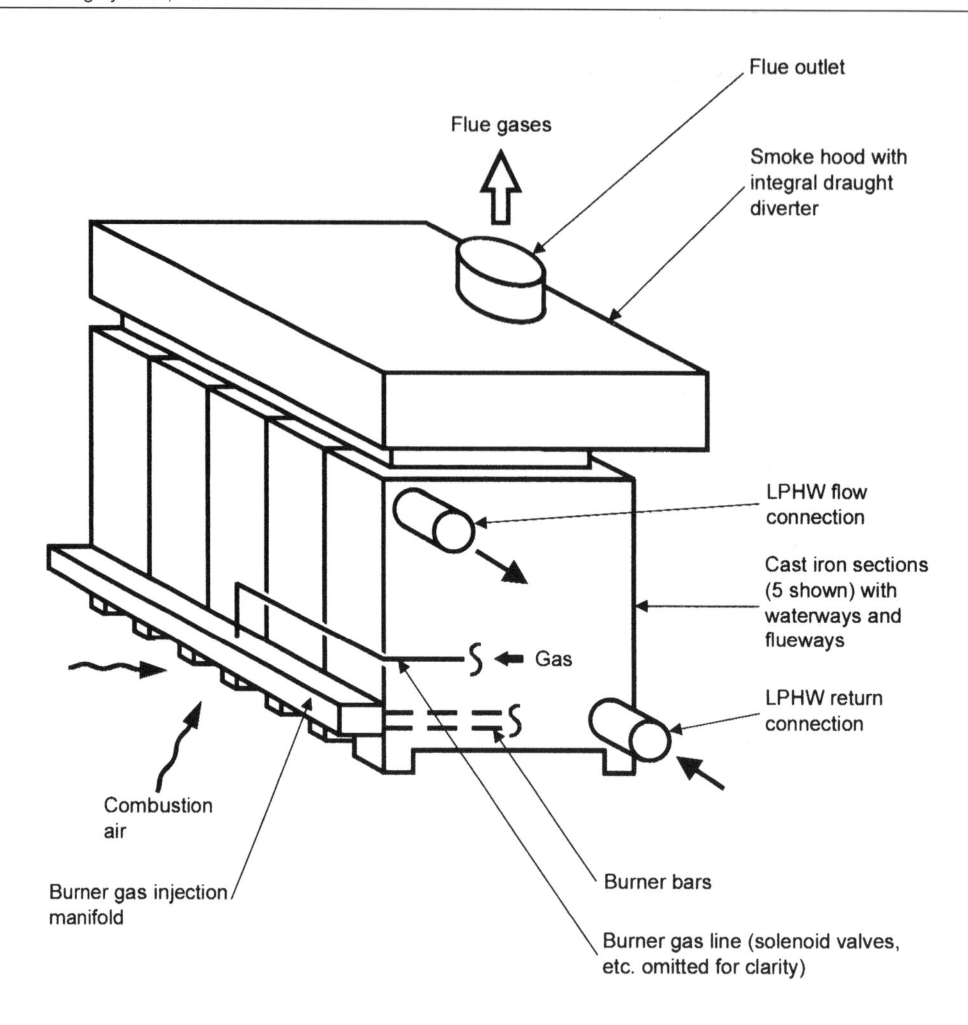

Figure 3.2 Layout of a typical vertically-sectioned cast-iron boiler with atmospheric natural gas burner.

In this a horizontal cylindrical combustion chamber is located in the centre of the boiler, and on leaving the combustion chamber the flue gases pass through a 'nest' of *fire tubes or smoke tubes* which terminate in the smoke box at the rear of the boiler. In modern shell-and-tube boilers the combustion flame may reverse along the length of the combustion chamber before the combustion gases enter the fire tubes. The whole assembly of the combustion chamber and fire tubes is contained within a water-filled shell to which the boiler flow and return water connections are made (typically at the top of the boiler 'shell'). The fire tubes are supported within the volume of this water-filled shell in such a manner that they are surrounded by the water, by means of *tube plates* at the front and rear of the boiler. The return water connection to the boiler shell will typically incorporate some form of

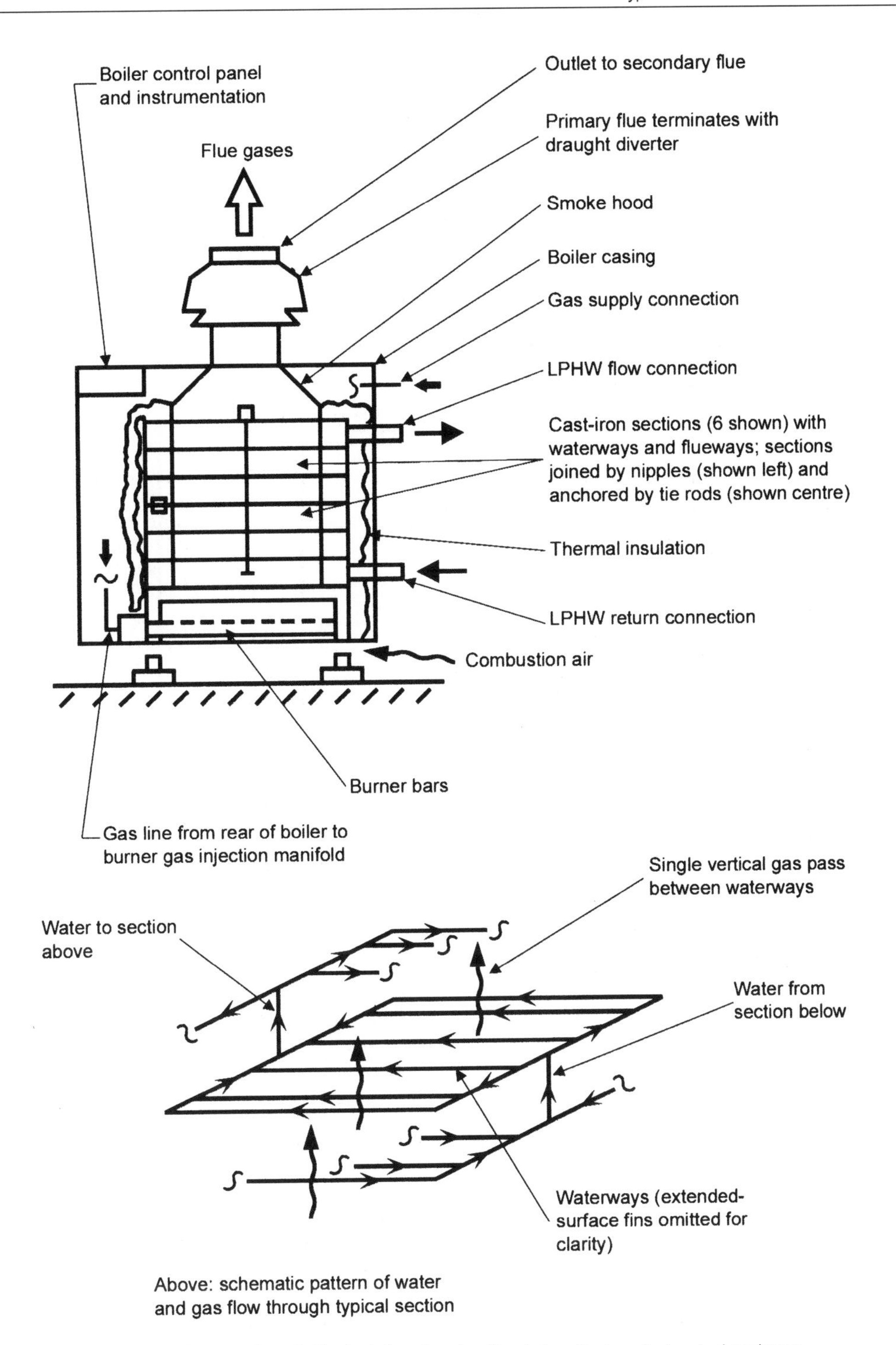

Figure 3.3 Layout of a typical horizontally-sectioned cast-iron boiler with atmospheric natural gas burner.

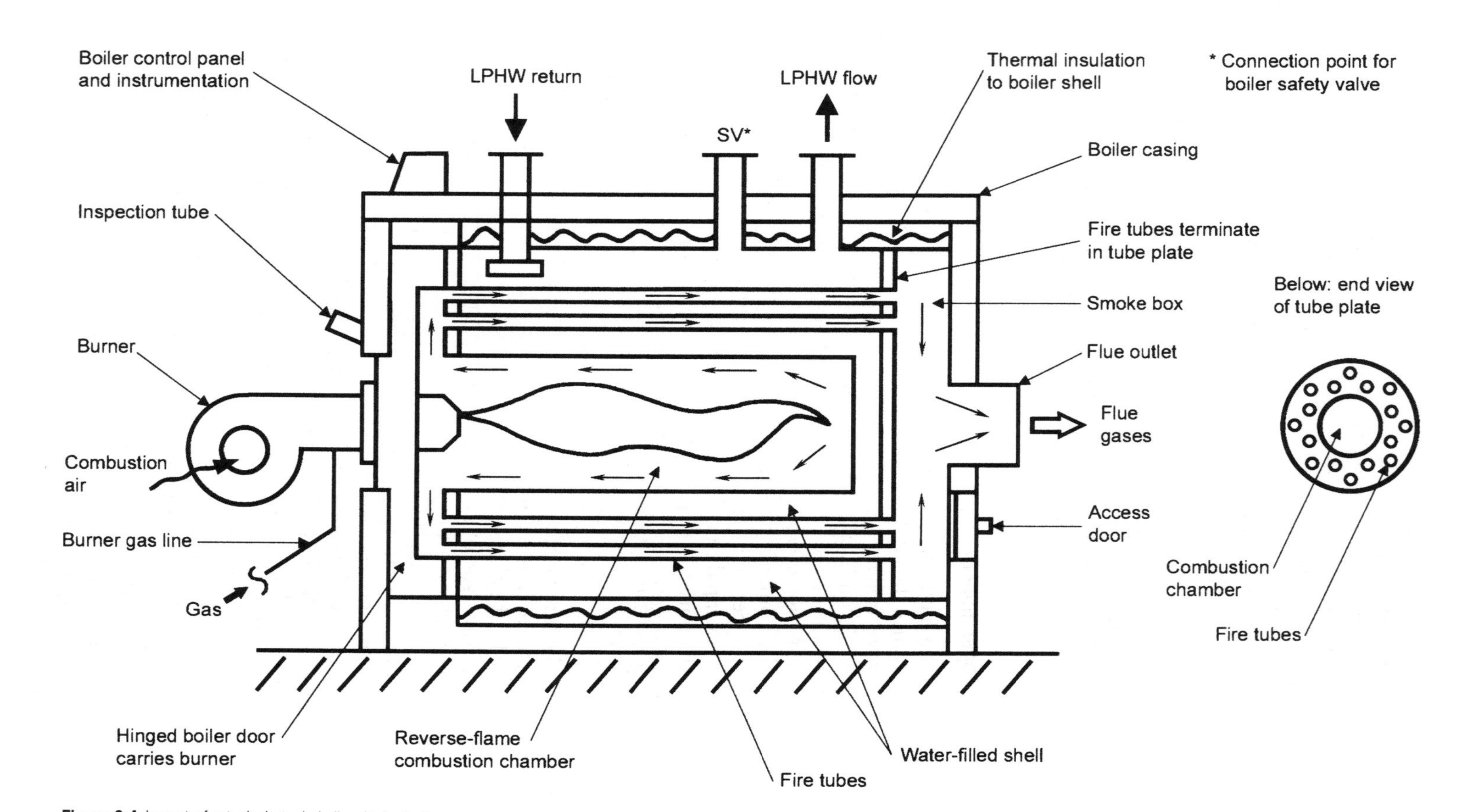

Figure 3.4 Layout of a typical steel shell-and-tube boiler.

distributor to ensure that water is delivered evenly around the circumference of the cylindrical shell.

The front of the boiler is taken up by a door that gives access to both the combustion chamber and the fire tubes, and which again provides a mounting for the burner and an opening into the combustion chamber for the burner 'blast tube'. Over the area of the combustion chamber and the fire tubes the inner surface of the boiler door is typically provided with a *ceramic* thermal insulation. The water shell is wrapped in a thermal insulation 'blanket', which may again be high-density fibreglass or other appropriate material. The structural integrity of the boiler is provided by a steel framework, which will include a substantial baseframe. The whole framework is finally encased in a plastic-coated or stove-enamelled steel casing. The boiler will naturally include provisions to allow a full 'drain-down' of the water contained within the shell, while provision is typically also made to allow the direct connection to the shell of a boiler safety valve.

3.3.3 Copper boilers

As we have already seen, the term 'copper' boiler is in some ways misleading, since the amount of copper actually used in the boiler may be quite small. Copper boilers typically employ atmospheric natural gas burners, with the flow of combustion gases rising vertically upwards from the burner bars to pass through a heat exchanger composed of a series of horizontal *copper-finned copper tubes* carrying the boiler water, and which is located at the top of the combustion chamber (Figure 3.5). The high thermal conductivity of the copper and the low volumetric content of the water tubes typically result in both rapid initial heat-up and fast response to changes in the heat output of the burner. On passing through the gas-to-water heat exchanger the combustion gases are collected in a flue hood located at the top of the boiler. This will typically incorporate an integral draught diverter, and may include directional vanes or 'baffles' to ensure that the combustion gases are directed cleanly past the draught diverter and straight into the flue outlet in the top of the flue hood. The whole assembly will typically be built up from a steel frame or *chassis*, and enclosed within a casing that may be fabricated from panels of galvanised sheet steel to which a painted finish is applied.

In this type of boiler the combustion chamber is *not* surrounded by water. The sides of the combustion chamber are therefore lined with refractory panels, and the base on which the boiler stands must be protected against downward radiant heat. Additional thermal insulation may be incorporated into the boiler construction and casing to minimise heat loss. Openings must be provided at low level around or in the base of the combustion chamber to allow the induction of combustion air to the atmospheric gas burner. The latter will typically comprise a number of parallel burner bars which are assembled within a tray or trolley that can be withdrawn from the boiler for inspection and maintenance. The parallel disposition of a number of burner bars typically means that the width of the boiler is substantially greater than its depth, and this may need to be taken into account when planning the layout and space requirements for a boiler plant comprising units of this type. Versions of this type of copper boiler provide rare examples of a *modulating* atmospheric natural gas burner, and in this case the supply of combustion air may be split between primary and secondary routes into the combustion chamber. As with all types of boiler construction,

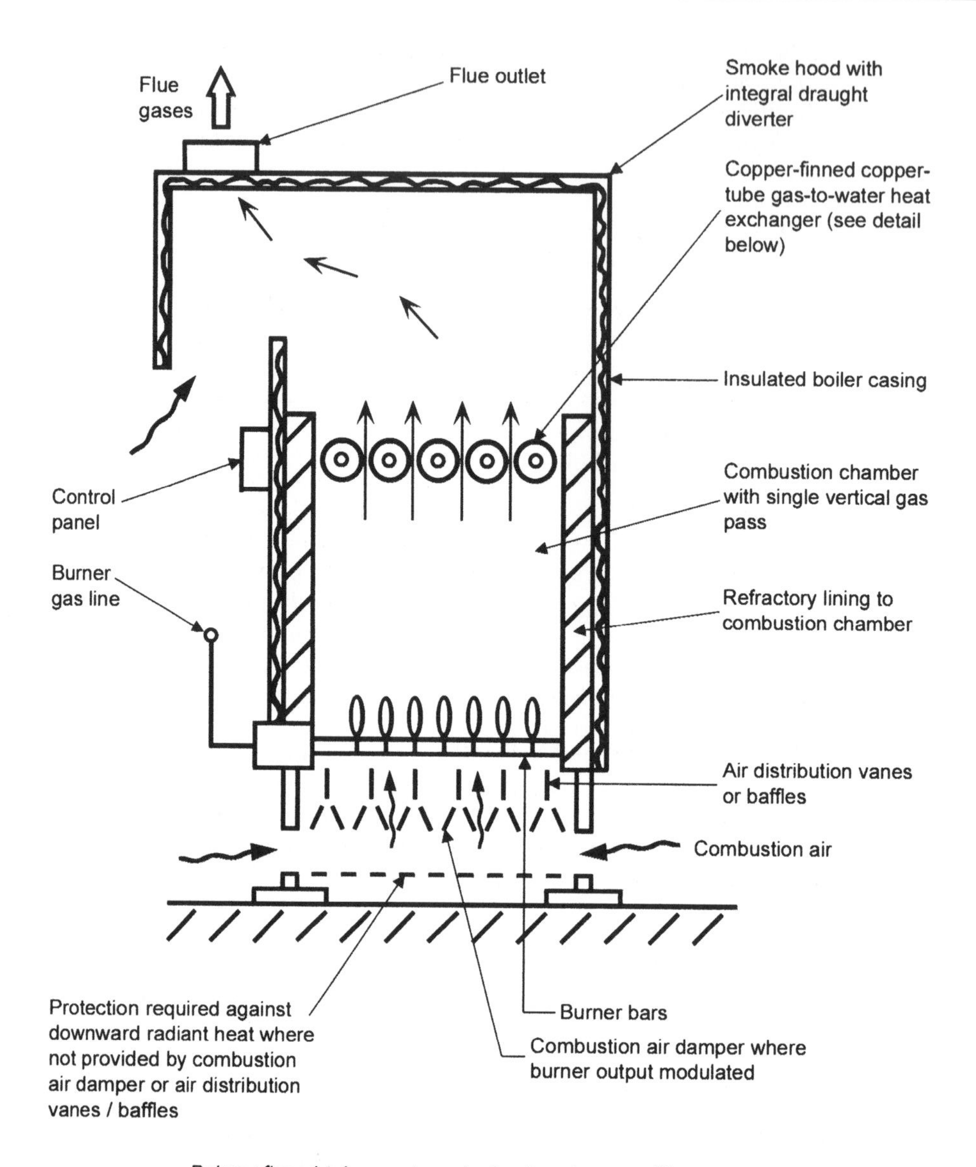

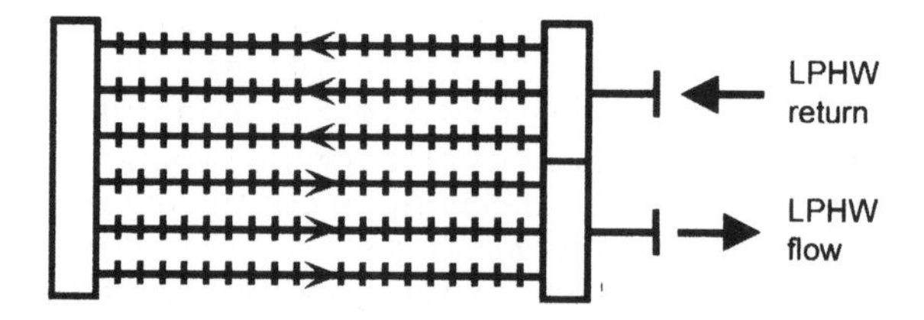

Figure 3.5 Layout of a typical 'copper' water-tube boiler with atmospheric natural gas burner.

the boiler will include provisions to allow a full 'drain-down' of the water within the heat exchanger.

Modular boilers typically also employ copper water tubes within the heat exchanger modules, although this type of boiler is rarely categorised as a 'copper' boiler.

3.4 Fire-tube and water-tube boilers

Boilers in which the combustion chamber and flue gas passages are surrounded by water are termed *fire-tube* boilers. The converse of this is the *water-tube* boiler, in which the water is contained within tubes with the combustion process going on around them. Cast-iron sectional boilers are inherently of the fire-tube type, with the passages of the combustion gas 'passes' representing the 'fire tubes' (Figure 3.1). Fabricated steel boilers may be of either fire-tube or water-tube type, although the shell-and-tube boilers typically associated with building services applications are of the former type (Figure 3.4). Water-tube boilers are typically associated with very large outputs (in excess of 10 MW), and with the kind of difficult operating conditions that arise when such boilers are fired on 'heavy' oil and waste-type fuels. The exceptions to this are the 'copper' boilers used in building services applications, since these are inherently of the water-tube type (Figure 3.5).

3.5 Modular boiler installations

3.5.1 Types of modular boiler installation

Modular boiler installations provide a cost-effective and potentially space-conscious way to provide (or even improve on) the flexibility and load-matching performance of a multiple-boiler installation where a low-to-medium installed boiler capacity is required. There are two principal approaches to the design of this type of installation. The first approach is, strictly-speaking, a *modular boiler system*, and is based on cast-iron sectional techniques and atmospheric natural gas burners. The second approach is represented by a true *modular boiler*, and is typically based on premix natural gas burners.

3.5.2 Modular boiler systems

A modular boiler system comprises a number of standard-sized boilers that are close-coupled in a modular arrangement, piped in parallel and controlled in sequence by a remote boiler sequence controller. Each boiler is completely self-contained and may in fact be installed as a 'stand-alone' boiler in its own right. The construction of this type of boiler module has been considered in an earlier section (3.3.1) of this chapter. Each boiler is provided with an integral gas line. While overall depth and height of the casing tend to be limited to approximately 1000 mm or so, the width is typically minimised to allow boilers to be close-coupled side-by-side at centres that typically do not exceed approximately 600 mm. In this way compact multiple-boiler installations can be built up (Figure 3.6). In some instances the spacing between module centres is minimised by the provision of a multi-module casing. Water and gas pipework connections are uniformly made to the rear of each boiler module, allowing access to the burner assembly, gas line components and boiler controls from the front of each module. Manufacturers typically offer prefabricated pipework headers for ease of on-site installation.

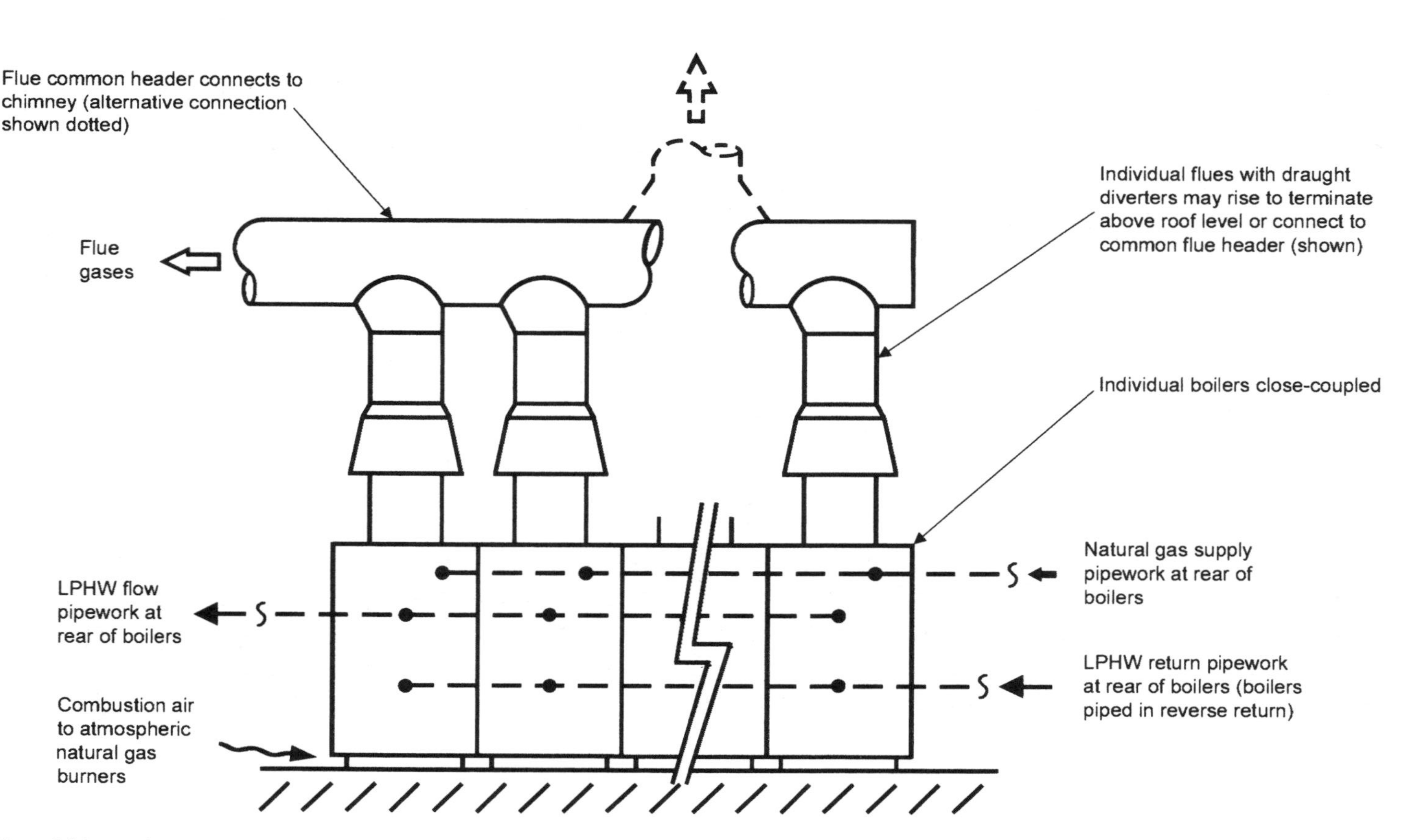

Figure 3.6 Layout of a typical modular boiler system.

Individual boilers may be of a single standard capacity, although the sectional technique also allows a range of capacity to be provided by some manufacturers. 50 kW is a common value for a single standard boiler capacity in modular boiler systems, since it is probably the largest capacity step that is acceptable for the range of commercial space heating applications that require total installed capacities in the region of 150–250 kW, and will also allow many smaller applications to be served by one or two boilers of this size. Where a range of boiler capacity is provided, it is unlikely to extend significantly below 50 kW or significantly above 100 kW. Nevertheless, where up to six boilers can be close-coupled in a modular configuration, a wide range of configurations and installed capacity is possible. The absolute maximum number of boilers that can be close-coupled in a single modular grouping (which is sometimes, and perhaps confusingly, referred to as a 'module' of boilers) is probably ten, although limits of six to eight are more likely to be recommended by most manufacturers. However, it is in principle possible to build up installed capacity in more than one such grouping, if suitable flueing and sequence control arrangements may be made. Except where space limitations may necessitate more than one smaller grouping or module of boilers, the practical benefit of doing so may be questionable. Condensing boilers are available that can be close-coupled in the same modular configuration, and some manufacturers provide units of both types in a compatible physical and dimensional format that allows mixing of both types within the same modular grouping.

In each individual boiler, control of the burner is typically on/off although types are available that offer 'high/low/off' burners. As in the case of conventional multiple-boiler installations, firing of individual burners (or burner firing stages) remains under the local control of their own boiler control thermostats, but is enabled or disabled in a sequence that is determined by the remote boiler sequence controller. The boiler and burner technology employed by modular boiler systems is conventional and well-proven, and individual boilers may be isolated for maintenance or repair. Where capacity of the individual boilers is greater than 50 kW, it is also a space-conscious approach. While the modular boiler system will require multiple valves and other pipework components, these will inherently be small and relatively low-cost items.

Modular boiler systems probably give way to conventional multiple-boiler installations somewhere in the range from 500–1000 kW of installed capacity, although there are no clear-cut rules regarding the application of either approach, and each designer or specifier of boiler installations will have his or her own preference and view on the subject. Ultimately, where either approach is viewed as an option for a particular installation, the life-cycle performance of each should be analysed against a chosen criterion of either minimisation of life-cycle energy consumption, minimisation of life-cycle cost or minimisation of life-cycle carbon emissions.

3.5.3 True modular boilers

A true *modular boiler* in the strictest sense of the term incorporates a number of identical *individual heat exchange modules*, each with its own premix natural gas burner (Figure 3.7a). The general form of these was described in Chapter 2 (section 2.4.6). The combustion gases from each heat exchange module are discharged into a common boiler casing, which provides an outlet

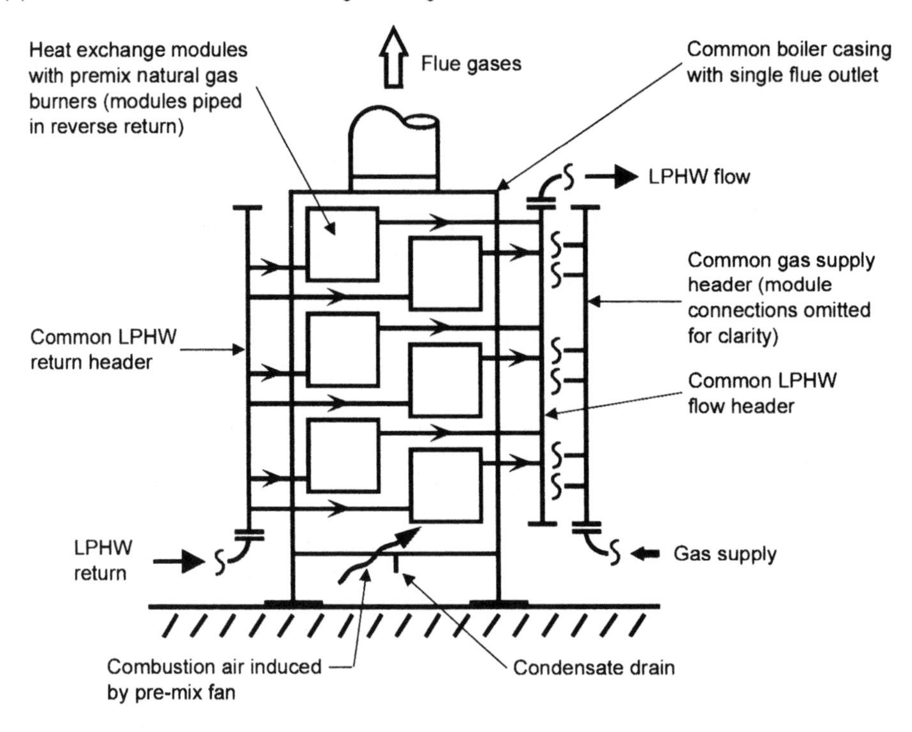

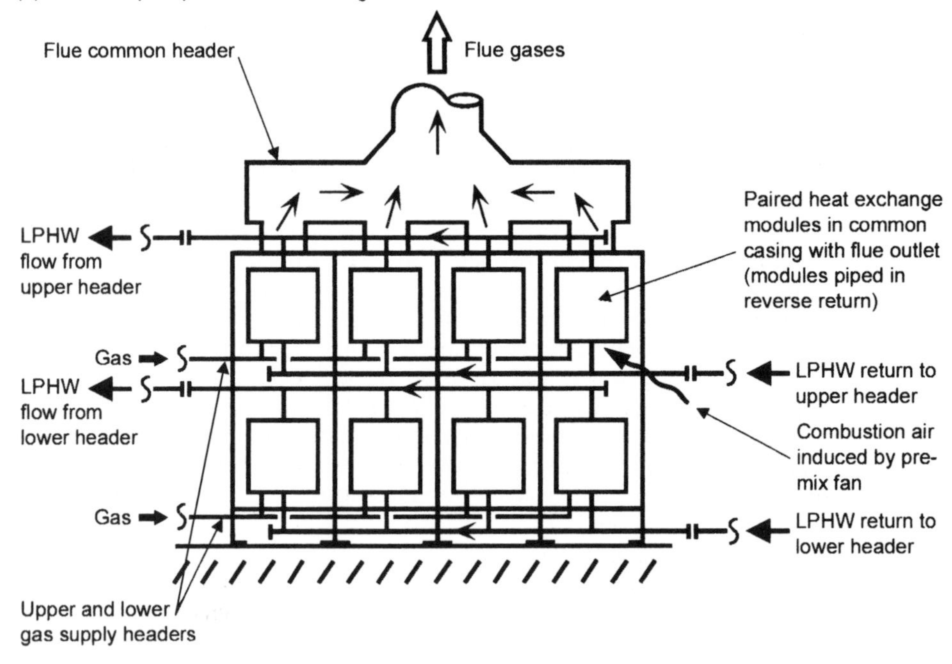

Figure 3.7 Configurations of typical modular boilers.

connection to the single boiler flue. Each heat exchange module has independent water flow and return and gas line connections to common pipework 'headers' that form part of the modular boiler itself.

Modular boilers have been around for many years now. The original concept standardised on heat exchange modules with a single capacity rating of 50 kW, with individual modular boilers typically providing up to six or twelve such modules disposed in a combination of horizontal and vertical banks, with unused module positions blanked off. While a module capacity of 50 kW probably remains most common, most manufacturers of this type of boiler also offer heat exchange modules with a capacity of 100 kW. Since it originally appeared on the UK market, the modular boiler concept has naturally developed to offer a range of variants. One such variant is effectively a *hybrid* of the modular boiler and the modular boiler system (Figure 3.7b), in that it builds up an installation by close-coupling what are in effect 'twin' modular boilers that each comprise a pair of vertically-superimposed heat exchange modules sharing a common casing and flue outlet. Close-coupling up to six such units allows installed capacities between 50 and 600 kW to be achieved. Again where a heat exchange module is not needed to achieve the required installed capacity, it may be blanked off. Such an arrangement, while for practical purposes qualifying as a modular boiler, does differ from this in its strictest sense in that the individual flue outlets from each pair of heat exchange modules must be connected to a single common flue by a flue header arrangement. Nor is this the only variation on the modular boiler theme, with individual manufacturers offering other groupings and combinations of heat exchange modules. In some instances modular boilers built up from paired heat exchange modules may provide a particular application for series-connected boilers in which the heat exchange modules *within each pair* are piped in series, to achieve a high temperature rise, but the close-coupled *pairs of modules* are piped in parallel. While the premix burners in heat exchange modules have traditionally fired on natural gas, modular boilers are available for use with LPG. Whatever the approach, twelve individual heat exchange modules typically remains the maximum for any individual modular boiler. Where a higher installed capacity than this permits is required, two or more modular boilers may in principle be installed, although (as noted for modular boiler systems) the practical benefit of doing so may be questionable.

Where heat exchange modules have a capacity of 50 kW, control of individual burners is typically on/off. Modules with a capacity of 100 kW may employ high/low/off burners. Each heat exchange module typically has an individual control system, including control and high-limit thermostats, and operation is fully automatic. In the strict modular boiler sequence, control of the burners in each heat exchange module has traditionally been achieved by appropriate setting of the module control thermostats, which sense the return water temperature to each module. Where paired or triple heat exchange modules are close-coupled to form an installation, sequence control of the heat exchange module burners is typically provided by a remote boiler sequence controller. However, particularly where a relatively small number of modules is involved, integral control packages are sometimes offered that include not only basic sequence control of the heat exchange module burners but also more-or-less comprehensive control functions that allow 'stand-alone' operation and automatic control of the complete heating system.

Use of finned-copper-tube heat exchangers typically results in a significant reduction in weight over conventional cast-iron sectional boilers, for the same installed capacity. Values between one-half and one-third of the weight of an equivalent arrangement of multiple cast-iron sectional boilers may be possible for some modular boiler installations. The heat exchange modules of modular boilers are designed for efficient heat transfer and to respond rapidly to a demand for heat. They have a low water content and are therefore potentially sensitive to any reduction in water flow rate through the heat exchanger when the burner is firing. Minimum acceptable water flow rate may be as high as 80% of the design value at a temperature rise across the module of 11 K. It is thus very important that the mode of operation of the remote heating system controls does not inherently result in any significant reduction in the water flow rate through the modules. The heat transfer characteristics of modular boilers may also pose specific requirements on the boiler flues, and this is considered in Chapter 5.

Modular boilers typically offer similar options to the designer or specifier of a heating system as does a modular boiler system. The two are, after all, variations on a common theme. Water-side isolation of individual heat exchange modules is typically not provided, although isolation of the gas supply to each module obviously is. Although individual heat exchange modules may be spaced at centres of less than 500 mm, it is perhaps more questionable as to whether a modular boiler installation always represents a *space-conscious* approach to providing the flexibility and load-matching performance of a multiple-boiler installation, at least where the capacity of the modules is limited to 50 kW. The technique has, however, spawned offspring in the form of heat exchange modules up to a capacity of 200 kW, which in single or vertically paired form can indeed provide a very small footprint for the installed capacity. As to the question at what point, if any, modular boilers should give way to conventional multiple-boiler installations, somewhere in the range from 500–1000 kW of installed capacity would again seem most likely. Again there are no clear-cut rules and there are many different opinions from designers and specifiers of boiler installations. Manufacturers, being natural optimists, frequently suggest that their products of this type are suitable for installed capacities up to 2000 kW. As ever, the solution for a particular application should take into account the appropriate life-cycle performance of the various options available to the designer or specifier. It is worth noting that, whether fairly or not, modular boilers have acquired a reputation among some designers and specifiers for poorer reliability than other types where maintenance regimes have been less than ideal.

3.6 Heating, hot water service and combined applications

The fundamental problem with using a boiler plant for both space heating and hot water service (HWS) arises where requirements for the latter are low or even minimal, as is the case in perhaps the majority of ‘office-type’ commercial applications. In such circumstances the boiler plant must continue to run outside the space heating season purely to satisfy an HWS heating load that represents only a small percentage of the installed boiler capacity. This has a detrimental effect on the seasonal (or in this case *annual*) efficiency of the boiler. The effect is worst where a single boiler with on/off burner is employed, while modulating burners and the modern practice of using multiple and modular boilers is certainly beneficial. Alternatively separate space heating and HWS boilers may be provided, allowing the former to be shut down outside the months of the heating season.

Since the *combination boilers* ('combi-boilers') that have been widely applied for domestic use are typically only applicable at the level of 'bed-sits' or small 'flats', their use in even the smallest commercial heating installations is inappropriate. This type of boiler will therefore not be considered here. On the commercial scale, packaged equipment is available that combines an LPHW space and HWS heating boiler and a 'close-coupled' HWS storage calorifier within a common casing (Figure 3.8). The calorifier is typically mounted above the boiler and may overhang it. The 'primary' (LPHW) flow and return connections to the calorifier are taken from the main boiler flow and return water connections, and the package typically includes a circulating pump for the primary LPHW circuit through the calorifier and an integral (HWS) control thermostat. Units of this type are typically offered as variants of a standard boiler range. With such equipment the intention is to reduce the cost and space requirements of a conventional combined space heating and centralised HWS installation.

The modern trend towards *separating* the space and HWS heating loads in commercial building services applications has led to the development of a range of direct gas-fired *hot water generators* that are specifically designed to serve HWS heating loads. In essence the approach takes the 'traditional' concept of a dedicated HWS boiler and removes the need for the associated storage calorifier. Hot water generators are considered further in Chapter 8.

3.7 Condensing operation

3.7.1 Principles and benefits

The principal constituents of the common boiler fuels – natural gas, LPG and the various grades of fuel oil – are carbon and hydrogen. When any of these fuels are burnt in the combustion chamber of a boiler, the hydrogen combines with the oxygen in the combustion air to produce water vapour (H_2O). In Chapter 2 we saw that it is an inherent characteristic of each of these hydrocarbon fuels that there are more hydrogen atoms in each molecule than there are carbon atoms. Natural gas is almost entirely (over 90%) methane, which has a molecular formula of CH_4 (i.e. four hydrogen atoms for every carbon atom). This is in fact the highest ratio of hydrogen atoms to carbon atoms in the common boiler fuels, and leads to substantial water vapour production. At the high temperatures of the combustion chamber this water vapour is superheated.

As the combustion gases are cooled on their passage through the flueways of the boiler block their temperature drops, since the process is one of sensible heat transfer. In theory, the more that the combustion gases can be cooled in their passage through the boiler, the better. As far as most of the combustion gases are concerned (the carbon dioxide produced by combustion, the excess oxygen remaining from the combustion air and the nitrogen of the combustion air), no matter how far they are cooled in the boiler, they will still enter the flue and ultimately be discharged to atmosphere as gases. However, if the water vapour in the combustion gases is cooled sufficiently, it will *condense* from water vapour to liquid water, releasing its latent heat during the process. This condensation occurs at a constant temperature, after which the liquid water or condensate may continue to cool by further sensible heat transfer, which will again be associated with a reduction in its temperature.

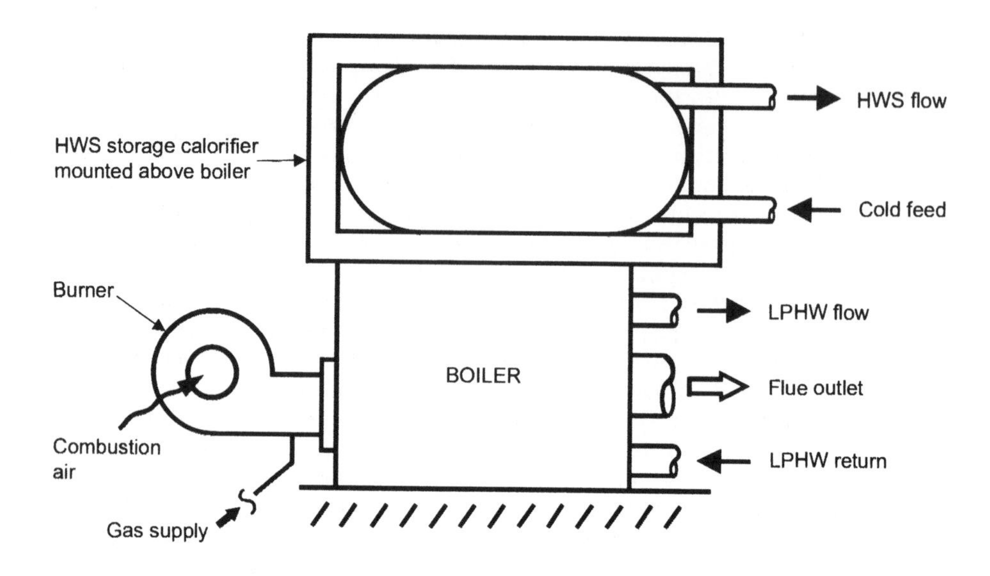

Below : rear elevation

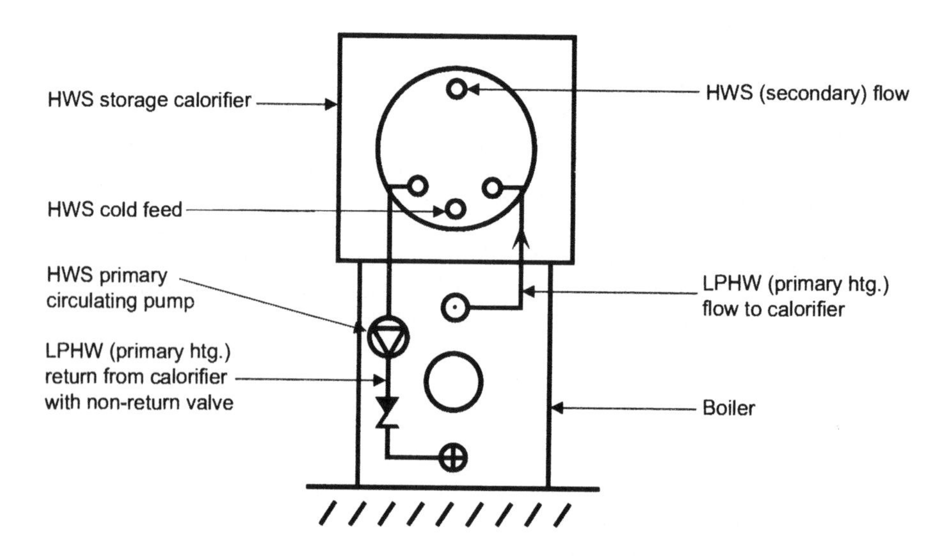

Note: Calorifier open vent, drain and secondary return connections omitted for clarity.

Figure 3.8 LPHW boiler with close-coupled HWS storage calorifier.

The condensation of water vapour releases a large amount of heat; 1 kg of water vapour condensing at 55°C releases approximately 2370 kJ of heat[2], and the combustion of 1 m^3 of natural gas will produce in the region of 1.5–1.6 kg of potential condensate. A conventional (non-condensing) natural gas boiler with a capacity of 100kW heat output to water at an efficiency, based on gross calorific value, of 85% will consume approximately 11 m^3 of natural gas for each hour that the burner is firing continuously at full output. Hence if all the water vapour produced by the combustion process could be continuously and steadily condensed (a situation that is, of course, unachievable in practice), approximately a further 11kW of heat would be released, or 11% of the boiler output. By comparison, a further steady and continuous cooling of the liquid condensate by 10K would release less than 0.2kW of heat, a negligible effect.

In conventional boilers the water vapour in the combustion gases leaves the boiler and passes into the flue still as a vapour. In some instances condensation of the vapour may occur during its passage through the flue to its discharge to atmosphere, and in any event its occurrence on exit from the flue will frequently be very visible as a *condense plume* drifting away from the top of the flue. Condensation occurring within the boiler flue, while still releasing latent heat, does not of course contribute this usefully to anyone (although in principle it may contribute to the buoyancy of the flue gases). It has been the convention to positively seek to avoid the occurrence of condensation within the flueways of the boiler, since exposing the metal of the boiler block to continuous or repeated heavy condensation will typically lead to accelerated corrosion and early failure of the boiler. To make matters worse, the condensate that forms is inevitably *slightly acidic*, since it will contain a dissolved quantity of the oxides of nitrogen that are generated as an unwanted by-product of the combustion process, together with other substances generated by the presence of traces of impurities in the natural gas. The latter may include trace substances added to the gas as an odorant to aid in the detection of leakage. The slightly acidic nature of the condensate (typically pH 3–5 for condensate formed from the combustion of natural gas) naturally increases its corrosive action. Since the risk of condensation is naturally greatest where the combustion gases are at their coolest just prior to entering the flue at the rear of the boiler, this type of boiler corrosion has traditionally been referred to as *back-end corrosion*, and methods of protecting the boiler against it as *back-end protection*. This remains the situation for conventional boilers, which are designed, installed, operated and controlled to avoid an operating regime in which continuous or repeated heavy condensing occurs.

It is clear, however, that there is a useful benefit to be obtained if we can arrange for a boiler to continually condense as much water out of the combustion gases as possible, and add the heat released usefully to its 'conventional or non-condensing' heat output to water, *without requiring early replacement of the boiler*. This is the purpose and principle of operation of a *condensing boiler*. We have seen that there is no worthwhile benefit to be gained from cooling the liquid condensate, so this is just something to be disposed of safely, conveniently and economically. Condensing operation does not come free of charge. Condensing boilers come at a substantial cost premium over conventional non-condensing types, although this is expected to be recovered from the saving in fuel cost that results from its improved efficiency. *Simple payback periods* of between two and five years are considered typical, although this may of course be affected by the characteristics of a particular installation and the actual unit cost that will be paid for the boiler fuel.

Two further benefits inherently come with condensing operation. The first of these arises since the combustion gases must be cooled further to cause condensation of the water vapour. They thus enter the boiler flue at a lower temperature. This inevitably means that more sensible heat has been extracted from them, with less heat being carried to waste 'up the boiler flue'. Since more of the heat generated by the burner has been usefully extracted, the efficiency of the boiler must have been further improved. For boilers fired on natural gas a reduction of 20 K in the temperature at which the combustion gases are discharged into the flue will typically reduce the associated sensible heat loss (and therefore increase boiler efficiency) by c.1% *in absolute terms*. There is, however, an implication for the flue, since the temperature difference between the combustion gases within the flue and the ambient air is the fundamental factor in generating the buoyancy or natural draught that maintains the flow of the gases themselves through the flue to atmosphere. The second additional benefit arises from the lower boiler flow and return water temperatures associated with condensing operation, resulting in a reduced mean temperature for the water content of the boiler. The surface temperature of the boiler block will reach lower values when the burner is firing, and less heat will be stored within the mass of the block. Heat loss from the surface of the boiler block to its immediate surroundings will therefore be reduced while the burner is firing, while the amount of stored heat lost will also be reduced during periods when the burner has ceased firing (the so-called 'standing' losses).

Since any of the common boiler fuels noted above contain hydrogen, any of them can in principle be used for condensing boilers. However, since natural gas has the highest ratio of hydrogen atoms to carbon atoms, it gives the greatest potential benefit from condensing operation. For natural gas approximately 10% of the gross calorific value (GCV), or 11% of the net calorific value (NCV) is theoretically available if complete condensing can be achieved. The potential benefit to be gained with condensing operation using LPG is less than with natural gas. Approximately 7–8% of the GCV of the fuel, or 8–9% of its NCV, is theoretically available. The least potential benefit arises from condensing operation with oil. Only approximately 5–6% of the GCV of light fuel oil (c.6% of its NCV) is theoretically available. The extent to which the theoretical benefit that is available, if complete condensing takes place, is actually realised depends not only on the design of the particular condensing boiler but also on the design of the heating system that it serves. The latter factor is considered in the following section. At the time of writing, from 50–80% of the potential benefit noted above may be achieved in practice by natural gas condensing boilers, if the heating system can provide what the boiler requires from it for condensing operation. For a particular design, smaller models in the range tend to achieve a higher proportion of the theoretical condensing benefit. A *condensate index* is sometimes used to indicate a measure of the extent to which the theoretical condensing potential is achieved, and is simply the ratio between the amount of condensate that is actually formed during condensing operation and the amount that is theoretically available due to the chemical structure of the fuel.

In the UK at least, condensing boilers are invariably *natural gas* condensing boilers. Condensing operation with even the lighter grades of fuel oil is not favoured, since the increased level of impurities in these fuels, and in particular of sulphur (which leads to the formation of sulphur dioxide in the combustion gases), leads to an *aggressively acidic condensate*. While in principle materials are available that can offer resistance to such condensate (in particular, specific grades of stainless steel), and although their application may only be necessary in certain parts of a condensing boiler, their use is typically uneconomic. Condensing operation with LPG has all of the cost premium associated with a natural gas condensing boiler, but less of the benefit, and the viability of its application, except in particular circumstances, must therefore be questionable. A special case of condensing operation may exist where 'dual-fuel' burners are employed. In such installations natural gas is invariably the first-choice fuel, and the burner is typically only fired on oil in the event of an interruption in the supply of natural gas. Hence condensing operation with oil is potentially infrequent and short-lived, and some leeway may be acceptable. Manufacturers will be best-placed to make recommendations regarding their own products, and should always be consulted in such cases.

For a given fuel the ratio of water vapour to carbon dioxide produced during complete combustion is a property of the fuel (the ratio of hydrogen atoms to carbon atoms). If we follow an analogy with which HVAC engineers will typically be familiar, that of mixtures of air and water, we can appreciate that, once the water vapour in the combustion gases stops behaving like a gas and becomes a condensible vapour, the temperature at which it will start to condense, *its dew-point temperature*, depends on how 'saturated' the 'dry' components of the combustion gases are with the water vapour. Since the ratio of water vapour to carbon dioxide produced during complete combustion is a fixed property of the fuel, the only additional factor that will affect this is the amount of excess air supplied for the combustion process. For a particular fuel, the percentage content *by volume* of carbon dioxide (CO_2) in the combustion gases gives a measure of this, and hence of the water vapour dew-point temperature for that fuel under the particular combustion conditions. The percentage content by volume of CO_2 in the flue gases is a parameter that has traditionally been measured to assist with, or subsequently check, the burner set-up in automatic boilers, providing an indication of the degree to which excess air is being supplied for combustion. For natural gas at c.9.5% CO_2 content the water vapour dew-point temperature is approximately 55°C. This reduces to approximately 40°C at c.4–5% CO_2 content, and to less than 25°C at 1% CO_2. The relationship is not linear but is a reasonably close approximation of this between 10% and 4% CO_2 content. The combustion air itself will not of course be dry. With light fuel oil the water vapour dew-point temperature will be lower for a given CO_2 content, from approximately 13 K at c.9.5% CO_2 to approximately 10 K at c.4.5% CO_2. While oil-fired boilers typically operate with a higher CO_2 content (c.12–13%) than natural gas boilers, the difference in the water vapour dew-point temperature is still of the order of 10 K. We shall see that this allows the requirements for condensing operation to be achieved more easily with natural gas. While it may seem obvious that it is the *water vapour dew-point* that we are referring to when considering condensing boilers, it is important to always specify this clearly. This is because the water vapour dew-point temperature is not the only dew-point temperature referred to for boilers. Irrespective of condensing operation, an *acid dew-point* also exists where, as in the case of oil-fired boilers, the fuel contains sulphur that burns to form sulphur dioxide (SO_2). Some of this is further oxidised to sulphur trioxide (SO_3), which readily combines with the water vapour in the combustion gases to form a vapour of weak sulphuric acid (H_2SO_4). This vapour will itself condense if the combustion gases are cooled below the acid dew-point temperature, at which they become *saturated* with the acid vapour. This acid dew-point temperature is considerably higher than the water vapour dew-point temperature.

3.7.2 Requirements of condensing operation

There are four basic requirements associated with condensing operation. First and foremost is a requirement for a boiler that is *designed for* condensing operation. Second is a steady feed of water returning to the boiler from the heating system *at a temperature that will cause condensation to occur.*

While for natural gas boilers condensation is likely to occur when water returns to the boiler at a temperature of approximately 55°C or less, this temperature has more relevance as a lower limit to be avoided to prevent (or at least limit) condensation in conventional boilers. It is not a design return water temperature to *achieve* condensing operation. The further the combustion gases are cooled below their dew-point temperature, the greater the quantity of condensate that will be formed. So in principle the lower the temperature at which water is returned to the boiler, the better. However, there is a limit to just how low the temperature of return water provided by a practical heating system (or part of it) can be. As far as space heating is concerned the return water temperature must maintain a minimum temperature differential above the temperature of the heated spaces if the full heat transfer surface of the heat emitters is to be usefully employed. This therefore probably puts a practical lower limit of c.30°C on the temperature at which return water can be provided to a condensing boiler in a space heating application. Special considerations may apply if process heating loads are served, but this is atypical of most building services applications. Manufacturers typically recommend that return water temperature should be in the range 30–45°C for condensing operation with natural gas. When it is remembered that heating systems in the UK have traditionally been designed for nominal boiler flow and return water temperatures of 82°C and 71°C respectively, it can be seen that the achievement of condensing operation has significant implications for system design and control, and that some forms of heating system will be more suited to the use of condensing boilers than others. The implications of condensing operation are considered further in Part B of this book. It is sufficient here to note that in achieving condensing operation a boiler may serve a complete heating system or may be piped up to serve a specific load circuit.

The third requirement of condensing operation is that a means must be available of disposing safely and conveniently of the slightly acidic condensate. In the UK it is typically permitted to discharge this directly into the public drainage system, although some European boiler manufacturers do offer treatment systems to increase the pH of the condensate prior to disposal. These treatment systems are typically based on bringing the condensate into contact with a suitable chemical neutralising agent. Due to the acidic nature of the condensate, copper tube is not suitable for condensate drainage pipework, and hard plastic (uPVC or polypropylene 'ABS') pipework should be used. Condensate drainage pipework must naturally be run to an adequate fall (gradient) to ensure a continuous and steady removal of the condensate to waste. If the pipework is run externally where freezing of the condensate could occur, or through relatively unfrequented plant spaces where a blockage could go unnoticed, any failure to clear the condensate continuously to waste may result in it backing up into the flue gas passages of the boiler block.

Since this is clearly undesirable, it is common practice in UK installations to arrange for the condensate from the boiler to discharge into the main run of condensate drainage pipework via a simple 'tundish'. In the event of any blockage occurring, the result will merely be a 'puddle' of greater or lesser proportions on the plant room floor (which will in any case typically be provided with some emergency drainage provisions of its own). Manufacturers of condensing boilers may, however, specify their own requirements for condensate discharge connections to their boilers. Since condensate may not be formed throughout the potential heating season, care should be taken either to avoid the possibility of trapped connections to the building drainage installation drying out (leading to the possible ingress of odours), or to ensure that they are regularly checked as part of the operating and maintenance regime for the installation. The condensate drain from the boiler itself should not, of course, allow the escape of combustion gases from the boiler into the plant room.

Finally, where a condensing boiler and one or more conventional boilers are part of a multiple-boiler installation, the condensing boiler should always be the 'lead' boiler. There are several reasons for this. First we have seen in the preceding section that, even if it is not condensing, the lower flue gas temperature that is an inherent part of the design of a condensing boiler will typically give it a higher efficiency than the conventional boilers in the installation. It is therefore sensible to make the boiler with the highest efficiency the 'lead' boiler, supplementing its capacity as necessary from the lower-efficiency units. Secondly, it is naturally desirable to maximise the hours of condensing operation in a heating season. Finally, in many heating systems it will be possible to reduce the flow water temperature to all or part of the system under those part-load conditions that will require the capacity of only one of the boilers in the installation, and thus 'force' the return water temperature to a level that will allow condensing operation. This is typically achieved by employing *weather compensation or 'scheduling'* of heating system flow water temperature. This subject is considered further in Chapter 10.

3.7.3 Condensing boilers

For reasons that have already been noted in the preceding sections, our consideration of condensing boilers will be restricted to natural gas condensing boilers. In the UK at least, few if any alternatives will be encountered in practice. Furthermore, as we have seen for their conventional, non-condensing, brethren, there are a range of approaches to the design of condensing boilers and it will not be possible to consider them all in detail here.

Cast-iron, carbon steel and copper are the materials most widely used to form the block or heat exchanger of a boiler. However, *none of them* are suitable to withstand the rigours of condensing operation. Both of the most commonly available materials that *are* able to do so, namely some grades of stainless steel and aluminium, may only be employed at a cost premium that is more or less substantial relative to the base material (copper, after all, comes at a cost premium relative to cast-iron). The cost premium is not only a function of the cost of the material itself but also of the techniques necessary to form it. Of the two available materials, stainless steel carries the highest cost premium. Add to this that only part of the boiler needs to be able to withstand the rigours of condensing operation, namely the part closest to the outlet to the boiler flue, and we can see one of the fundamental characteristics of condensing boiler construction – that the whole boiler is *not* made to withstand condensing operation. It would be absurdly uneconomic to do so. Therefore condensing operation is achieved by the incorporation of an additional, secondary, heat exchanger into what is essentially a conventional, non-condensing boiler. The design of the primary and secondary heat exchangers are naturally optimised in their relationship to each other, but the principle remains true. Hence, for example, cast-iron may continue to be employed for the primary heat exchanger or boiler block, while the secondary or condensing section may employ aluminium as the primary construction material. Strictly-speaking, there is no problem on the water side of the secondary heat exchanger; it is only the surfaces exposed to the acidic condensate that need to be of an enhanced durability.

The principle of the primary/secondary heat exchanger arrangement is illustrated in Figure 3.9. The physical relationship of the secondary to the primary is a function of the design of each individual boiler, and Figure 3.9 shows examples where the secondary is above, below and behind the primary. Almost without exception, however, the secondary heat exchanger comprises an array of water tubes with fins that provide extended gas-side heat transfer surfaces, and it owes more to the design of compact gas/liquid heat exchangers than to conventional cast-iron or steel boilers. In any event it is essential that the flow of the combustion gases should not hinder the continuous and steady removal of condensate from the surfaces of the secondary heat exchanger, for which gravity is the principal driving force, and should preferably aid it. Condensate that is trapped or cleared only slowly may acquire an increased level of acidity that the heat exchanger is not designed to withstand, and accelerated local corrosion may result. Small quantities of particulate matter, in the form of chemical salts, may be deposited with the condensate and left behind as it drains away. Periodic brushing of the surfaces of the secondary heat exchanger may therefore be required to prevent a build-up of material that could both reduce the condensing performance and again result in local corrosion.

There are two fundamentally-different approaches to the way in which a condensing boiler is connected to the load circuit that it serves. In the first approach the primary and secondary heat exchanger are connected *in series* with each other. Where, as is typically the case, the series connection is made within the boiler, common flow and return water connections are employed (Figure 3.10a). Where it is made external to the boiler, separate flow and return connections are made to the primary and secondary heat exchangers (Figure 3.10b). In the second approach the primary and secondary heat exchangers are connected *in parallel*, and here again separate flow and return connections are made to each (Figure 3.10c). This is sometimes referred to as a *split system*.

Condensing boilers tend to employ atmospheric or premix natural gas burners. The secondary heat exchanger inevitably presents an additional resistance to the flow of combustion gases through the boiler. Where atmospheric gas burners are employed, it is therefore common practice for an *induced-draught fan* to be provided on the outlet side of the secondary heat exchanger to ensure that adequate draught is provided to draw the necessary air into the combustion chamber, and to draw the combustion gases through both the primary and secondary heat exchangers. This fan typically also provides adequate pressure to expel the combustion gases through the boiler flue

(a) Secondary heat exchanger above primary

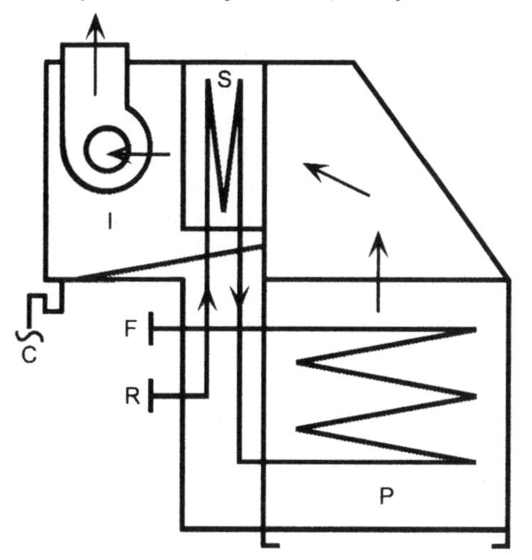

C Trapped condensate drain
F LPHW flow from primary heat exchanger
I Induced-draught fan
P Primary heat exchanger/boiler block (burner not shown for clarity)
R LPHW return to secondary heat exchanger
S Secondary (condensing) heat exchanger (shown connected in series with primary)

(b) Secondary heat exchanger below primary

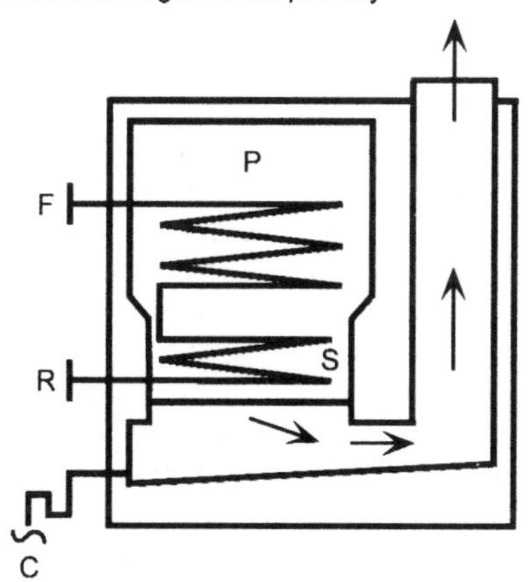

Note : Configuration employed by boilers with pre-mix burners. Burner fan provides mechanical draught for flue.

(c) Secondary heat exchanger behind primary

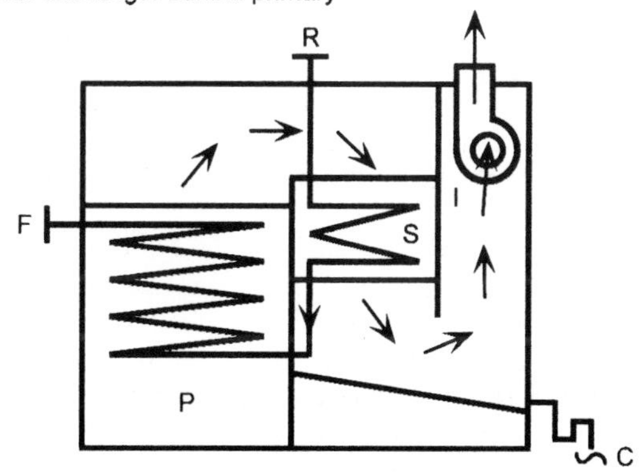

Figure 3.9 Configurations for primary and secondary heat exchangers in condensing boilers.

(a) Primary and secondary connected in series internally

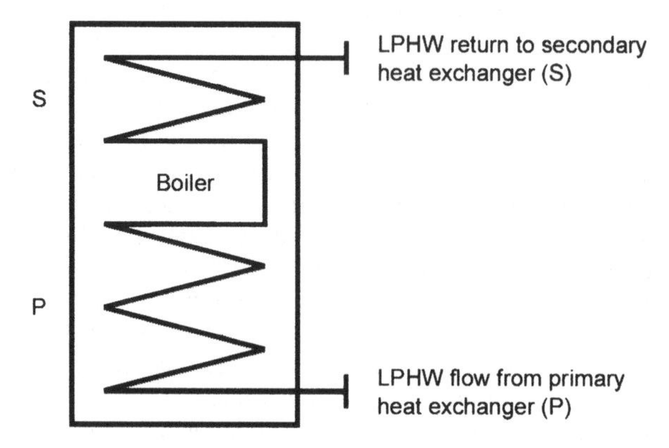

(b) Primary and secondary connected in series externally

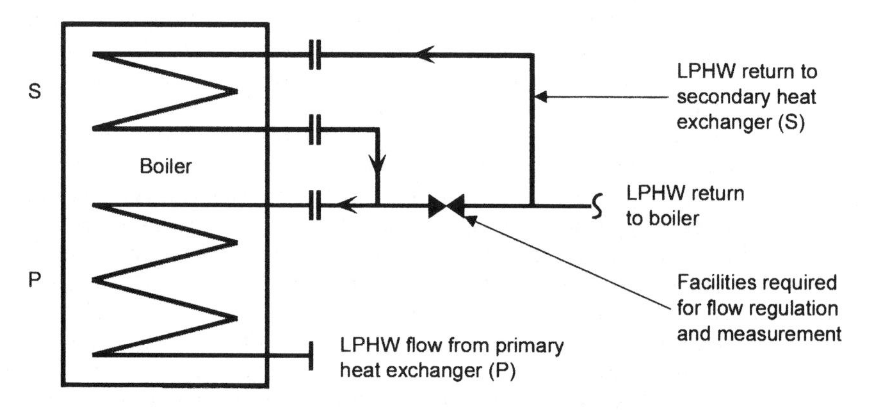

(c) Primary and secondary connected in parallel

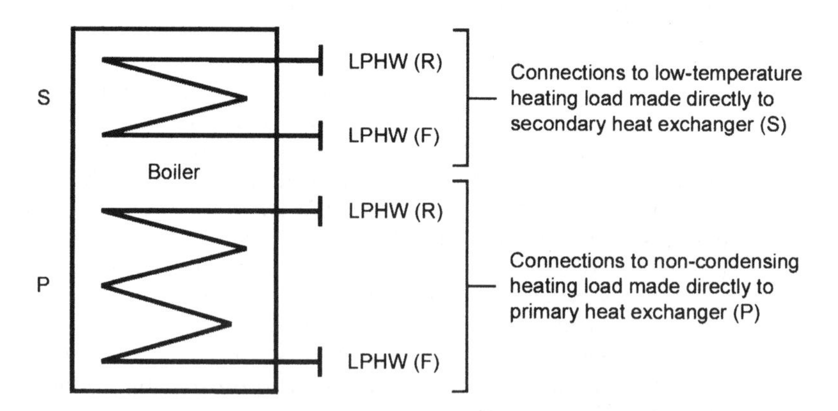

Figure 3.10 System connection options for condensing boilers.

without reliance on the weakened natural draught available from the low-temperature gases. Although provision of the induced-draught fan is common practice, it is not universal, the need for it (or its benefit) depending on the design of the particular boiler. For practical manufacturing purposes, an induced-draught fan will need to be able to cope with a range of flue pressure requirements in an energy-efficient manner, and this is typically achieved by a fan with an adjustable speed range that may be set to the requirements of the particular flue configuration during the on-site commissioning of the boiler. Strictly-speaking the electrical energy consumed by the induced-draught fan should be taken into account in analysing the energy, cost and carbon emission performance benefits of the condensing boiler. The energy consumed by the combustion fan in a fan-assisted or forced-draught burner should, of course, also be taken into account when considering the efficiency of conventional (non-condensing) boilers, but typically it is not. In any event, the use of an induced-draught fan will inevitably place the boiler flue under *positive pressure*, and the standard of its design, manufacture and installation must therefore be such as to prevent any significant leakage of combustion gases into the plant room. In the case of premix burners the premix fan or common combustion air fan provides a forced draught and operates in the more benign environment on the plant room side of the boiler.

In a condensing boiler the addition of the secondary heat exchanger connected in series with the boiler block or primary heat exchanger typically increases the water-side pressure drop across the boiler in comparison with a conventional (non-condensing) boiler of equivalent capacity. Condensing boilers may typically be employed as one or more units in parallel with conventional types in a multiple-boiler installation. It is therefore important that their *different hydraulic characteristics* are taken into account in the design of such installations, and adequate provision is made to ensure that the correct design water flow rates are achieved through each boiler. In such situations it will typically not be adequate simply to pipe all of the units in a parallel 'reverse return' arrangement. Flow measurement and regulating valves will be essential to all non-condensing boilers, and a local shunt pump may be required for each condensing boiler. Condensing boilers may also be piped in series with other condensing or non-condensing boilers. In the latter case the condensing boiler will be the first in the series, providing back-end protection for the non-condensing unit. Where the primary and secondary heat exchangers of condensing boilers are piped in parallel with the heating system load circuits, manufacturers may recommend minimum water flow rate or load criteria for the secondary heat exchanger. In any event the water-side pressure drop across the secondary heat exchanger must be accounted for by the circulating pump in the appropriate load circuit of the heating system. In such situations it may be possible to connect the secondary heat exchangers of more than one condensing boiler in series. However, this is only likely to be of practical utility where the boilers employ modulating burners with a wide range of modulation, so that there is no (or minimal) penalty to be paid, in terms of the efficiency of the installation, for potentially having the burners in each boiler modulating at reduced output.

Condensing boilers may employ on/off, high/low/off or modulating natural gas burners. With the latter two types of burner it is important for condensing operation that the combustion air flow is adjusted to avoid any significant increase in the level of excess air, either at the burner low firing rate or throughout its modulating range respectively, since this would reduce the dew-point temperature of the combustion gases. The control of individual boilers and their sequencing in multiple-boiler installations is generally the same as for conventional (non-condensing) types. We have already noted the principal exception to this, which is that the 'lead' boiler in a 'mixed' multiple-boiler installation should typically *always be a condensing boiler*.

While it is typically not possible to convert an existing non-condensing boiler to a condensing one, there are of course exceptions to every rule, and the authors are aware of at least one manufacturer with parallel designs of conventional and condensing boilers that do permit the on-site conversion of an existing boiler from the conventional range to an equivalent model in the parallel condensing range.

3.7.4 Condensing economisers

A condensing *economiser* provides a means of achieving condensing operation without employing a specially-designed condensing boiler. It may thus be used in conjunction with a conventional cast-iron sectional or steel boiler with forced-draught natural gas burner. The condensing economiser is an *independent* secondary gas-to-water heat exchanger that is close-coupled on the gas side to the boiler flue outlet connection (Figure 3.11). Condensation therefore occurs *outside the boiler*. The heat exchange surfaces of the condensing economiser typically comprise a battery of extruded aluminium finned tubes, and all surfaces that are in contact with the condensate are similarly of aluminium construction. However, in some instances stainless steel construction is employed, and the heat transfer surfaces then take the form of a battery of smooth stainless steel tubes. The flue gases may flow downwards across the heat exchanger to aid the gravity flow of condensate and promote its steady and continuous removal. A condensate collector is provided, with a suitable connection to allow disposal of the condensate to an external drain line. Construction of the economiser should provide access for periodic inspection and cleaning of the heat exchanger. The whole unit is enclosed in a thermally insulated steel casing, which is typically free-standing on steel legs.

The water side of the condensing economiser may be connected in series or in parallel with the 'host' boiler. When series-connected, the economiser pre-heats water returning from the heating system before it enters the boiler block. The boiler minimum return water temperature must be respected. With the simple piping connections shown in Figure 3.11 the minimum return water temperature to the condensing economiser must be limited to a defined offset from that to the boiler. The value of this offset will naturally depend on the design of the condensing economiser but may typically be expected to have a maximum value of 2–3 K. This simple arrangement is therefore only suitable for condensing operation if the host boiler, while not actually designed for condensing operation, can tolerate return water temperatures substantially below 55°C. However, more complex circuiting and control arrangements are possible that allow the return water temperature to the economiser to reach low values without detriment to the boiler.

The temperature of the flue gases from the conventional host boiler are high, and the 'condensing' economiser will recover a substantial amount of sensible heat at higher (non-condensing) return water temperatures. The high gas entering temperature also means that the economiser will typically require protection against the occurrence of an excessively high flow water tempera-

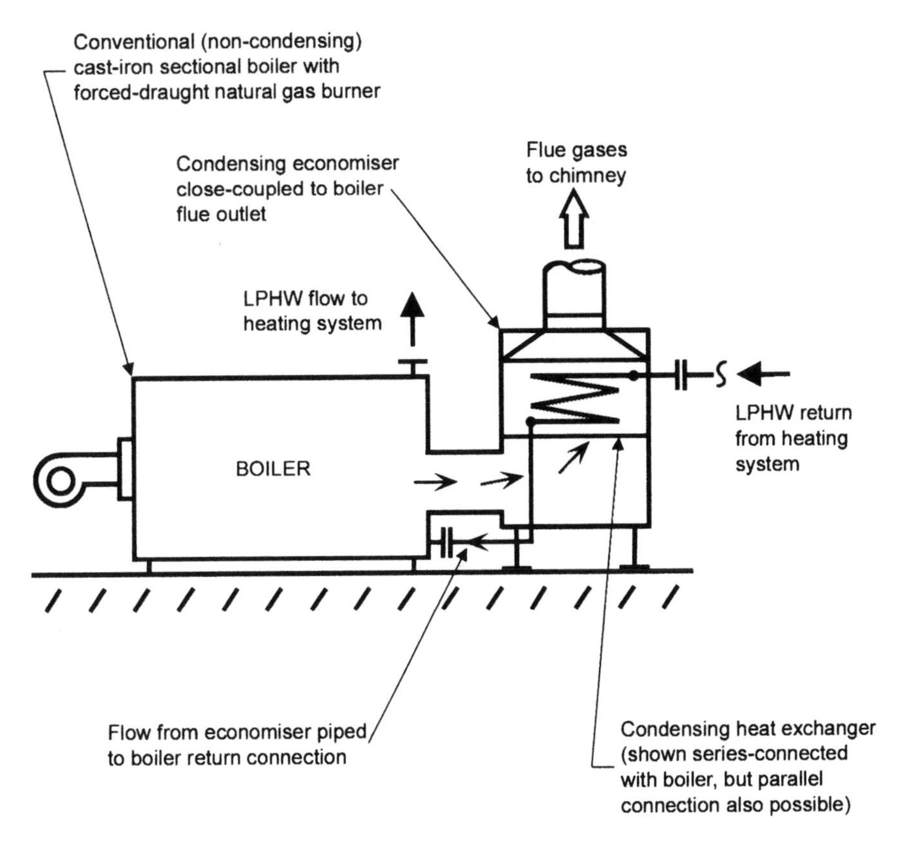

Figure 3.11 Layout of a cast-iron sectional boiler with condensing economiser.

ture. This may be achieved by ensuring that a minimum water flow rate is maintained through the unit and by a safety interlock to shut down the burner of the host boiler. A safety valve will be required if the water-side connections of the condensing economiser may be isolated while the burner of the host boiler may commence or continue firing. The additional gas-side pressure drop associated with a condensing economiser must be made good by the host boiler's forced-draught combustion fan and the design of the boiler flue. This may put the condensing economiser, together with part or all of the flue, under positive pressure. Appropriate standards of design, manufacture and installation will therefore be necessary to prevent any significant leakage of flue gases into the plant room. While condensing economisers are typically designed for operation with natural gas boilers, their use may be acceptable where 'dual-fuel' burners are employed.

In some instances a condensing economiser may be installed in a 'bypass' off the common flue header of more than one boiler (Figure 3.12). In this situation the economiser is equipped with a fan to draw the flue gases from the header and return them to it without reliance on the combustion air fans of the individual boilers. It is clear that the installation must be designed to avoid the escape of flue gases, and that it must be sized or controlled such that at any time it may not draw from the flue header any more combustion gases than are being generated by the boiler or boilers that are firing at that

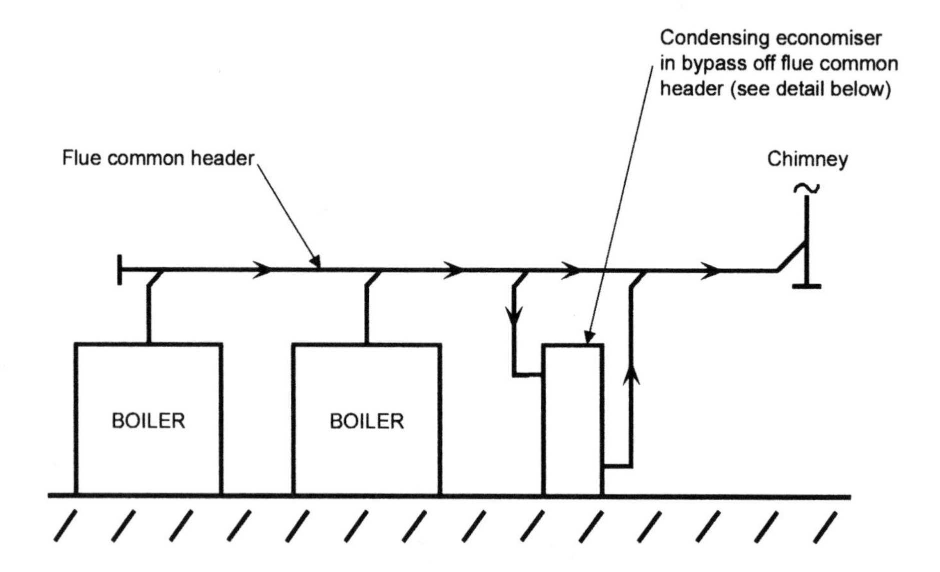

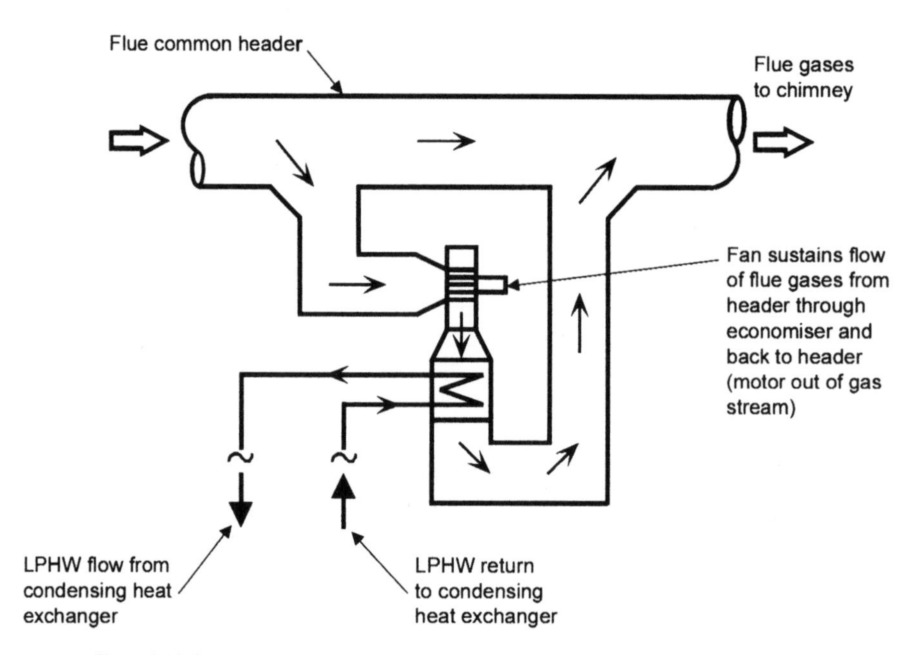

Figure 3.12 Schematic of a condensing economiser in a flue 'bypass'.

time. The flue header bypass fan must be suitable for the continuous handling of high-temperature flue gases, and the fan motor will inevitably be located out of the gas stream. The installation on the discharge side of the fan will be under positive pressure. Installing the condensing economiser in a flue header bypass does, however, offer the possibility of taking it out of operation if a 'dual-fuel' boiler installation must operate on oil firing for any sustained period of time.

3.8 Boiler efficiency

3.8.1 Choosing the correct definition

Unfortunately the term 'efficiency' is frequently used rather loosely with respect to boilers, and an unqualified boiler efficiency may therefore be interpreted in several ways. Add to this that any such boiler efficiency may be based on either of *two different calorific values* for the boiler fuel, and we have significant potential for confusion.

3.8.2 Basis of the definition

At its simplest any efficiency term may be represented by a very basic formula:

$$\text{Efficiency (\%)} = 100\left(\frac{\text{Output quantity}}{\text{Input quantity}}\right) \tag{3.1}$$

As far as the efficiency of gas-fired and oil-fired boilers is concerned, the input and output quantities are typically expressed in terms of MJ or kWh of heat energy. While MJ is the more scientifically 'rigorous' unit, kWh is a very convenient one, since both natural gas and electricity are typically sold by the kWh. (Multiply kWh by 3.6 to obtain MJ, and vice versa.) It is also frequently convenient to substitute *rates* of input and output for the absolute quantities of equation 3.1, and for boiler efficiency the rates of heat input and output are expressed in kW.

The boiler fuel has a potential to be converted into heat by combustion that is expressed in terms of its *calorific value*. This is defined in terms of the amount of heat released when a unit mass of the fuel is completely burnt under closely specified conditions (which include the reference and end states and the thermodynamic process that occurs in-between) in a specific piece of apparatus called a calorimeter. The techniques for determining the calorific values of gaseous and liquid fuels are laid down in British Standards. As far as HVAC engineers are concerned, calorific values that have been determined in accordance with the appropriate standards are declared for natural gas, LPG and the various grades of fuel oil. It is these values that are applied to the consideration of boiler efficiency. Calorific values are expressed in $\mathrm{MJ\,m^{-3}}$ for gaseous fuels, including LPG, and in $\mathrm{MJ\,kg^{-1}}$ for liquid fuels. However, it is again often convenient to convert these values to $\mathrm{kWh\,m^{-3}}$ for natural gas and LPG, and to $\mathrm{kWh\,l^{-1}}$ for fuel oil. While HVAC engineers are typically used to expressing rates of flow in terms of $\mathrm{m^3\,s^{-1}}$ (gas) and $\mathrm{l\,s^{-1}}$ (oil), it will again be convenient here to express them in terms of $\mathrm{m^3\,h^{-1}}$ and $\mathrm{l\,h^{-1}}$ respectively. The rate of heat input to a gas-fired boiler may therefore be expressed as:

$$\begin{array}{lcl} \text{Rate of heat input} = & \text{rate of fuel input} \times & \text{calorific value of fuel} \\ (\mathrm{kW}) & (\mathrm{m^3\,h^{-1}}) & (\mathrm{kWh\,m^{-3}}) \\ (\mathrm{kW}) & (\mathrm{m^3\,s^{-1}}) & (\mathrm{kJ\,m^{-3}},\ \text{or}\ \mathrm{MJ\,m^{-3}} \times 1000) \end{array} \tag{3.2}$$

For oil-fired boilers the rate of fuel input and calorific value refer, of course, to litres.

3.8.3 Gross and net calorific value

The *gross calorific value* (*GCV*) of a fuel is defined on the basis that all of the water vapour produced by its complete combustion is subsequently converted into the liquid state. The heat output from a boiler per unit of fuel input to the burner can never exceed the GCV, and boiler efficiencies referring to the GCV can therefore never exceed 100%. Furthermore, in the definition of calorific value the reference state for fuel and combustion air and the end state for the products of combustion are both 25°C. In any practical heating boiler the products of combustion will leave the boiler at a temperature above this. Therefore, even with condensing operation efficiency will always be *less than* 100% when referred to GCV.

The *net calorific value* (*NCV*) of a fuel is defined on the basis that all of the water vapour produced by its complete combustion remains in the vapour state. If some of this vapour is subsequently condensed, and the latent heat released is usefully added to the heat output of the boiler, it is possible for the heat output per unit of fuel input to the burner to exceed the NCV. The condition for this is, of course, that the heat released by condensing of the water vapour (which is over and above the NCV) exceeds that lost both to atmosphere with the flue gases and from the surface of the boiler to its surroundings. *Boiler efficiencies referred to the NCV may therefore exceed 100%*. Only condensation that contributes to useful heat output counts, however. Condensation that takes place in the flue does not.

Traditionally GCV has been used for the calculation of boiler efficiencies in the UK, while European practice has been to use NCV. However, in recent years this division has become blurred, and it is therefore very important that the basis of any boiler efficiency should be unequivocably defined. The difference between GCV and NCV depends on the ratio of hydrogen atoms to carbon atoms in the fuel molecules, and is thus fuel-specific. The difference is greatest for natural gas. In approximate terms:

$$\text{Efficiency referred to GCV} \times 1.11 = \text{efficiency referred to NCV} \quad (3.3)$$

while

$$\text{Efficiency referred to NCV} \times 0.9 = \text{efficiency referred to GCV} \quad (3.4)$$

The approximate factors required in equation 3.3 when other fuels are considered are 1.09 for LPG (propane), 1.08 for LPG (butane) and 1.06 for light fuel oil. Typical average values for GCV are:

Natural gas	$38.6\,\text{MJ}\,\text{m}^{-3}$,	$10.72\,\text{kWh}\,\text{m}^{-3}$
LPG (propane)	$96\,\text{MJ}\,\text{m}^{-3}$,	$26.67\,\text{kWh}\,\text{m}^{-3}$
LPG (butane)	$122\,\text{MJ}\,\text{m}^{-3}$,	$33.89\,\text{kWh}\,\text{m}^{-3}$
Fuel oil	$38\text{–}41.22\,\text{MJ}\,\text{l}^{-1}$,	$10.56\text{–}11.45\,\text{kWh}\,\text{l}^{-1}$

The values for fuel oil vary according to the grade of oil, increasing from the lighter to the heavier grades. Under UK statute the average calorific value for natural gas must not fall below its declared value over a quarter, although it is permitted to fall substantially for short periods, provided that a drop of 5% is not exceeded in any two-hour period. Corresponding approximate values of NCV may be derived using the factors from equations 3.3 and 3.4. Values are quoted for general reference in the context of this text only. For specific applications, and where the results of calculations may be sensitive to input values, precise calorifc values should always be confirmed with the relevant boiler manufacturer or the fuel supplier. This is particularly so in the case of LPG and fuel oil.

3.8.4 Overall efficiency

The overall efficiency of a boiler, which is frequently referred to simply as the *boiler efficiency*, is defined by the ratio of the total heat output to water to the heat input to the burner. When the burner is firing continuously we may employ rates of heat output and fuel input to derive an expression for overall efficiency, i.e.

$$\text{Overall efficiency (\%)} = 100\left(\frac{\text{total rate of heat output to water}}{\text{rate of fuel input} \times \text{calorific value of fuel}}\right) \quad (3.5)$$

where all quantities have the units described in section 3.8.2 above.

This definition of overall or boiler efficiency naturally assumes that steady-state conditions of boiler operation exist on both the gas side and the water side of the boiler block. For boilers with on/off burners it will only be applicable under full-load conditions, i.e. when the heating load is equal to the rated capacity of the boiler. For boilers with high/low/off burners it will also be applicable when the heating load is equal to the boiler capacity at the burner's low firing rate. For boilers with modulating burners it will be applicable throughout the modulating range of the burner. Where burner cycling will occur under reduced heating load, the rate terms must be replaced by total quantities aggregated over a number of burner operating cycles. The heat output is specified as 'total', since it must include the output of any secondary heat exchanger that provides condensing operation, and which may be piped in parallel with the boiler block or primary heat exchanger. Strictly-speaking, a condensing economiser increases the overall efficiency of the *boiler installation*, rather than the overall efficiency of the boiler itself.

Determination of the overall efficiency of a boiler depends on accurate measurements of fuel input and water-side performance (flow rate and temperature rise through the boiler) but it is neither practical nor economic to perform these on boilers in an actual installation. Furthermore, to allow any meaningful comparison to be made between different types and models of boiler, it is necessary for the measurements to be made under standard conditions and according to a standard methodology. Overall efficiency is therefore determined by testing carried out at laboratory or factory test facilities, and in accordance with prescribed British and European Standards. Such testing is carried out only on typical examples of a particular model of boiler, as part of its design and development to production status.

As far as HVAC designers are concerned, manufacturers may declare values of overall or boiler efficiency for their products, or (more typically in the UK), these may simply be deduced from stated values of heat output to water and heat input published in the manufacturer's technical and performance data for a particular boiler. Such data is typically provided for continuous operation at full burner output, and additionally for continuous operation at the low firing rate where a high/low/off burner is fitted. Except in some instances where modulating burners are employed, manufacturers do not typically provide curves of overall efficiency against heating load *normalised* as a proportion of the rated boiler capacity, and which would indicate the effect of burner cycling under part-load conditions.

Modern conventional (non-condensing) boilers fired on natural gas may typically be expected to have an approximate full-load overall efficiency in the range 78–85% when referred to GCV, which equates to 86–94% when referred to NCV. Modular boilers with premix natural gas burners would be expected to be well-represented towards the upper end of this range, with most conventional cast-iron sectional types occupying the middle third. Where high/low/off or modulating natural gas burners are employed, overall efficiency may actually be higher at the burner low firing rate, or with the burner modulating at an intermediate firing rate, than at its full output.

When modern condensing boilers fired on natural gas are considered in *non-condensing* operation, a minimum value of full-load overall efficiency of approximately 85% may be expected when referred to GCV (94% when referred to NCV). Maximum values of nearly 88% when referred to GCV (c.97% when referred to NCV) may be reached by some designs. In *condensing* operation, a full-load overall efficiency in the range 92–96% may be achieved, when referred to GCV, which equates to 102–107% when referred to NCV. The implications of boiler efficiency are discussed further in Chapter 10.

3.8.5 Combustion efficiency

The definition of *combustion efficiency* regards the output term in the efficiency equation as equal to the rate of heat input to the burner less the rate at which heat is wastefully carried away from the boiler by the combustion gases as they enter the flue. Hence:

$$\text{Combustion efficiency (\%)} = 100 \times \left[1 - \left(\frac{\text{rate of heat removal by flue gases}}{\text{rate of fuel input} \times \text{calorific value of fuel}}\right)\right] \tag{3.6}$$

where again all quantities have the units described in section 3.8.2 above.

The heat carried away from the boiler when the combustion gases enter the flue is also termed the '*stack*' *(or chimney) loss*. Under steady-state conditions of boiler operation the only other factor accounting for a difference between the rate of heat input to the burner and the rate of heat output to water is the rate of heat loss from the outside surface of the boiler block to its surroundings. The latter heat loss occurs by convection and radiation, although it is frequently termed the *radiation loss*. We would therefore expect the combustion efficiency to be greater than the overall efficiency, and the difference to be a direct measure of the extent of the radiation loss. For modern boilers fired on natural gas this may typically range from less than 0.5% to approximately 3%, when referred to GCV. In this instance, and for practical purposes, the values will be the same whether referred to GCV or NCV. An approximate value for combustion efficiency may be determined, for a non-condensing boiler, from site measurements of flue gas temperature and percentage CO_2 content. Reference will typically be required to graphic or tabular data for the specific fuel used to fire the boiler, although empirical relationships may also be applied. Equation 3.7 shows a formula due to Siegert, which expresses the 'stack' loss for a conventional (non-condensing) boiler as a percentage of the *NCV* of the fuel:

$$\text{Stack loss (\%)} = K\left(\frac{\text{flue gas temperature (°C)} - \text{combustion air temperature (°C)}}{\%CO_2}\right) \quad (3.7)$$

where K = a fuel-specific constant, for which values of 0.38 (natural gas) and 0.56 (fuel oil) are typically quoted[3].

The flue gas temperature should represent as closely as possible the temperature of the combustion gases *leaving the boiler*, while the combustion air temperature is the temperature at which combustion air enters the boiler (effectively the plant room temperature in most installations). An increase in CO_2 content, resulting from a reduction in the amount of excess air supplied for combustion, will typically reduce the magnitude of the 'stack' loss. Equipment and procedures for the on-site sampling and analysis of boiler flue gases are prescribed by British Standards. However, measurements should take place where the flue gases are well-mixed, avoiding areas where stagnant pockets of gas could occur (such as bends) or where leakage could occur into a flue under negative pressure. In the latter case any leakage will be of 'cold' plant room air into the flue, which may affect measurements or remain as a discrete layer adjacent to the wall of the flue.

3.8.6 Seasonal efficiency

The seasonal efficiency of a boiler represents its efficiency taken over a complete *heating season*. It is effectively the mean or average efficiency of the boiler during all of the hours of the heating season in which the burner is firing, and the average efficiency with which every kWh of heat input is converted to heat output to water. It is the seasonal efficiency that determines the energy consumption and fuel cost of the boiler installation over a complete heating season, and its carbon emission. For the present purposes we can represent seasonal efficiency simply as:

$$\text{Seasonal efficiency (\%)} = 100\left(\frac{\text{seasonal heat output}}{\text{seasonal fuel input} \times \text{calorific value of the fuel}}\right) \quad (3.8)$$

where seasonal heat output is in kWh; seasonal fuel input is in m^3 (gas) or litres (oil); calorific value is in $kWh\,m^{-3}$ (gas) or $kWh\,l^{-1}$ (oil).

Since calorific value is a term in equation 3.8, any given value of seasonal efficiency must be referred to either GCV or NCV, and the factors given in equations 3.3 and 3.4 will allow conversion between values referred to either calorific value.

The value of the overall efficiency that is typically available to the designer or specifier of a boiler installation is the *full-load value*, which assumes that the boiler is operating continuously at full burner output. In an actual installation, however, this will represent boiler operation for a relatively small proportion of the potential heating season (and for virtually none of it, excluding pre-heat periods, in an oversized single-boiler installation). The seasonal efficiency takes into account any effects that result because a boiler installation spends almost all of its operating time annually satisfying a heating load that ranges from being *substantially less* than the design value to *very substantially less*, and which may vary widely over this part-load range from day to day or even hour to hour. These effects will particularly include the cycling of burners. Typically the seasonal efficiency will be influenced to greater or lesser extent by boiler oversizing, type of burner control and the configuration of the boiler plant, i.e. the number and size of the capacity steps that it offers. Sequenced multiple boilers, modular boiler systems, modular boilers and modulating burners offer HVAC designers the potential to maximise seasonal efficiency by matching the heat output from the boiler(s) as closely as possible to the heating load at all times. Boiler sizing and the control techniques for matching boiler output to demand are considered further in Chapter 9.

3.9 Carbon intensity

From 1 April 2002 new central boiler plant installed in England and Wales is typically required by legislation to meet a new performance criterion, that of *carbon intensity (CI)*. The definition of 'new' boiler plant refers to the boiler plant and not to the application. Hence the replacement of existing boiler plant will typically also be subject to this requirement. Carbon intensity is measured in kg of carbon per kWh of rated boiler output ($kgC\,kWh^{-1}$), and is defined by the expression:

$$CI = \left[\frac{1}{\sum \text{boiler rated outputs}}\right] \times \sum\left[\frac{\text{boiler rated output} \times \text{CEF}}{\eta t}\right] \qquad (3.9)$$

where CEF = carbon emission factor for the boiler fuel, defined as the carbon emitted in kg per kWh of delivered fuel consumed ($kgC\,kWh^{-1}$); η_t = gross thermal efficiency of the boiler, expressed as a decimal fraction (i.e. kWh of heat output per kWh of delivered fuel).

The carbon intensity of the boiler plant must be assessed under two conditions of boiler output. The first of these is at 100% of the combined rated outputs of all boilers in the installation. The second is at 30% of their combined rated outputs. The limiting values of carbon intensity that must be met or bettered in each case are:

- At 100% of the combined rated boiler outputs, $CI \leq 0.068$ ($kgC\,kWh^{-1}$) for firing with natural gas, $CI \leq 0.091$ ($kgC\,kWh^{-1}$) for firing with other fuels
- At 30% of the combined rated boiler outputs, $CI \leq 0.065$ ($kgC\,kWh^{-1}$) for firing with natural gas, $CI \leq 0.088$ ($kgC\,kWh^{-1}$) for firing with other fuels. These values represent a c.5% reduction on the full-load case.

In both cases the value of η_t is that for continuous operation at the full rated output of the boiler. In the simplest case when a new or replacement boiler plant comprises several identical boilers the value of CI is equal to CEF/η_t. With a prescribed value of $0.053\,kgC\,kWh^{-1}$ for the CEF of natural gas, a minimum value for η_t of 0.78 will satisfy the requirement for a maximum CI of $0.068\,kgC\,kWh^{-1}$ at the full output of the plant, while a value of 0.815 will satisfy naturally both this *and* the requirement for a maximum CI of $0.065\,kgC\,kWh^{-1}$ at 30% of full output. Where all the boilers in the installation cannot achieve a value of 0.815 or better for η_t, the CI criterion at 30% output may be met by selecting one or more boilers of higher efficiency (such as a condensing type, for example) to provide lead boiler capacity equal to at least this output.

In some instances where the carbon intensity criterion cannot be met, the opportunity may exist to offset an increased carbon emission per kWh of output from the boiler plant by a reduction in the total number of kWh required. This is achieved by reducing the thermal transmittance of the building fabric by the ratio (limiting value of CI/actual value of CI). This may not of course be practical where the replacement of existing boiler plant is concerned.

3.10 The needs of the boiler installation

Boilers have a variety of needs which must be met if they are to function effectively and efficiently as heat generators for the space and water heating systems they serve. Principally these needs concern:

- Hydraulic stability
- Return water temperature
- Pressure in the boiler circuit
- Fuel supply
- Ventilation of the boiler plant room
- Water treatment.

Each of these needs is considered in the remaining sections of this chapter. Since satisfying the boiler's needs for hydraulic stability and the maintenance of a minimum return water temperature rely heavily on the use of appropriate circuiting and control techniques (subjects dealt with in Chapter 10), they will only be briefly introduced at this stage.

3.11 Hydraulic stability

The low water content of modern boilers makes them relatively sensitive to changes in the rate of water flow through them. While reductions or restrictions in flow rate are typically the greatest cause for concern, manufacturers typically also caution against allowing flow rates that are significantly *in excess of* the design value. Major reductions in the water flow rate through a boiler may ultimately risk thermally overstressing the block or heat exchanger, while smaller changes may have undesirable effects on its control or the control of the heating system it serves. *Boilers like a constant water flow rate*, but in the absence of this *they must have a minimum water flow rate at all times.*

The boiler's needs in this respect should be dealt with at two levels. At the first level is the use of appropriate circuiting and system control techniques, and these are considered in Chapter 10. It is worth noting that this must include managing intentional shut-downs of water flow through a boiler, since no change can be bigger and this is one type of major change that all boilers will be subjected to. At the second level are the actual piping arrangements to individual boilers. For single boilers these are straightforward. Where multiple boilers are piped in parallel the *reverse return* arrangement is the preferred means of connecting the boilers into whatever circuiting arrangement is adopted. While such an arrangement adds slightly to the extent of the boiler pipework, it should obviate the need to provide flow measurement and regulating facilities to each individual boiler *where these have identical hydraulic characteristics.* Where, for any reason, multiple boilers may not be piped in a reverse return configuration, individual flow measurement and regulating facilities will be necessary to ensure that design water flow rates are achieved through each boiler.

3.12 Return water temperature

It is typically undesirable for a conventional (non-condensing) boiler to operate for substantial periods with a return water temperature below approximately 55°C for firing with natural gas, and below approximately 60°C for firing with oil. If it is allowed to do so, accelerated rates of corrosion and deterioration of the boiler may be expected. These temperatures are, of course typical. Manufacturers should always be consulted for guidance concerning their own particular products. Particular designs may in some instances have less stringent requirements. Indeed there are instances where manufacturers state that there is *no* requirement for a minimum return water temperature. Undoubtedly a suitable construction to achieve this will carry a cost premium.

The boiler's needs in this respect are typically dealt with by the use of appropriate circuiting and control techniques, which are considered in Chapter 10. For the purposes of the present discussion it is simply worth noting that potentially the greatest risk of prolonged boiler operation at low return water temperature exists during start-up of the installation, and that this period will be greatly extended if the boilers must gradually bring both themselves and the whole heating system up to temperature at the same time. It is therefore desirable for the boilers to first heat *themselves* up as quickly as possible, in order to minimise the time during which they are exposed to the formation of condensate within the boiler 'block' or heat exchanger. Mixed installations of condensing and non-condensing boilers may also require specific circuiting and/or control arrangements to protect the non-condensing units. This aspect is again considered in Chapter 10.

3.13 Pressure in the boiler circuit

In the UK, considerations of health and safety in the workplace require that a minimum temperature difference of 17K is maintained between the temperature at which the boiler control thermostat operates to stop firing of the burner and the temperature of saturated steam corresponding to the pressure in the heating system pipework at its highest point above the boiler. This description does not purport to match exactly the wording of the guidance from the UK Health & Safety Executive[4], but it is nevertheless a practical implementation of it that is typically made by HVAC designers.

The approach typically adopted in the design of heating systems is to fix the design flow and return temperatures for the boiler according to the general design strategy adopted for the application, and then to *pressurise the installation* to achieve the level required for the 17K temperature margin. The latter is commonly referred to as the *anti-flash margin*, since it is designed to prevent the inadvertent 'flashing' of the boiler flow water into steam as a result of a reduction in its pressure. The risk of this occurring will typically be greatest at the highest point in the installation. The pressurisation will typically be achieved either by employing the system feed and expansion tank to provide an appropriate static head of water (in an 'open' heating system), or by the use of a pressurisation set (in a sealed heating system). For the purposes of the following discussion it will be assumed that we are dealing with a boiler plant located in a roof plant room, where the flow connection to each boiler *is* the highest point in the heating system pipework.

In the case of a single-boiler installation the gauge pressure to which the system should be pressurised is determined at a temperature equal to (design boiler flow temperature + 17K). The setting of the boiler high-limit thermostat will typically be 6–7K less than this. A glance at steam tables[2] will show that design boiler flow temperatures up to 83°C will not generate a positive gauge pressure to be applied to the system according to this criterion, and minimum values of 0.2 bar (gauge), or 2m static head, are therefore applied. Design boiler flow temperatures above approximately 88°C will yield requirements above these minimum values.

In the case of multiple-boiler installations the gauge pressure to which the system should be pressurised is determined at (design *mixed flow temperature* in the boiler primary circuit + 17K + the design temperature rise across each boiler). The boiler high-limit thermostat will again be set 6–7K below this temperature. The additional temperature difference term has been introduced since in multiple-boiler installations the control thermostats of each individual boiler will typically have an increased setting to allow a remote controller to sequence burner firing. This means that for a design mixed flow temperature of 82°C and boiler temperature rise of 11K, a minimum pressure of approximately 0.42 bar (gauge) will always have to be applied to the system. In the event that the water flow rate through each boiler can vary, the *worst-case temperature rise* should be used.

Up to this point our consideration has been limited to the maintenance of a *minimum* pressure in the boiler circuit. The risk of *excessively high* pressures occurring must also be mitigated. Two provisions are typically made in this respect. The first provision differs according to whether the boiler is installed in an open or sealed system. If the former is the case, each boiler that can be individually isolated is provided with an 'open vent' connection that comprises a section of pipe that branches off the flow pipework from the boiler and rises vertically to discharge over the heating system feed and expansion (f & e) tank. In the event of inadvertent isolation of the boiler or total failure of the control and high-limit thermostats, an escape path is available for any steam generated to discharge harmlessly to atmosphere over the f & e tank. The open vent connection must naturally be made between the boiler and any valve that can isolate the flow from the boiler. In multiple-boiler installations each boiler may be connected to a common vent pipe; via a *three-way vent cock*. The three ways for fluid into and out of this device (one in and two out) may be configured in *two positions*. One of these connects the boiler flow to the common vent pipe; the other isolates it from this, while opening an alternative vent path to atmosphere local to the boiler (typically piped to low level adjacent to the boiler). In any event the open vent pipe must rise a sufficient distance above the surface of the water in the f & e tank to ensure that heated boiler flow water is not wastefully discharged into the tank under any normal operating condition. Naturally in the case of sealed heating systems no boiler open vent is possible. In lieu of this a *safety circuit* provides for the pressurisation set to shut down the boiler on sensing high-limit and low-limit pressures (the former indicating a limiting over-temperature condition, while the latter is indicative of serious water leakage from the installation), and to inhibit further firing of the burner until this safety circuit has been manually reset.

The second provision is required irrespective of whether the installation is open or sealed, and requires that each boiler that can be individually isolated is provided with a *safety (pressure relief) valve*. As its name suggests, the purpose of this device is to relieve excessive pressure in the boiler, and it is the last line of protection against explosion in the event that a boiler has been isolated while the open vent connection is blocked for any reason, and yet is still allowed to fire. Safety valves typically lift against the pressure of a spring, which is either pre-set and non-adjustable or set and subsequently locked to prevent casual or inadvertent tampering. Modern safety valves typically employ non-stick valve seats, and those in larger installations typically include a manually-operated quick-release. The safety valve may be installed off the flow pipe from the boiler, but again it must not be possible to isolate the boiler from the safety valve. It is fairly self-evident that the discharge from a boiler safety valve should be made to a location that is both safe and visible. The requirements for, and sizing of, both boiler open vents and safety valves are prescribed by British Standards, while guidance based on these is given in the *Guide* published by the Chartered Institution of Building Services Engineers (CIBSE)[5].

The maximum pressure to which the boiler may be subjected during normal operation, as a result of the effects of static head of water, system pressurisation and pump pressure, must also fall within its design envelope. Many cast-iron and steel boilers are designed for a maximum operating pressure in the range 4–6 bar (400–600 kPa) gauge.

3.14 Fuel supply

3.14.1 Natural gas

Where boilers are fired on natural gas the requirements for fuel supply are typically straightforward. Beyond the provision of an incoming supply of gas to the installation that is of adequate capacity and suitably metered, the pipework carrying the gas from the meter to the boilers must be sized to constrain the pressure drop along its length (up to the connection to the burner gas line) to a maximum of 1 mbar (100 Pa). Strictly-speaking, this '1 mbar' requirement applies to a *low-pressure supply* of natural gas that is metered at a design pressure of 21 mbar, which probably represents the majority of commercial building services applications. Where the supply is normally metered at a design pressure in excess of 21 mbar, the limiting pressure drop is 10% of the design pressure. All pressure drops are, of course, taken at the maximum volume flow rate of gas in the pipework. It is not within the scope of the present text to consider in detail the design of natural gas pipework installations, which are naturally subject to other constraints (for example, maximum velocities for the flow of gas in pipes and venting requirements for ducts, voids and risers that may form a route for pipework), statutory regulations, recommendations and considerations of good practice[6].

Additional requirements will arise, however, where the pressure drop constraint cannot be met either using economic pipe sizes or within the available space. Under such circumstances the pressure available at the outlet from the gas meter must be raised or 'boosted' to permit a practical pipework installation to be achieved. The supply pressure is raised by a *gas booster* that is installed between the gas meter and the boilers (Figure 3.13). In modern building services applications the gas booster is typically supplied as a fully-functional packaged unit or *gas booster set*. The heart of the booster set is a

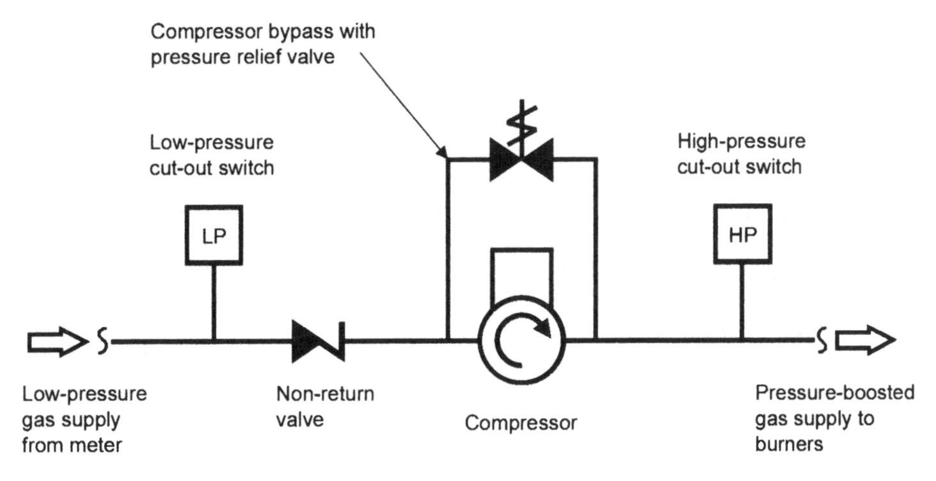

Figure 3.13 Schematic of a basic gas booster set.

compressor driven by an electric motor, while other essential features include a bypass around the compressor (incorporating a pressure relief valve) and a non-return valve that is located upstream of the compressor and its bypass. The latter ensures that pressure surges or pulses are not transmitted back towards the meter or to any non-boosted supplies taken from the installation upstream of the booster set. Where fluctuations in the outlet pressure from the gas booster set may cause problems to the burners, or where the changing supply requirements of the burners may pose problems for the booster set itself, the basic scheme of Figure 3.13 may be extended by incorporating a gas receiver on the high-pressure side of the compressor. This high-pressure receiver should be provided with a safety (pressure relief) valve that discharges to atmosphere outside the building, a pressure gauge and a condensate drain cock.

Control of the gas booster set is typically effected via pressure (cut-out) switches located on either side of the set. The pressure switch on the upstream (meter) side of the booster set is arranged to shut down the compressor if a specified low-limit value of the gas supply pressure is reached. Manual reset of this control circuit will typically be required. While the pressure switch on the upstream side of the booster set is an essential feature of the installation, that on the downstream (boiler) side is intended to prove that adequate gas pressure is available for operation of the burner's own automatic firing control, and is therefore recommended rather than mandatory practice. As is the case generally when packaged equipment is employed, it is incumbent on the designer or specifier of the installation to confirm that the equipment complies with all necessary standards, statutory or other approvals and good practice recommendations. It is also particularly important to ascertain any items that the packaged equipment may not provide, and yet which may be essential to achieve a fully-functioning installation.

Where a gas booster handles most or all of a low-pressure gas supply, additional provisions may be required to avoid repeated nuisance cut-outs that may be caused if a transient low supply pressure condition is induced by the starting of the compressor itself. The manufacturer should always be advised of the nature of the environment in which the gas booster set will be required to work, and this should include both the supply pressure available from the meter and the *proportion of the total incoming gas supply* that is to be handled by the set. The utility responsible for operation of the gas supply network should also be consulted regarding any proposed use of a gas booster.

Where a gas booster is employed, the constraint on pressure drop between the meter and the boilers is then applicable to the extent of pipework between the meter and the booster set. Boosting of the gas supply pressure may also be required if there are special requirements for the gas pressure to the burners in a particular installation. Since this is most likely to occur in very large installations, which are likely to have a correspondingly high annual consumption of natural gas, it may be both possible and preferable to negotiate a gas supply at an appropriate design pressure.

The other principal component of a natural gas supply to a boiler installation is an emergency gas shut-off and isolation valve or *fire safety shut-off valve*. This is installed in the main gas supply pipe serving all the boilers, typically at a point within the plant room, and is in addition to all requirements for the manual isolation of the gas supply to the boiler installation as a whole and to individual boilers. The gas emergency shut-off valve is typically a solenoid type that fails closed if the power supply to the valve is interrupted for any reason. Closure of the valve is activated by heat detectors located above the burner of each boiler, by emergency manual 'knock-off' or 'panic' buttons at each point of entry/exit from the plant room, and on activation of the building fire alarm system. It should be borne in mind that the pressure drop associated with the gas emergency shut-off valve may represent a significant proportion of the 1 mbar available for the sizing of a non-boosted gas supply. Formerly gas emergency shut-off valves were also employed that were mechanically closed by the release of a substantial 'deadweight'. In normal circumstances this type of valve was kept open by a taut steel wire that ran above the burners of each boiler. Links of a low-melting point metal alloy joined sections of this wire together above each burner. In a fire situation one or more of these alloy links would melt or fuse under the heat (hence they were termed *fusible links*), releasing the 'deadweight' to close the emergency shut-off valve. However, this type of valve tends not to be favoured for modern boiler installations since at the crucial time of operation the mechanical action may be adversely affected by friction or corrosion that has passed unnoticed. Additionally this type of emergency shut-off valve is inherently of the *normally-open type*, whereas solenoid valves are of the *normally-closed type* that is typically preferred in modern design practice. Where the gas emergency shut-off valve is located outside the plant room in which the boilers are installed (as, for instance, closely downstream of a gas meter that serves the whole building), any nuisance activation of the valve by the boilers may adversely affect other users of gas in the building. This will particularly be true if a gas supply used for catering loads is also shut down. It is a situation which has sometimes been encountered in speculative office developments that are subsequently fitted-out to a tenant's requirements that include the introduction of staff catering facilities.

In the UK an enhanced approach towards the metering of energy consumption within buildings is embodied within revisions of the Building Regulations which were introduced on 1 April 2002. This approach is likely to require that, where the gas supply to a building serves loads other than the heating boilers, the gas consumption of the boilers can nevertheless be independently metered or deduced. This is, in any event, good practice in respect of monitoring and managing the energy use of a building. Techniques for monitoring and targeting ('M & T') of building energy use are considered in Chapter 11.

3.14.2 LPG

Where boilers are fired on LPG (propane or butane) the requirements for fuel supply are considerably more extensive than for natural gas. Most fundamentally, bulk storage of LPG is required, and a minimum capacity for six weeks' operation at maximum load is recommended to allow an adequate reserve against an extended period of cold weather[7]. LPG installations are likely to remain a relatively rare occurrence for most HVAC designers dealing with commercial building services applications. In the UK the installation must comply with a range of health and safety, gas safety and building regulations, and the requirements of the LPG supplier. The latter should always be consulted at an early stage if an LPG installation is considered, and their recommendations and advice both listened to and heeded.

We have already seen that LPG has a higher calorific value than natural gas, and hence lower volume flow rates of gas will be required for a given burner heat output. Low-pressure supplies of LPG are typically regulated to a higher design pressure than for natural gas, nominally 28 mbar for butane and 37 mbar for propane, with a pipework pressure drop constraint of 2.5 mbar.

LPG may be chemically reactive with some materials. In the gaseous state it is flammable at low concentration (c.2–11% of gas by volume), and any accumulated leakage represents a risk of fire and explosion. LPG installations within buildings should therefore be *gas-tight*. When handling LPG, the 'acceptable' leakage is *zero*. Since LPG is denser than air, any leakage will accumulate at low level until physically dispersed. Moreover, the gas will sink down through any gaps or openings into low-lying spaces and voids, such as ducts or drains. Hence *gas detection systems* should be fitted in LPG boiler installations, with detectors positioned both adjacent to the burner and at low level in the plant room. On the other hand, if LPG is employed in a 'high-rise' application, the *effect of altitude* (Chapter 2, section 2.4.4) must be taken into account when sizing the gas pipework and determining appropriate test pressures.

LPG storage tanks are pressure vessels. These are typically cylindrical in form (horizontal or vertical), and in commercial heating applications will generally form a static installation. While LPG storage tanks will incorporate safety (pressure relief) valves, venting of any tank contents is both highly uneconomic and environmentally unacceptable, and such valves are strictly for 'last-resort' emergency protection. It will typically be advantageous for inspection and maintenance (although more expensive) if the required storage capacity is provided by two smaller tanks in lieu of a single larger tank. It must therefore be possible both to isolate individual tanks and to positively confirm when they are empty. LPG storage tanks are never completely filled, and this must therefore be taken into account when determining the amount of space required for the tankage necessary to provide the desired storage capacity of LPG. *Filling ratios* are specified by British Standards and are defined as the ratio of the maximum *mass* of LPG allowed within the tank to the mass of water that would fill it at 15°C. In small installations individual portable LPG vessels may be connected into a static site installation that basically comprises a multi-tank pipe manifold incorporating an auto-changeover device to ensure that the supply of gas is not interrupted when the contents of an individual vessel are exhausted.

LPG tanks are positioned away from the building to avoid the risk of overheating in the event of a building fire. Increasing the temperature of the LPG within the tanks will naturally raise its pressure. Ambient temperatures in the UK are typically high enough to cause vaporization of the LPG once it leaves the storage tanks. The larger the capacity of the tanks, the further away from the building they must be sited and the greater must be the spacing between tanks (minima of 7.5 m away from the building and 0.9 m between tanks for tank capacities in the range 2.25–9 m^3, while these values increase to 15 m and 1.5 m respectively for tanks that exceed this capacity[7]). LPG pipework, which may typically be of small diameter, will generally need to run below ground between the external storage tanks and the building. Since any leakage of the heavier-than-air gases into recessed or underground pipe ducts will not be self-venting, it will be preferable that the pipework is simply buried at an appropriate depth and suitably protected against corrosion if and as necessary. The LPG supplier should be consulted regarding the use of appropriate materials for pipework, fittings and jointing (although the zero-tolerance to leakage of LPG will naturally favour as few joints as possible). The storage installation must provide the pressure reduction necessary to suit the requirements of the burners.

In bulk storage installations LPG is delivered by road tanker, and it is essential not only that the storage tanks are closely accessible to the size of vehicle that the LPG supplier will employ, but that the delivery driver can keep his vehicle in sight at all times during the filling of the tanks. Since in the gaseous state LPG will tend to accumulate and 'hang' at low level, it is typically not the practice to 'bund' LPG storage tanks. In the event of leakage from the tanks the bund walls may trap the heavier-than-air gases. Similarly, LPG tanks should not be placed within or beneath any man-made or natural enclosure, *including trees*. The temptation to improve the appearance of the site by enclosing the storage tanks with close-boarded fencing or dense shrubbery must therefore be avoided.

As in the case of natural gas, an emergency gas shut-off and isolation valve or fire safety shut-off valve should be installed in the LPG supply line within the plant room and at a point where it will shut off the supply of LPG to all the boilers served. Installation of the emergency shut-off valve is again in addition to all requirements for the manual isolation of the LPG supply to the boiler installation as a whole and to individual boilers. In modern installations the emergency shut-off valve is typically a normally-closed (fail-safe) solenoid type. While shut-off is generally activated as described for a natural gas installation, it should also occur on activation of the gas detection system.

Finally it is particularly important that operating and maintenance personnel should be familiar with all the characteristics of LPG. *If skin contact is made with leaking LPG at close quarters, 'freeze burns' are likely to result.* Anyone who has ever filled a gas cigarette lighter and felt the effect of the escaping liquid gas should have an appreciation of a relatively mild and harmless form of this effect.

3.14.3 Oil

Oil-fired boiler plant requires bulk storage of fuel, and the design of the installation depends on the grade of fuel oil that is to be used. These grades range from 'C' to 'H' for fuel oil derived from petroleum and are supplemented by various grades of coal tar fuel derived from the distillation of coal. The former are defined by British Standard BS 2869: 1983[8], while the latter are defined by British Standard BS 1469: 1962[9]. It is now unlikely that any grade of coal tar fuel will be encountered in a typical commercial building services application in the UK. Of the petroleum-derived fuel oils, grades 'C' and 'D' (kerosene and 'gas oil') are termed *distillate fuels*, while grades 'E' to 'G'(respectively 'light', 'medium' and 'heavy' fuel oil) are termed *residual fuels*. Grade 'H' is typically only applicable to heavy power generation or industrial applications. Grade 'D' fuel oil is also sometimes referred to as '35 sec oil', which relates to a former specification of its viscosity as *35 seconds Redwood*, and in the UK it is easily recognizable since it is dyed a red colour to the requirements of HM Customs and Excise. Since the sulphur content of the fuel oil *increases drastically* for grades beyond 'D' (by over four times for grade 'E', and in excess of four-and-a half times for grades 'F' and 'G'), the emission of sulphur dioxide that results from their combustion is similarly increased. If for no other reason than this, it is very unlikely that the use of any of these fuels would (or even *should*) realistically be given credence for even a very large modern commercial building services application in the UK. Since in general the 'heavier' the grade of fuel, the more specific are its storage and handling requirements, we will in general confine our considerations to the distillate fuels, grades 'C' and 'D'.

A minimum (actual) storage capacity of three weeks' consumption at maximum load or the content of a normal fuel delivery plus two weeks' consumption at maximum load, whichever is the greater, is recommended[10]. Oil storage tanks are typically constructed from welded mild steel and may be rectangular or cylindrical (vertical or horizontal). Tanks are typically supported by steelwork (joists for rectangular tanks, cradles for horizontal cylindrical types) that span low-height 'sleeper' walls. Standard types, construction methods and installation details for oil storage tanks are defined by British Standard, and for the interested reader the principal requirements are summarised in Section B13 of the *Guide* published by the Chartered Institution of Building Services Engineers[10]. In the case of oil storage tanks it is *essential* that the installation is 'bunded' to provide a 'catchpit' that is impervious to the passage of oil, and that this is both large and strong enough to contain the full holding capacity of the tank(s) in the event of catastrophic leakage. While the catchpit may not contain any permanent drainage, it must be possible to remove the accumulated rainwater that will inevitably occur in an external tank location in the UK climate. Adequate space should be provided to allow periodic inspection of the complete perimeter of all oil storage tanks. In very large installations oil storage tanks may be installed underground (within or external to the building), and may be of reinforced concrete construction with a suitable oil-resistant lining. Steel tanks should not of course be buried in direct contact with the earth, since corrosion is likely. In any event, the 'flotation' effect of ground water on a buried oil storage tank must be considered (especially when the contents of the tank are low), and suitable anchorages provided.

All oil storage tanks must be provided with a suitable vent pipe, which will terminate in a downward-facing bend with wire mesh grille to shield it from the potential ingress of rainwater, dirt and debris, or small birds and large insects. The tank vent pipe must rise continuously, and preferably steadily, to its termination without any 'dips' that could act as traps for liquid to either restrict or prevent the safe venting of oil vapour. While it is naturally desirable to dissipate fuel odours that may develop during filling of the tank, if the vent pipe must rise to an excessively high level to achieve this, the tank may be unduly pressurised by the resulting static head of oil should an overfill condition occur. In such circumstances, to avoid damage to the tank that might result in catastrophic leakage, it may be necessary to relieve a high-limit pressure by a safety discharge into the catchpit. Oil storage tanks should never be completely filled, in order to avoid a discharge of oil through the open vent as a result of surging or 'frothing' of the oil within the tank during the filling process, and the sizing of the tank should take this into account for the actual storage capacity required. The air gap that therefore exists at the top of the tank, when the contents gauge indicates that the tank is full, is termed the *ullage*. It is typically a minimum of 100 mm deep or the equivalent of 5% of the total contents of the tank, whichever is the greater. The oil supply outlet from the tank will similarly be taken from a minimum level (c.76 mm) above the base of the tank, in order to prevent water (which is of course heavier than oil) and any dirt or sediment that has settled to the bottom of the tank (the two will typically mix to form a sludge) from entering the oil supply line to the burners. Periodically this accumulation of sludge must be removed from the tank and disposed of *in an environmentally acceptable manner* (which is most definitely *not* into the local drainage system), and the tank must therefore incorporate an appropriately-sized and valved outlet for this purpose at its lowest point. Where oil is drawn from the storage tank by pump, this may be done via a 'foot valve' that is immersed at low level in the tank (but above the 'sludge line'). In the UK, grade 'C' and 'D' fuel oils are typically stored at ambient temperatures, and tank heaters are only required for the 'heavier' grades.

Fuel oil is again delivered by road tanker. Both the oil fill point and the vehicular access to it should be as close to the tank as is reasonably practical. In any event the delivery vehicle will require access to within a maximum of 30 m from the fill point. While it should be possible for the driver to keep the storage tank clearly in view throughout the filling operation, a visual and audible overfill alarm should in any event be provided to alert the operator when the tank is full.

Additional storage requirements arise where an oil-fired boiler plant is located in a roof plant room. In such circumstances the main storage remains at or below ground level, while the short-term needs of the boilers are taken care of by an additional local storage or *daily service* tank in the roof plant room. For safety reasons (fire risk) the maximum capacity of this high-level storage is typically limited, and oil must therefore be frequently transferred from the main low-level storage by a system of automatic pumping. Duplicate transfer pumps (run and standby) are essential, and the high-level oil tank must be provided with an overflow pipe that drops down the building to discharge into the main low-level storage tank. A catchpit that is impervious to oil must also be formed within the roof plant room that can accommodate the contents of the day tank in the event of catastrophic leakage.

The oil distribution system from the storage tank(s) is typically one of three principal types. The simplest and most straightforward of these is the *single-pipe, gravity-fed system* (Figure 3.14a), which is therefore to be preferred wherever its use is practical. The oil supply line will typically include both a filter to remove any sludge drawn from the tank and a non-return valve to

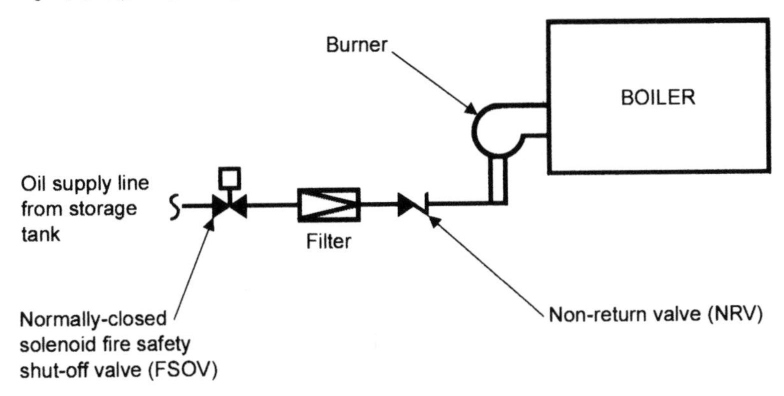

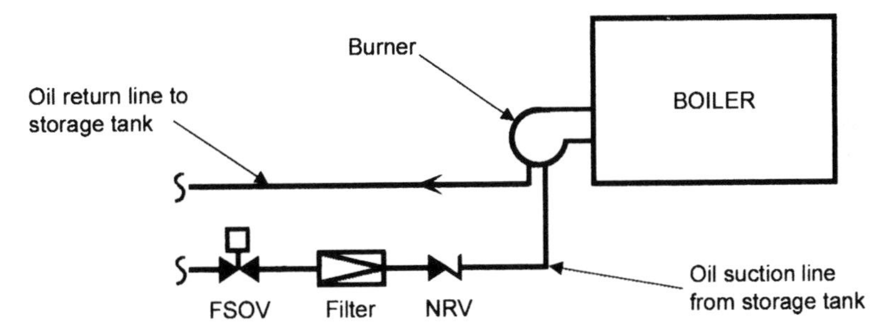

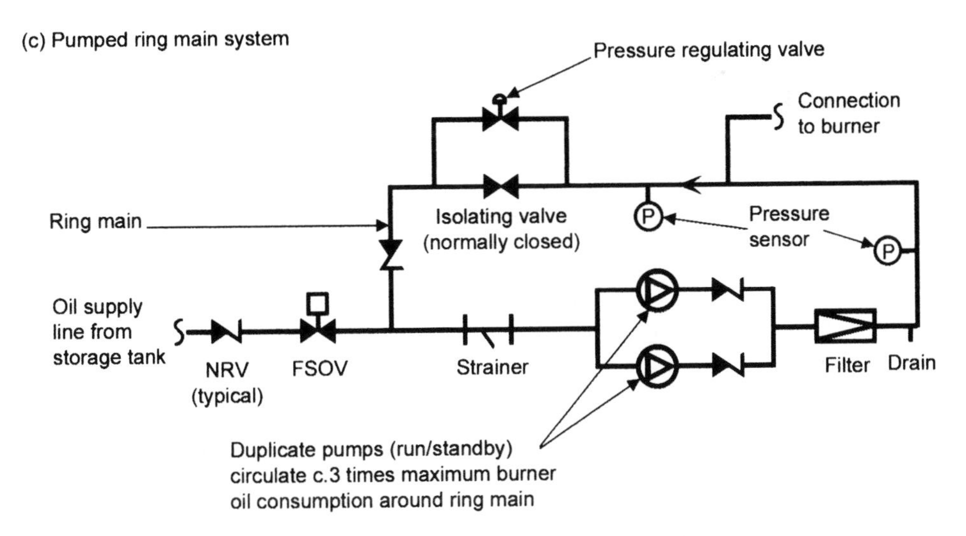

Note: In each system general isolating valves have been omitted for clarity.

Figure 3.14 Principal types of fuel oil distribution system.

prevent any tendency for oil to drain back from the burner (which will typically be a pressure-jet type with its own oil pump) when it stops firing. Some boiler manufacturers recommend that air venting of the oil line is provided between the filter and the burner, either by a dedicated vent valve or by a combined filter and vent valve. As in the case of natural gas, an emergency oil shut-off and isolation valve or fire safety shut-off valve should be installed in the oil supply line within the plant room, and at a point where it will shut off the oil supply to all the boilers served. Installation of the emergency shut-off valve is again in addition to all requirements for the manual isolation of the oil supply to the boiler installation as a whole and to individual boilers. In modern installations the oil emergency shut-off valve is typically a normally-closed (fail-safe) solenoid type, and shut-off is activated as described for a natural gas installation. Where oil-fired boilers are installed in a roof plant room, the emergency shut-off valve will be installed in the oil supply line from the daily service tank, and the pumps transferring oil to this tank from the main low-level storage should also be shut down both on closure of the emergency shut-off valve and on general fire alarm. The chemical characteristics of the fuel oil used may have implications for the materials employed in the oil supply line. Black mild steel pipework with either malleable cast-iron or wrought iron fittings is typically favoured. Copper tube may be suitable for grade 'C' fuel oil, but its use for other grades should be confirmed with the prospective fuel supplier since copper contamination may degrade some fuel oils. The use of brass and alloys of copper and zinc is not typically acceptable, and oil does not 'mix' with most natural or synthetic rubber products. Where the oil supply line must run external to the building it is essential to minimise the risk of leakage from pipework leading to contamination of soil, watercourses and the local water table.

Where oil cannot flow to the burner by gravity, a *two-pipe system* should be used (Figure 3.14b). In this case oil is drawn from the storage tank by the suction of the burner's oil pump. Boiler or burner manufacturers will advise on the maximum suction lift that can be accommodated where it is necessary for the storage tank to be located below the level of the burner. Problems may be experienced in two-pipe systems due to a loss of 'prime' by the burner oil pump, which leads to a delay in establishment of the combustion flame and a nuisance tripping of the burner's flame-failure protection circuit. Sizing of the two-pipe system is therefore important in achieving satisfactory operation. An undersized oil supply line will generate excessive suction pressures, while an oversized line will increase the quantity of oil subjected to suction conditions that tend to cause air and any other dissolved gases in the oil to come out of solution. The air and/or gases thus released from solution in the oil will ultimately tend to collect in the burner pump and again cause nuisance tripping of the flame-failure protection circuit when establishment of the flame is delayed. Since the oil supply line is under suction pressure, any minor leakage at joints and fittings, particularly where these are of screwed type, will tend to be of *air into* the oil line, rather than of *oil out of it*, and consequently will be very difficult to trace.

In large multiple-boiler installations employing the 'lighter' distillate grades of fuel oil (grades 'C' and 'D'), or where the boilers are remote from the storage tank, a *pumped ring main system* may be used (Figure 3.14c). In such systems oil is drawn from the storage tank into the circulating pipework ring, but unused oil is not returned to the storage tank. Where a ring main is employed the oil emergency shut-off and isolation valve will be installed in the oil supply line between the outflow from the storage tank and its junction with the ring main.

Where the 'heavier' grades of residual fuel oil (grades 'F' and 'G') are employed, a complex *'hot oil' ring main system* will typically be required in any situation, with the temperature of the stored fuel oil (c.40°C for grade 'G' oil) raised for pumping (to c.50°C for grade 'G' oil) on leaving the storage tank. The temperature of the oil may subsequently be further increased to near its burner atomizing temperature by 'line heaters' installed downstream of the pumps in the ring main. The pipework of the 'hot oil' ring main will be both thermally insulated and trace heated, and safety (pressure relief) valves will be required to relieve excess pressure in the event that pipework is inadvertently isolated while trace heating continues.

Where oil firing of boilers is proposed, the fuel storage and distribution installation should be discussed with both the prospective boiler (or burner) manufacturer and the fuel supplier at an early stage, and the approval of each should be sought for the final design of the installation.

3.15 Ventilation of the boiler plant room

Ventilation of the boiler plant room is necessary for two reasons. The first is naturally to *ensure that an adequate supply of combustion air is available* for the efficient and safe operation of the boilers themselves. The second reason is to *limit overheating of the plant room and its equipment.* This is particularly important in modern plant rooms that may contain, in addition to the boilers themselves and various other mechanical plant and equipment (pumps, fans, etc.), sophisticated control and electrical switch panels. Even in mild winter weather, the ambient temperature is at least relatively low, and the ventilation air can therefore pick up a fairly substantial amount of heat (approximately 12 kW per $m^3 s^{-1}$ of volume flow rate at a temperature rise of 10 K). In principle the worst case for the ventilation requirement may reasonably appear to be where the boiler installation must operate during summer months of high ambient temperature, for example to satisfy domestic hot water heating. However, in an efficiently-configured installation the proportion of the plant required to service this load should be substantially reduced, and with it the heat gain to the plant room. The mid-season (spring, autumn) condition may therefore require the greatest flow of ventilation air.

Combustion air requirements are ultimately based on the combustion chemistry of the common boiler fuels, although where boilers employ atmospheric natural gas burners, account must also be taken of the additional induction of plant room air that will occur at the draught diverters. The Building Services Research and Information Association (BSRIA)[11] notes the combustion air requirements for boiler plants with rated heat outputs in the range 100 kW–1 MW as varying from 90 $l s^{-1}$ per 100 kW rated output for an atmospheric natural gas burner employed with a draught diverter (60 $l s^{-1}$ per 100 kW without draught diverter), to 45 $l s^{-1}$ per 100 kW rated output for a fan-assisted or forced-draught natural gas burner and 50 $l s^{-1}$ per 100 kW for a pressure-jet oil burner (both without draught stabilisers). The ventilation requirement for limiting the maximum temperature reached in the boiler plant room to an acceptable level is based on standard HVAC design practice for determining both heat gains and losses and the sensible heat pick-up of a volume flow rate of air. In the UK the *Guide* published by the Chartered Institution of Building Services Engineers (CIBSE) gives recommendations on the supply of combustion air and a method for the detailed assessment of the ventilation requirement[12]. Within the UK, Building Regulations have typically only considered gas-fired boilers with rated heat inputs of up to 60 kW, and oil-fired boilers with rated heat inputs up to 44 kW. However, British Standard BS 6644[13] prescribes requirements for the provision of combustion and ventilation air to gas-fired hot water boilers with rated heat inputs between 60 kW and 2 MW, while Part 2 of British Standard BS 5410 considers the requirements of oil-fired boilers of rated heat input 44 kW and above[14]. The CIBSE method for assessing the ventilation requirement is based on a limiting temperature of 32°C at ceiling level within the boiler plant room, with temperatures in the range 21–27°C in the *working zone*, while BS 6644 recommends slightly higher limiting values of 25°C at low level (100 mm above floor level), 32°C at mid-level (1500 mm above floor level) and 40°C at high level (100 mm below ceiling level).

In principle, and when the boilers are firing, the air supplied to the boiler plant room for combustion leaves the plant room via the boiler flues (together with any induction via draught diverters). Hence only the ventilation air needs to be both supplied and extracted from the plant room. In the UK climate it is not typically necessary to pre-heat the air supplied to the boiler plant room, and it is therefore simplest, and generally to be preferred, if both air supply and extract are achieved by natural means (Figure 3.15a). In any event, air for both combustion and ventilation should be supplied at low level, with the ventilation air extracted at high level. Due to the difference between the supply and extract volume flow rates, the area of low-level openings will be greater than that of the high-level openings. For natural ventilation the air supply and extract openings should be permanent and fitted with purpose-designed grilles or louvres that cannot (as far as is possible in practice) be blocked inadvertently. Consideration should be given to the effects of wind forces when positioning the ventilation openings. The air supplied to the boiler plant room should be free of any significant sources of contamination, including natural or man-made dust, fibres and (in particular among industrial-type pollutants) halocarbons. The air within the boiler plant room should also not be subject to abnormal humidity levels. For natural ventilation with gas-fired boilers, the requirements of BS 6644 for the *free area* of ventilation openings are typically expressed as a base provision plus an incremental increase for each kW of input in excess of 60 kW. Hence the requirement for low-level openings is given as [540 + 4.5 (rated heat input – 60)] cm^2, and for high-level openings as exactly half this or [270 + 2.25 (rated heat input – 60)] cm^2. It is essential that the rated heat input in these expressions is based on the correct calorific value. The relationships used in BS 6644 were naturally derived at a time when the use of gross calorific value was prevalent in the UK, *and must therefore be based on GCV*. If values of rated heat input based on now-common values of net calorific value are used, the free areas that result will be only approximately 90% of those actually required. For particular boilers it may be possible to make a combustion air connection directly from outside the boiler plant room to a physical connection with the boiler itself.

Where the required rates of air supply and extract cannot be achieved by natural ventilation, the preferred strategy is to mechanically supply the combustion and ventilation air to low level, while retaining natural extract from high level (Figure 3.15b). There are two reasons for this. First, the boiler plant room *should not be placed under negative pressure*, since this may interfere

(a) Natural supply and extract

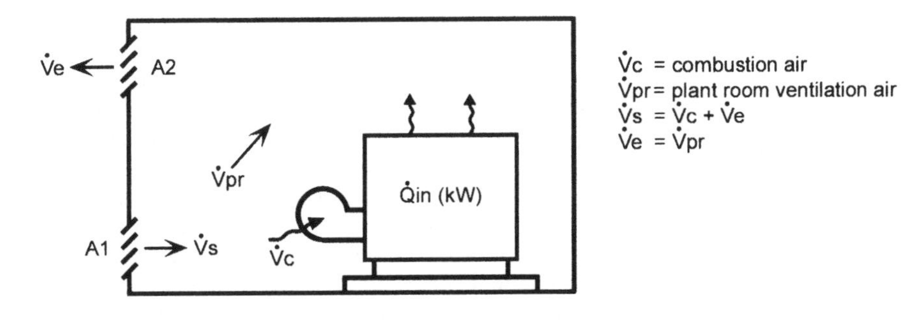

A1 (cm² free area) = 540 + 4.5 (Q̇in – 60); A2 = 0.5 A1

(b) Mechanical supply and natural extract

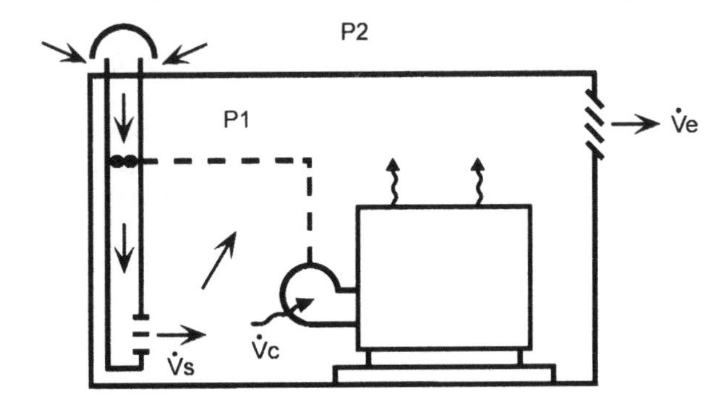

V̇s = 1.25 (V̇c + plant room ventilation air requirement); P1 ≥ P2

(c) Mechanical supply and extract

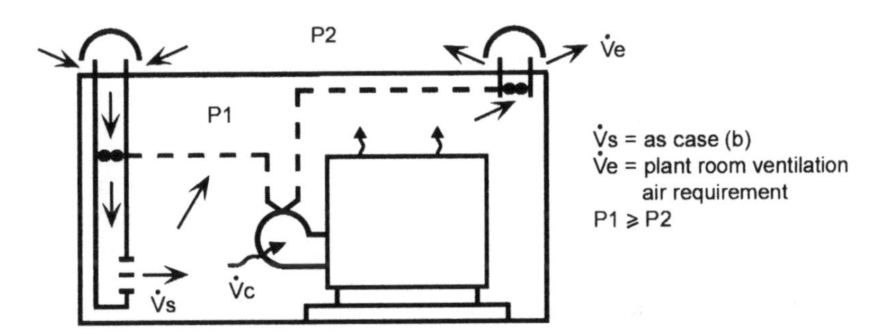

Figure 3.15 Options for the ventilation of boiler plant rooms.

with the safe and efficient operation of the burners. This is particularly so where atmospheric gas burners are employed, since a 'suction' that is established along any air path out of the plant room other than through the boiler will interfere with the induction of combustion air that is driven by the flue draught. While less susceptible than atmospheric gas burners, fan-assisted and forced-draught types will also be affected. Secondly, high-level natural extract takes advantage of the convection forces that inherently exist within the boiler plant room. To ensure that the plant room as a whole does not adopt a negative pressure relative to ambient conditions, it is recommended[12] that the mechanical supply is sized *in excess of* the combined combustion air and ventilation requirement by a minimum of 25%. A control interlock will be required to ensure that no burner can fire unless operation of the mechanical supply fan is proven.

Where even natural extract is not possible, a combined mechanical supply and extract ventilation strategy must be adopted (Figure 3.15c). Here also it is essential to avoid a situation in which the boiler plant room can adopt a negative pressure with respect to ambient conditions, and a mechanical supply sized at least 25% in excess of the combined combustion air and ventilation requirement is again recommended[12]. Interlocking of burner operation with that of the mechanical supply and extract fans is again essential to prevent the possibility of a burner attempting to fire in the absence of an adequate supply of combustion air.

The configuration and size of most boilers used in other than the smallest commercial heating applications has traditionally precluded the use of so-called 'balanced flues' that also provide a path for combustion air to the boiler. However, the balanced-flue principle has been developed into the *balanced-compartment system* that provides both flueing of the boilers and ventilation of the boiler plant room. This technique is considered further in Chapter 5.

3.16 Water treatment

In the UK the incoming mains water supply to a building, which provides both the initial fill for a heating system and its subsequent make-up requirements, typically contains dissolved mineral salts. These are not, of course, harmful to human beings and do not affect the cleanliness or potability of the water. However, while these mineral salts are dissolved in the water at the temperature of the incoming mains supply, when heated to temperatures above approximately 50°C, as in the familiar household kettle *or in a heating boiler*, these mineral salts are converted to an insoluble form and precipitate out of the water to form a *scale* on any surface in contact with the heated water. In a heating system there are an abundance of such surfaces, specifically including the waterside surfaces of the boilers. Furthermore, since these salts have a negative solubility constant (i.e. solubility decreases with water temperature), there will be a tendency for most scaling to occur on the hottest surfaces, which are naturally to be found on the waterside of the boiler block or heat exchanger.

The concentration of dissolved mineral salts in the incoming mains water supply to a building is measured, both qualitatively and quantitatively, on a scale of water 'hardness', and varies according to the geology of the region that is the source of the water supply. The principal dissolved minerals are salts of calcium and magnesium, and the former is the main source of problems in heating systems and boilers. Principal among the dissolved calcium salts is *calcium bicarbonate, Ca* $(HCO_3)_2$, which is sometimes also termed *calcium hydrogen carbonate*. This is an acid salt of the weak *carbonic acid* that is formed by the dissolution of carbon dioxide in water. On heating it is converted to *calcium carbonate, Ca* CO_3, which has a low solubility in water. This is the familiar white scale found in domestic kettles and, unfortunately, frequently in heating systems and boilers. Quantitatively, the scale of water 'hardness' is expressed in mg Ca CO_3 per litre of water or *mol* Ca CO_3 per m^3 of water. Since a 'mol' of a substance is its molecular weight in g, and the molecular weight of Ca CO_3 is 100, 1 mol Ca CO_3 per m^3 of water is equal to 100 mg Ca CO_3 per litre of water. Qualitatively, we are familiar with the terms 'hard' water and 'soft' water. Since calcium carbonate is 'chalk', water sourced from such regions is typically 'hard'.

The principal problem caused by scale is that it forms a layer on the water-side heat transfer surfaces of the boiler block or heat exchanger. The thermal conductivity of the calcium carbonate scale is approximately 1% of that of carbon steel, and slightly less than this for cast-iron[11]. Scale therefore acts as an insulating layer in the heat transfer path from the combustion gases to the boiler water. As a layer of scale builds up on the water-side surfaces, the rate of heat transfer is reduced and the fuel consumption of the installation increases (more fuel is consumed in a heating season for the same size and pattern of heating load). Important enough as this is, it may be more significant for the continuing operation and ultimate longevity of the boiler that the insulating layer of scale results in an increase in the temperature of the metal in the boiler block. If the build-up of scale is sufficient, it may result in thermal overstressing of the boiler metal and cracking of cast-iron sections, or even in a degradation of the metal characteristics in areas subject to prolonged overheating. The parts of the boiler block that are most sensitive to the effects of scale are areas that are subject to high thermal loading, in particular the surfaces of the combustion chamber that are exposed to radiant heat from the flame, or to its actual impingement (itself, of course, undesirable). Boilers designed for high heat transfer rates with low water content are therefore also more susceptible (e.g. modular boilers). Even if scale does not continue to build up in a layer, but rather breaks away into the water flow, it will subsequently tend to collect elsewhere in the boiler, typically at the lowest points. Scale formation decreases the cross-sectional area of the waterways, and substantial formation will therefore tend to reduce water flow rate through the boiler, further increasing the risk of overheating of the metal and also potentially affecting the operation of boiler sequencing controls. Cast-iron sectional boilers are typically considered to be more tolerant to some scale formation than fabricated steel boilers, in which cracking of tube plates and tube ends are the potential consequences of scale-induced overheating and deformation of the boiler metal. Calcium carbonate is not, of course, the only potential scale-forming deposit. Silicon-based compounds, such as calcium silicate, may have a thermal conductivity only approximately 0.33% of that of carbon steel or cast-iron (i.e. only one-third of that of calcium carbonate scale)[11]. Hence relatively thin layers of such deposits may cause significant overheating of the boiler metal, yet remain difficult to detect visually. Other mineral salts that may form deposits within a boiler include calcium sulphate and calcium phosphate, although the thermal conductivities of these are respectively c.5.3 and c.8.7 times that of calcium carbonate[11]. One manufacturer of cast-iron sectional boilers estimates that a layer of scale as little as 1 mm thick in the boiler can increase the fuel consumption of an installation by 10%, while a layer 3 mm thick may increase this to 25% *and* be sufficient to cause premature boiler failure due to local overheating. Even when reduced boiler flow and return temperatures are employed, the temperature of the *water boundary layer* (immediately adjacent to the metal surface of the boiler) will typically be high enough to cause scale formation.

Once the mineral salts within the heating system fill water have been precipitated out, scale formation from the initial fill of water will cease. All scale is not, of course, formed immediately on the first heating of the fill water, and the rate of scale formation may be expected to be rapid during the initial few hours of burner operation, trailing off after that in a more-or-less exponential fashion. However, every time new 'make-up' water is introduced into the heating system from the incoming mains supply, whether to replenish leakage or as a result of draining down and refilling associated with modification work or the inspection and maintenance of plant and equipment, it will bring with it additional mineral salts that will be subject to heating within the boiler. If the water fill and make-up supply to a heating system is individually metered, it is then possible to monitor water use by the installation, and to identify any instances of excessive use that might otherwise have passed unnoticed and to identify when additional chemical treatment of system water may be required.

The presence of mineral salts in the initial water fill and subsequent make-up water for a heating system is not the only cause of concern. Oxygen, from air dissolved in the fill or make-up water, can cause 'pitting' or 'grooving' corrosion of both cast-iron and steel and the formation of a *sludge* (much of which is a black substance called *magnetite*, Fe_3O_4, formed by secondary chemical reactions associated with rusting). If the quantity of this sludge becomes significant, it may coat the heat transfer surfaces of the boiler, restricting the flow of water and affecting the heat transfer characteristics, and potentially leading to local overheating and failure of the boiler. The air (and its oxygen) that is initially dissolved in the fill water is driven out of solution under heating. In an adequately vented installation it will therefore be removed after a relatively short period of operation. Unless substantial quantities of make-up water are repeatedly or steadily introduced into the heating system, the cause of which should in any event be investigated and remedied, the main avenues for oxygen to be reintroduced into the system are at the surface of a feed and expansion (f & e) tank, and via any leakage of air into the system in areas where the internal pressure within the pipework is below atmospheric pressure. While various proprietary types of *air separator* may be installed in the boiler circuit pipework, in modern HVAC design practice it is increasingly being seen as an advantage (although not simply for this reason) to remove the need for an f & e tank by 'sealing' the system and using a pressurisation set.

Acidity of the boiler water is also undesirable (a *pH value* of 7 is neutral, and a value of 6 or less indicates some acidity of the water, which will be greater the lower the pH value). Boiler manufacturers typically recommend that the water in the heating system should have a pH value in the range 8–9.5.

The formation of *biofilms* within pipework and heat exchanger tubes, as a result of bacteriological growth within the heating system, can seriously restrict the cross-sectional area of pipework available for the flow of water, and also the performance of heat emitters. While this is typically of greater relevance to the load circuits of heating systems than to the boilers themselves, in some installations restriction of the water flow rate returning from the heating load circuits may adversely affect water flow rate through the boiler. Also pipe failure resulting from bacteriological contamination of the heating system may lead to greater use of make-up water, with its introduction of 'new' mineral salts and oxygen. In the UK the potentially serious effects of the *pseudomonas* bacterium, which may be introduced into heating systems via the incoming mains water supply, have recently drawn considerable attention from HVAC engineers and building services operating and maintenance specialists.

All of the above may be addressed by the application of suitable chemical water treatment measures, which will typically include a scale and corrosion inhibitor, a biocide and a pH regulator. Periodic re-dosing with these chemicals will be required to maintain the necessary concentration levels, and this will be particularly important following any substantial make-up to the system. However, chemical treatment should be seen as complementary to sound initial design and good operating and maintenance practice, which will include monitoring of water use by the heating system. The application of any chemical treatment to a heating system should of course be closely monitored and logged, and should not be permitted to have any significant adverse health and safety implications for the occupants of the building served by the system.

Finally, and perhaps most fundamentally, modern boilers have a relatively low water content and a tendency towards increasingly complex configurations of heat exchange surfaces. They therefore need protection against blockage or damage by general dirt and debris from the heating system load circuits. In a new installation the risk of this should of course be minimised, both by good site practice during the installation period and by appropriate cleaning and flushing of pipework during the commissioning period. However, where a replacement boiler plant is being installed in an existing heating system of some age, the new boilers should typically be given specific protection. If a substantial part of the existing heating system is drained and refilled, quantities of sludge, loose scale and rust may be circulated through the boiler. Since the boiler will typically represent an area of lower water velocity, relative to the flow and return pipework, it will be a favoured site for the deposition of such material. Since the air venting capabilities of the existing heating system may also be of unknown performance, it may be prudent to provide both a suitable dirt or sludge trap and an air separator to handle the flow of return water to the boilers. Conventional pipeline-mounted strainers are typically not intended for this type of application. Various proprietary types of dirt or sludge trap are available to the designers or specifiers of boiler installations. These devices, which are mechanically simple, typically rely either on achieving gravitational settlement of solids in suspension at low water velocity, or on the effect of inertia when the direction of a flow containing solids in suspension is sharply changed. Perhaps most importantly from the practical perspective, it must ultimately be possible to get the accumulated sludge out of the trap and into a form suitable for convenient disposal without adverse health and safety implications for the operative carrying out the task, which in some installations may have to be carried out daily. In some cases it may be advisable to consider the use of a boiler that is more tolerant to the particular water-side environment in which it will be expected to operate.

The cost of providing suitable chemical water treatment and mechanical protection devices may initially 'test' a limited project budget. However, it is inherently considerably less than the cost of subsequently dealing with early failure of a boiler in service and its premature replacement. It is also worth noting that damage resulting from scaling and corrosion is invariably disclaimed by the product warranties on boilers.

References

1. Walker, D.C. (1990) *Investigating Boiler Failures*. A guide for building services engineers. Technical Note 2/90, p. 25. Building Services Research and Information Association, Bracknell.
2. Rogers, C.F.C. & Mayhew, Y.R. (1981) *Thermodynamic and Transport Properties of Fluids*, 3rd edn (SI units). Basil Blackwell, Oxford.
3. Chartered Institution of Building Services Engineers (CIBSE) (1986) Combustion Systems, Section B13 of the CIBSE *Guide*, p. 45. CIBSE, London.
4. Health & Safety Executive (HSE) *Automatically Controlled Steam and Hot Water Boilers*. Guidance Note PM5, HSE, London.
5. Chartered Institution of Building Services Engineers (CIBSE) (1986) Heating Systems, Section B1 of the CIBSE *Guide*. CIBSE, London.
6. Institution of Gas Engineers (IGE) *Gas Installation Pipework, Boosters and Compressors on Industrial and Commercial Premises*. Utilization Procedures IGE/UP/2, Communication 1598. IGE, London.
7. Chartered Institution of Building Services Engineers (CIBSE) (1986) Combustion Systems, Section B13 of the CIBSE *Guide*, p. 15. CIBSE, London.
8. British Standards Institution (BSI) (1983) BS 2869: Specification for fuel oils for oil engines and burners for non-marine use. BSI, London.
9. British Standards Institute (BSI) (1962) BS 1469: Coal tar fuels. BSI, London.
10. Chartered Institution of Building Services Engineers (CIBSE) (1986) Combustion Systems, Section B13 of the CIBSE *Guide*, p. 19. CIBSE, London.
11. Walker, D.C. (1990) *Investigating Boiler Failures*. A guide for building services engineers. Technical Note 2/90, p. 19. Building Services Research and Information Association, Bracknell.
12. Chartered Institution of Building Services Engineers (CIBSE) (1986) Combustion Systems, Section B13 of the CIBSE *Guide*, pp. 41–44. CIBSE, London.
13. British Standards Institution (BSI) BS 6644 Specification for installation of gas-fired hot water boilers of rated input between 60 kW and 2 MW. BSI, London.
14. British Standards Institution (BSI) Code of practice for oil firing. BS 5410: Part 2 Installations of 44 kW and above capacity for space heating, hot water and steam supply purposes. BSI, London.

4 Alternative Means of Heat Generation

4.1 Scope of the alternatives

Thus far we have considered gas-fired or oil-fired boilers as *the* means of generating the heat that will be used by a commercial space heating system. While these are certainly the most common type of heat generating plant, there are alternatives which may be worth considering. These fall into three categories. Firstly there are alternatives to the boiler as a means of generating LPHW. Secondly there are alternatives to the use of gas and oil as boiler fuels. Finally there are alternatives to LPHW heating itself. In each case our consideration will be confined to the context of providing space heating, since in commercial applications it will still frequently be advantageous to de-couple the space and domestic hot water heating requirements and serve the latter independently. This is considered further in Chapters 8 and 11.

There are probably only two alternatives to the boiler as a means of generating LPHW. These are combined heat and power (CHP) and the heat pump. Alternative fuels include the so-called 'bio-fuels' and certain types of waste. Although these are not as convenient to handle as gas or oil, the desire to reduce consumption of the latter in favour of more sustainable fuels has increased interest in these alternatives in recent years. Their use is briefly considered at the end of this chapter. Electricity may be used for commercial space heating, but it is the premium fuel both in cost and environmental impact. Even if generated sustainably, it is generally better employed for purposes other than space heating. However, an exception to this is the role of electricity as the preferred fuel for heat pumps. Electric heating will not be considered in this book outside this context. A range of alternative heating systems are available to the HVAC designer. These are typically based on decentralised heat generation and include various forms of electric heating (both direct and employing thermal storage), direct gas-fired warm air and radiant-tube systems, and air-to-air heat pumps. This list is by no means exhaustive. However, all of these heating systems lie outside the scope of this book and none of them will be considered further.

4.2 Combined heat and power units

4.2.1 The concept

Combined heat and power or CHP units are undoubtedly the most significant alternative to a boiler for generating heat in an LPHW heating system. As the name suggests, a combined heat and power system produces both electrical power and heat. The generation of electrical power is inherently accompanied by the production of a considerable amount of 'waste' heat. Large-scale power generation employing the latest combined-cycle gas turbine plant may achieve electrical conversion efficiencies of 50% or more of the gross calorific value (GCV) of the natural gas fuel, compared with less than 40% for the best conventional coal-fired power stations. Small-scale power generation employing reciprocating internal combustion engines may achieve a conversion efficiency in the range 28–35%. In all cases the rest of the GCV of the fuel appears as heat, and where electricity generation is the only interest this is wastefully rejected to atmosphere. It is the concept of a CHP system to both generate electrical power and to recover as much as possible of the available heat for use. This significantly increases the overall conversion efficiency from the GCV of the fuel to useful power and heat. In building services applications the obvious use for this heat is in the provision of space and water heating. From the preceding discussion it can be seen that CHP systems invariably produce substantially more heat than power.

4.2.2 Application to building services

CHP itself is a well-established technology, and CHP systems generally may be based on conventional steam boilers and turbines, gas turbines and waste heat boilers or internal combustion engines with heat recovery systems. The scale of both electrical power and heat generation therefore varies widely. The largest schemes, which can in principle serve large commercial or industrial complexes or whole areas of a town or city, employ steam and gas turbine systems. These require specialist operating skills and are most suited to continuous operation over extended periods. Systems based on internal combustion engines are more suited to the intermittent operation that will be associated with the daily operating cycle of many buildings. In building services applications CHP systems will typically be of the type referred to as *small-scale packaged CHP*. These will almost invariably be based on internal combustion engines and will be supplied as one or more complete packaged CHP units.

The size of a CHP unit is designated in terms of its nominal *rated electrical power output in* kW_e. Small-scale packaged CHP units are typically available across a size range from a minimum output of approximately 25 kW_e up to a maximum output of 500 kW_e. The ratio of heat output to electrical output (it is typically referred to as the *heat ratio*) depends on the type of CHP system. Steam systems can provide heat ratios in the range between 3:1 and 10:1. Basic gas turbine CHP systems may achieve a heat ratio of approximately 2:1, while the addition of *supplementary firing* may increase this to as much as 5:1. The heat output from a small-scale packaged CHP unit based on an internal combustion engine, as likely to be employed in a building services application, will typically be approximately twice the electrical output, although the ratio may vary somewhat according to the size of the unit, with larger units potentially offering an increased electrical conversion efficiency (and hence a slightly reduced heat ratio).

Considerable experience of the use of packaged CHP exists in the UK, both in building services and industrial applications, with many hundreds of systems installed since the mid-1980s. However, in commercial heating applications packaged CHP units will never be installed as a *complete alternative* to boilers, but will rather always be part of a 'mixed' heating plant that combines one or more CHP units with one or more boilers.

4.2.3 Principal elements of a small-scale packaged CHP unit

The principle elements of a typical small-scale packaged CHP unit are shown schematically in Figure 4.1. Such a unit is built around a water-cooled reciprocating internal combustion engine that is coupled to an electrical generator. Engines are typically spark-ignition types fuelled on natural gas. For the

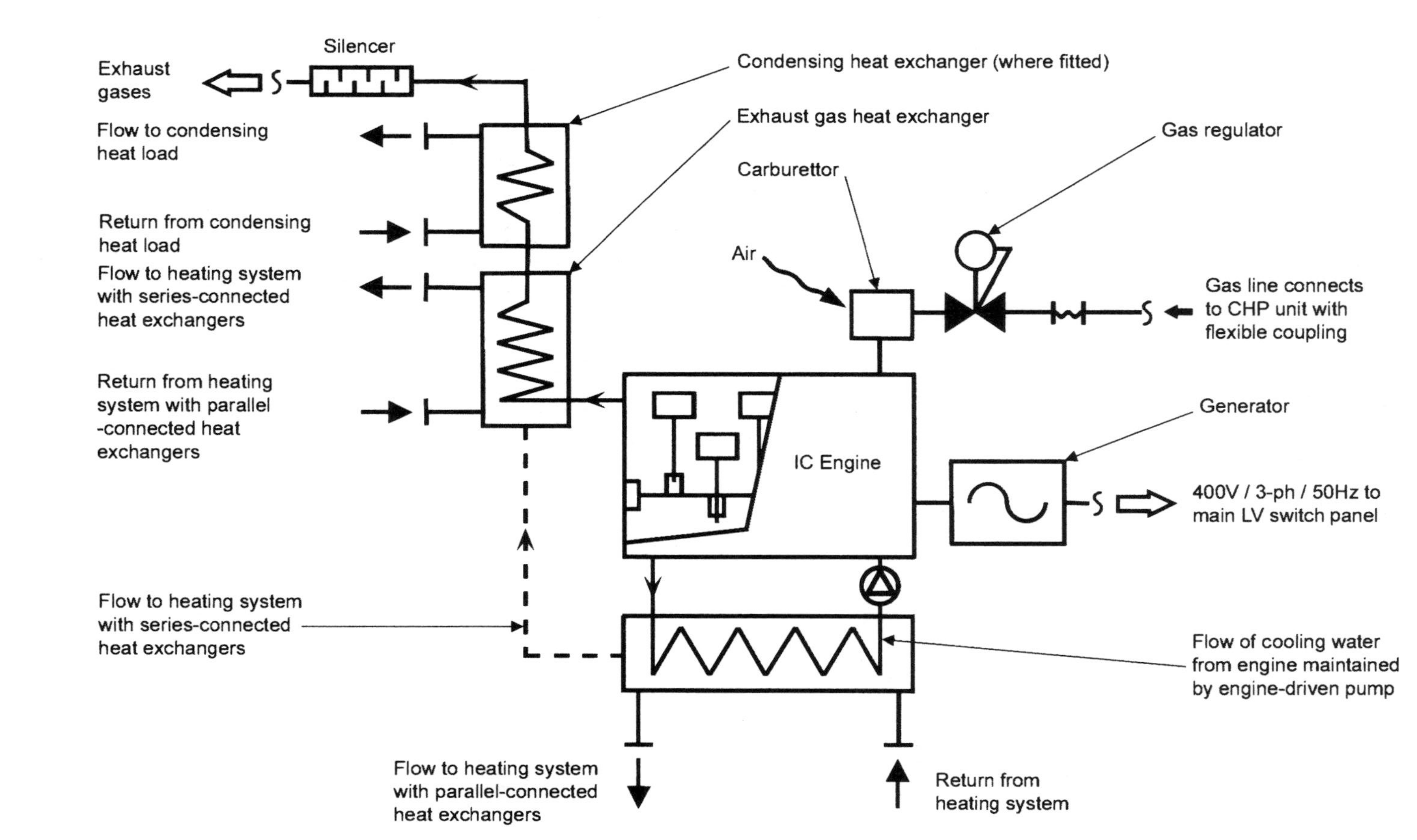

Figure 4.1 Principal elements of a typical small-scale packaged CHP unit.

smaller packaged units with rated output in the range up to c.200 kW_e engines are most commonly derived from automotive use (but heavy goods vehicles, rather than cars), although in CHP use such units typically run relatively slowly (c.1500 rpm) and at constant speed. Engines may also be based on 'stationary' industrial engines originally designed for pumping or primary power generation, and this will typically be the case for units with rated outputs in excess of 200 kW_e. Due to their lower production volume, industrial engines tend to have higher capital and spares costs than automotive-derived units, but are typically designed for low wear and ease of maintenance in unsophisticated operating environments.

The unit incorporates a *heat recovery system* that comprises a minimum of two heat exchangers, via which heat can be transferred to the building heating system. The first allows transfer of the heat picked up by cooling water circulating through the engine cooling jacket and oil cooler, while the second allows recovery of heat from the engine exhaust gases. A physical separation is maintained between the engine's water cooling system and the building heating system in order to avoid any risk of the latter causing corrosion or thermal shock to the engine. Hence the transfer of heat recovered by the engine cooling water to water from the building heating system takes place across a water-to-water heat exchanger. The exhaust gas heat exchanger is a gas-to-water type that allows heat to be transferred directly from the engine exhaust gases to water from the building heating system. Since the exhaust gas heat exchanger can allow hot water to be generated at temperatures in the MPHW range, it may in some applications be desirable to connect the building heating system to the two heat exchangers in parallel, although in straightforward LPHW applications a series-connection is all that is necessary, with single flow and return connections to the heating system. CHP units may employ an intermediate heat exchanger to provide a common source of heat for the heating system to interface with.

Where the CHP unit is fuelled with natural gas and a suitable source of low-temperature return water is available from the heating system, a third heat exchanger may be added to provide *condensing operation*, piped in parallel with the main flow and return water connections to the CHP unit. This heat exchanger mirrors the role of the secondary heat exchanger in a condensing boiler, and typically requires a source of water returning from the building heating system at a temperature in the range 30–40°C to ensure that substantial and sustained condensation occurs. The condensing heat exchanger must include appropriate provisions for the drainage of condensate without escape of exhaust fumes.

The exhaust from the CHP unit is discharged to atmosphere via an exhaust system that mirrors the flue of a boiler but is typically of smaller diameter and will incorporate appropriate silencing. The CHP unit comes complete with integral controls, with the user interface provided by a local control panel. All elements of the unit are mounted on a common base frame with anti-vibration mountings, and in some instances the whole unit may be enclosed within a casing that functions as an acoustic enclosure.

4.2.4 Typical energy balance for a small-scale packaged CHP unit

If approximately 28% of the GCV of the fuel input to the CHP unit may be delivered as electricity, approximately a further 55% may be recovered at a high enough temperature from the engine cooling jacket, oil cooler and exhaust gas heat exchanger to allow its transfer to water from a heating system operating at conventional LPHW flow and return temperatures of c.82°C and 71°C respectively. Approximately 60% of this recovered heat (c.33% of the GCV of the fuel), may be expected to derive from the engine cooling jacket and oil cooler, with the remaining 40% (c.22% of the GCV of the fuel), deriving from the exhaust gas heat exchanger. Since the exhaust gases enter this at a temperature of up to 650°C, and are cooled to a temperature of the order of 120°C, this heat may potentially be transferred to a heating load circuit operating at flow and return temperatures in the MPHW range or used to generate steam. A further 10% of the GCV of the fuel input to the CHP plant may be recovered in a second, condensing exhaust gas heat exchanger, both by the generation of condensate and the increased cooling of the exhaust gases. This is, of course, provided that a suitable heating load circuit exists. In any event the exhaust gases from the engine still retain some heat, even during condensing operation, and this part of the GCV of the fuel is lost from the system when the exhaust gases leave the last heat exchanger. This is equivalent to the 'stack' or chimney loss from a boiler. Some heat is also lost, by radiation and convection, from the body of the CHP unit, and this is equivalent to the 'radiation' loss from a boiler.

This suggests a nominal *overall conversion efficiency* to useful energy output of c.83% (non-condensing) and c.93% (condensing), based on the GCV of the fuel. However, the 'boiler' efficiency, or conversion efficiency to useful heat output, is only c.55% (non-condensing) and c.65% (condensing). Since even if there were *zero* heat loss via the exhaust gases and from the body of the unit, these values would only increase to c.62% and 72% respectively, we can conclude that heat alone may always be generated more efficiently by a boiler than by a CHP unit. This statement is certainly not to diminish the potential benefits of packaged CHP units, and its significance will become clear in a subsequent section of this chapter. For the present Figure 4.2 summarises the energy balance described above. For a given CHP unit the exact energy balance may vary with the size of unit and its particular design characteristics.

4.2.5 Fuels for CHP

Natural gas is typically the first-choice fuel for boilers in commercial heating space and water heating applications. Since a small packaged CHP unit will invariably run in conjunction with conventional boilers, whether in an existing installation or a new one, natural gas is most likely to be used as its fuel. However, CHP units may also run on liquified petroleum gas (LPG), grade 'A' fuel (DERV or 'diesel') or grade 'D' fuel oil ('gas oil'). Fuel storage will be required for all three of these alternatives to natural gas, and the additional cost of providing this must be taken into account. However, a choice of either LPG or gas oil is only likely to be made if the boilers in the heating plant are similarly fired. The effect of the CHP unit on the required fuel storage and supply system may therefore be minimal. Where DERV or gas oil is used, the engine of the CHP unit will run on the diesel (compression-ignition) cycle which typically results in higher conversion efficiencies for electrical power generation, albeit at the expense of increased emissions from the engine and the formation of deposits of unburnt carbon in the engine ('sooting'). The former may affect the location of the exhaust termination, while the latter may increase maintenance costs. Set against this, diesel

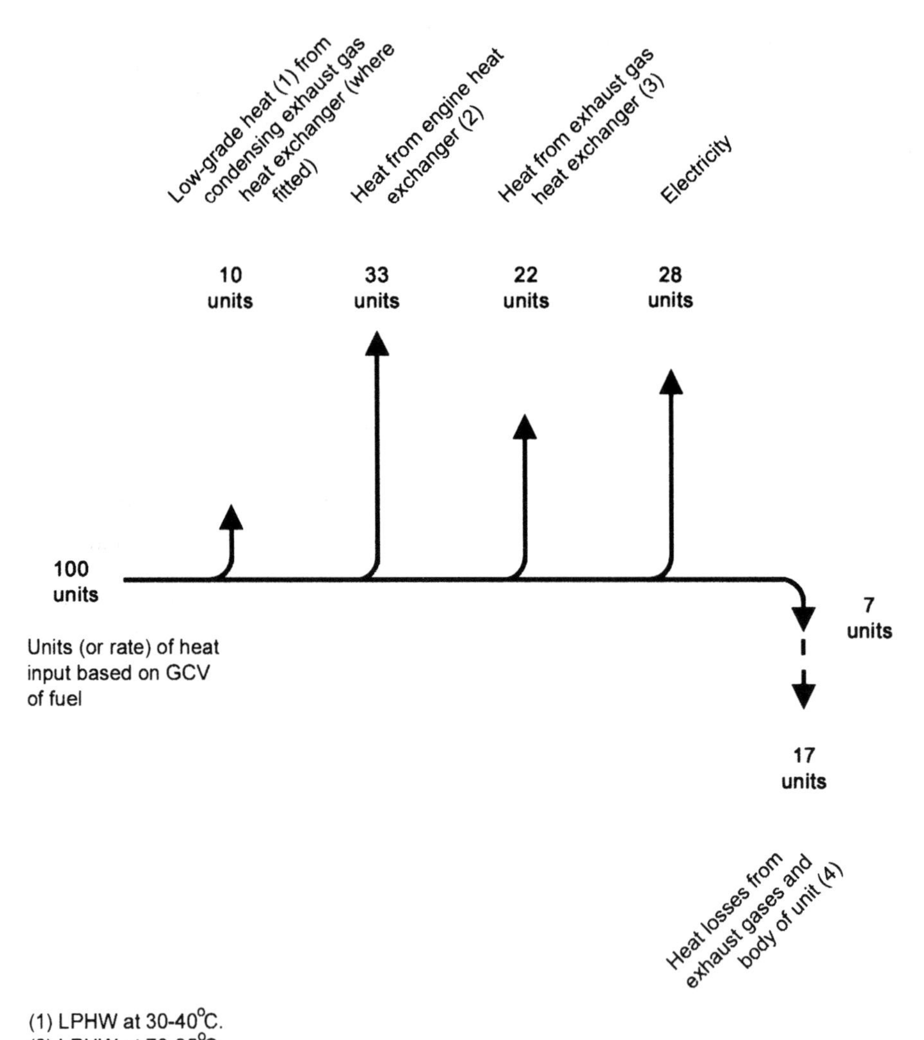

Figure 4.2 Typical energy balance on a small-scale packaged CHP unit.

engines are a standard type, are widely used in standby generator sets, and the skills necessary for their maintenance are widely available.

Small packaged CHP units may also run on 'biogas', which is methane gas produced by the decay of biological waste and waste foodstuff. Methane itself, CH_4, is of course the principal constituent of natural gas. Biogas is only likely to be available in adequate quantities for fuelling a CHP unit under specific circumstances, in particular at sewage treatment plants, at some landfill sites, and possibly in some agricultural or horticultural applications. However, in these situations the biogas is typically produced by natural biological action, whether required or not, and must in any event be disposed of safely. Since methane is also a strong greenhouse gas, with a global warming potential of 21 times that of carbon dioxide, taken over a 100-year period, there is an obvious benefit in using it beneficially, rather than simply venting it to atmosphere or wastefully burning it off. However, this does not necessarily mean that the use of biogas is 'free', since the presence of traces of hydrogen sulphide (H_2S) in the gas, together with excess water, may both require provisions for its use that increase the capital cost of the unit and result in increased maintenance costs. The presence of hydrogen sulphide also implies that the use of a condensing exhaust gas heat exchanger will not be appropriate, since the condensate formed will be aggressively acidic. The manufacturer of the prospective CHP unit should be consulted as early as possible where operation on biogas is an option under consideration. There are also, of course, practical issues to be solved in the collection and supply of the biogas to a CHP unit. Except where environmental considerations are given prominence, the use of biogas is unlikely to be relevant to most typical building services applications of small-scale packaged CHP.

4.2.6 The economic case

In terms of its capital or 'first' cost, a small-scale packaged CHP unit is *very substantially more expensive* than a conventional (non-condensing) gas-fired boiler. Published general guidance to HVAC designers suggests that the installed cost of a packaged CHP unit may be of the order of £600–850 kW_e^{-1} (i.e. per kW of electrical output), or from 1.7 to 3 times the cost of a standby generator of equivalent electrical output.[1] Incremental cost, in £ kW_e^{-1}, may typically be expected to be greatest for small units, dropping off sharply over the first 150–200 kW_e and ultimately 'bottoming out' for large installed outputs. Unfortunately most building services applications of small-scale packaged CHP will typically fall within the range of relatively high incremental cost. At a heat ratio of approximately 2:1 this is equivalent to an installed cost of the order of £300–425 *per kW of heat output*. By comparison, the installed cost of a conventional natural gas boiler of equivalent heat output may be expected to be approximately 10% of this for the smallest boilers (assuming a minimum CHP unit size of c.25 kW_e, with a corresponding heat output of c.50 kW), falling to 6–7% of it as boiler size increases to give heat outputs beyond approximately 200 kW (i.e. a CHP unit of approximately 100 kW_e). While this will continue to decrease to approximately 5% as boiler heat output increases, it is likely that the size of packaged CHP units employed in building services applications will typically be *considerably less than 500* kW_e.

Small packaged CHP units are more sophisticated than boilers, and require specialist maintenance expertise. Maintenance requirements must include both routine servicing and long-term overhaul and rebuild costs. While the latter may be infrequent, they are major cost items that must be spread over the whole life cycle of the CHP unit. The maintenance of a small-scale packaged CHP unit is therefore considerably more expensive than that of a boiler of equivalent heat output. Building owners or operators will typically not have such skills 'in-house', and maintenance will therefore need to be contracted to an outside organisation. Most manufacturers of CHP units will be able to offer comprehensive maintenance of their products following installation. For general guidance and initial scheme assessment a guideline figure of approximately 0.7p kWh_e^{-1} (i.e. per kilowatt-hour of electrical output, based on the *full* output of the unit) is considered typical,[2] although this will not necessarily represent the cost of an 'all-inclusive' maintenance contract including an emergency call-out service.

It has sometimes been considered that the heat output from a small-scale packaged CHP unit is 'free'. This is typically *not the case*, however, since heat alone may always be generated more efficiently in a boiler. The concept of *total marginal cost (TMC)* or *marginal generating cost (MGC)* balances the cost of fuel and maintenance for the CHP unit against the cost of generating its *useful* heat output in an alternative boiler plant. Expressed in pence per kilowatt-hour of electrical output ($\mathrm{pkWh_e^{-1}}$), TMC is the *effective* cost to the CHP operator of generating one kWh of electricity.

$$\mathrm{TMC} = \left[\frac{I_{CHP}F_{CHP} - \left(\dfrac{100 \times H_O \times F_{Boiler}}{\eta_{Boiler}} \right)}{E_O} \right] + m \quad (4.1)$$

where I_{CHP} = fuel input rate to the CHP unit (kW); F_{CHP} = unit cost of fuel for the CHP unit ($\mathrm{pkWh^{-1}}$); H_o = *useful* heat output rate from the CHP unit (kW); F_{Boiler} = unit cost of fuel for the boiler ($\mathrm{pkWh^{-1}}$); η_{Boiler} = overall efficiency of a boiler that *could otherwise be used* to generate the heat (%); m = maintenance cost for the CHP unit ($\mathrm{pkWh_e^{-1}}$); E_o = electrical power output of the CHP unit ($\mathrm{kW_e}$).

The electricity generated by the CHP unit substitutes for imported mains electricity, and the unit purchase price (tariff) of the latter may therefore be compared directly with the value of TMC. Provided that the TMC is less than the applicable electricity tariff, an overall cost saving will be made from running the CHP unit. Annual savings are determined as the summation of the difference between the tariff for imported power and TMC for every kWh of electricity generated by the CHP unit during the course of the year. This calculation may be straightforward if a single electricity tariff applies and the full heat output of the CHP unit may always be usefully employed. However, the situation is more complex where electricity tariffs are not constant, as with a *seasonal time of day (STOD)* tariff, and where TMC varies (as when all of the heat output of the CHP unit cannot be usefully employed). In any event it is essential that a careful analysis is carried out based on both electrical and heat load profiles for the application. Ideally an hourly analysis should be carried out for at least a 'typical' operating day in each month. While a detailed discussion of the operating strategy for CHP is beyond the scope of this book, such an hourly analysis will identify periods during which TMC exceeds the electricity import tariff (and the CHP unit should be switched off), together with those during which either the electrical or heat output of the unit exceeds the site requirements. When mains electricity is available at 'off-peak' rate, it will typically be preferable to utilise such a supply, supported by a boiler for heat output, in preference to operating the CHP unit. While surplus electricity may be exported (or transferred to other sites), it is rarely cost-effective to do so. The electrical output of some CHP units may be modulated down to c.50% of their rated output. However, efficiency reduces with load, and hourly maintenance costs must be spread over the reduced quantity of electricity generated (the value of m used in equation 4.1 must be adjusted to reflect this). Hence the value of TMC increases.

It is important to note that equation 4.1 is based on the heat output from the CHP unit that may be *usefully employed*. It is clear that if this reduces for a constant rate of fuel input to the unit, TMC will increase and savings will reduce. While the CHP unit is running, all the heat output *must be dissipated*. If it cannot be usefully employed, it must be rejected to waste (typically to atmosphere) using dedicated heat rejection equipment. The provision and control of such a dedicated heat rejection facility increases the capital cost of the installation. Furthermore, where heat is rejected to atmosphere the heat rejection equipment will typically comprise one or more dry air coolers. There will thus be a *parasitic power consumption* associated with the rejection of unwanted heat, although the cost of this will typically be small in comparison with the fuel cost for unwanted heat. The rejection of unwanted heat to a thermal store, for subsequent use, is not generally considered to be economic.

In view of the above it will typically only be economically viable to install a small-scale packaged CHP unit where there is a substantial year-round demand for heating. Good examples of applications that meet this requirement are hospitals, hotels, sports and leisure centres. All of these provide high year-round hot water service heating loads. Other potential applications include college and university (residential) campuses, sheltered residential accommodation for the young or elderly, military bases, 'institutional' accommodation (prisons and residential detention centres) and any application that includes a swimming pool. Industrial process heating may also be eminently suitable for small-scale packaged CHP units, particularly where the process is continuous. The scope for CHP units in commercial applications may be increased by the use of water chilling plant that operates on the *absorption cooling cycle*, which is a heat-operated cycle. However, such plant typically requires either hot water at temperatures in the MPHW range or the availability of steam, and only a proportion of the heat output of some small-scale packaged CHP units may be available in this form.

The optimum size of a small-scale packaged CHP unit may only be determined from an appropriate hourly analysis of the electrical and heating load profiles for the particular application. Too large a machine will result in long periods of shut-down. Too small a unit may achieve high running hours, yet forego potential savings. However, it may be possible to subsequently add a further unit if the potential for further savings is great enough. Hence undersizing is typically regarded as by far the lesser of the two evils. Unfortunately the sizing process is a time-consuming and complex exercise. In the past simple rules of thumb have been used to size CHP units using the concept of a *base load* and minimum annual operating hours. One such technique that is applicable at an early stage of scheme planning is set out in a publication from BSR1A[2]. However, computer-based techniques are now common and range from manufacturers' proprietary sizing software to application-specific models based on load profiles determined from actual measurements.

Two further factors have recently been introduced into the economic assessment of small-scale CHP schemes in the UK. From 1 April 2001 the cost of all forms of non-domestic energy use became subject to a climate change levy (CCL) that initially amounts to an increase in the cost of electricity, natural gas and LPG of 0.43p$\,$kWh^{-1}, 0.15p$\,$kWh^{-1} and 0.07p$\,$kWh^{-1} respectively. Energy users pay the CCL through the bills received from their energy suppliers, much in the same way that they would pay value added tax, to which the CCL is also subject. However, 'good quality' CHP installations in building services applications will be exempt from the effects of the climate change levy if they meet threshold performance criteria that will be specified by a scheme of *CHP quality assurance (CHPQA)*. The definition of 'good quality' basically requires that most of the recovered heat is usefully employed, and that minimum values of annual conversion efficiency to electrical power are met. Furthermore, such CHP installations will also be eligible for a scheme of *enhanced capital allowances (ECA)* that will typically allow 100% capital allowance against tax during the year in which the scheme is installed. Qualification for the benefits of ECA and exemption from the climate change levy will depend on both calculation procedures and monitoring of the performance of an installation. Hence the necessary provisions should be considered at an early stage in the planning of a small-scale packaged CHP scheme, and early input from the prospective manufacturer or supplier of the unit will in most cases be beneficial. While exemption from the CCL and the benefit of an ECA will undoubtedly improve the economics of employing a small-scale packaged CHP unit, it is unlikely that meeting the qualification requirements for these benefits *per se* will obviate the basic design considerations inherent in achieving a successful (i.e. economically viable) implementation.

The capital cost of implementing a small-scale packaged CHP scheme is inevitably high and this can prove a barrier to its adoption in some of the very applications in which it would exhibit favourable economic potential. For 'capital-strapped' clients there are options whereby the responsibility for some or all of the capital outlay, operating and maintenance costs over the life of the CHP unit are taken over by the manufacturer or supplier of the unit. Naturally in return for taking over these costs, the manufacturer or supplier of the unit also takes over a greater or lesser proportion of the savings generated by the unit over its life cycle. Such arrangements vary from *supplier finance* to full *contract energy management*.

4.2.7 The CHP unit as part of a boiler installation

A CHP unit may be installed *in parallel with one or more boilers* (Figure 4.3a), or *in series with parallel-connected boilers* using a bypass off the common primary return to the boilers (Figure 4.3b). Installation in parallel is recommended where the CHP unit may supply a significant proportion of the heating load, and in new boiler installations generally. Installation in series may be advantageous where the CHP unit is introduced into an existing boiler plant. In any event the water-side pressure drop across the CHP unit is likely to be greater than exists across the boilers, with each at its respective design water flow rate, and the CHP unit will therefore typically require a dedicated 'shunt' pump. Since the CHP unit will typically need to operate during the summer months in order to generate the critical number of running hours for a viable installation, it must be possible to maintain the design water flow rate through the CHP unit even if all boilers are shut down. This is specifically relevant to the 'series' installation of a CHP unit. Where bypass loops are not employed around individual boilers, an alternative means of bypassing them will be required to maintain flow in the CHP circuit.

The pipework connections and gas line to a typical small-scale packaged CHP unit are illustrated schematically in Figure 4.4. Since the CHP unit is built around a reciprocating engine, final connections between the unit and the pipework of the building heating system should always be made via

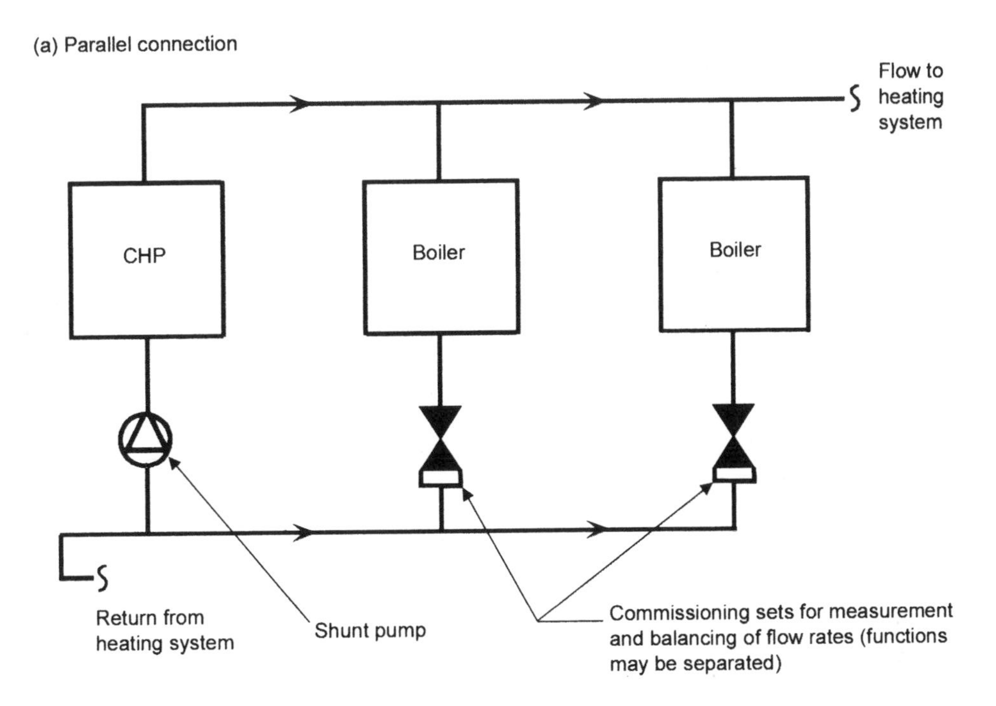

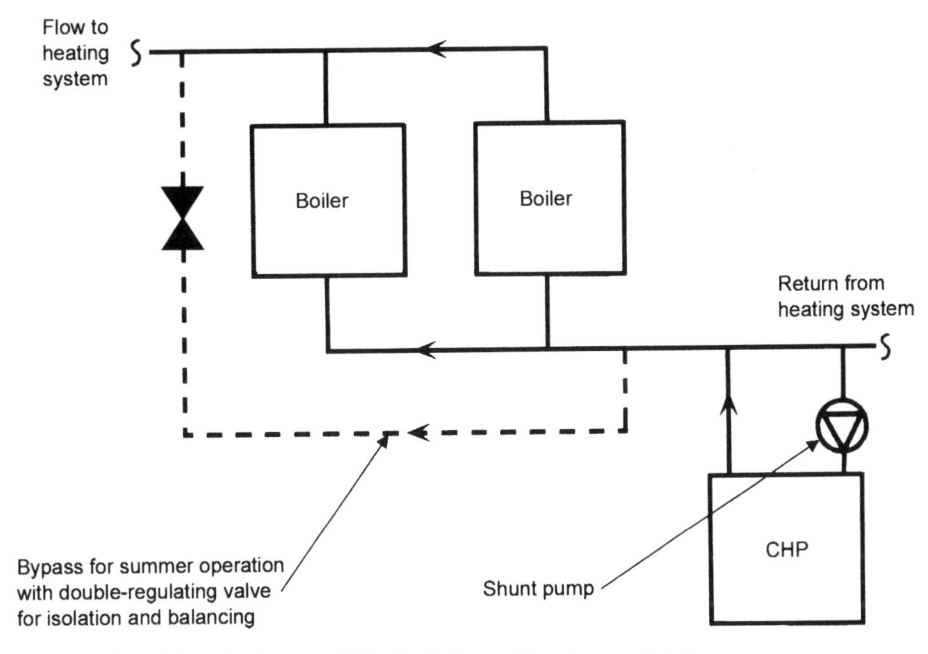

Figure 4.3 Small-scale packaged CHP units piped in parallel and in series with boilers.

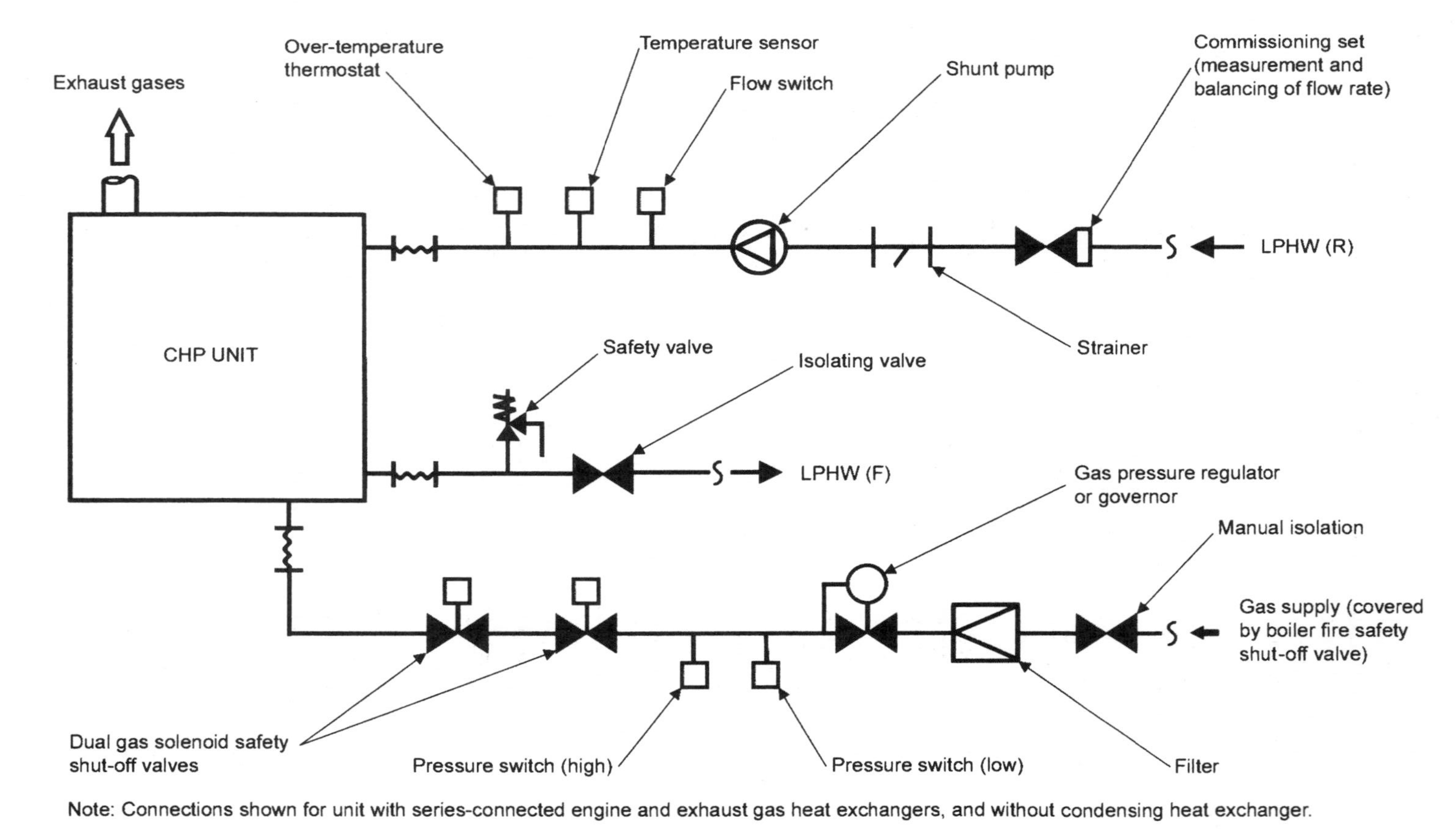

Figure 4.4 Pipework connections and gas line to a typical small-scale packaged CHP unit.

flexible couplings. Failure of water flow to the CHP unit can be *catastrophic*, resulting in a risk of major and expensive damage (such as cracked cylinder heads and damaged heat exchangers). Flow switches should therefore be employed in the return water pipework from the heating system to any individually-connected heat exchanger (including a condensing exhaust gas heat exchanger), with an additional *high-limit or 'over-temperature' thermostat* (manually reset) in the main return water pipework to the unit. Strainers should in any event be fitted to remove any debris from the heating system that could block a heat exchanger, and operation of the CHP unit should be interlocked with the relevant heating system circulating pumps.

Condensate must be drained, both where a dedicated condensing exhaust gas heat exchanger is fitted and from the exhaust system generally if such a heat exchanger is not fitted. In any event condensate must not be allowed to drain back towards the CHP unit (and the engine). Condensate formed will be acidic and will discolour concrete and other surface finishes. This may need to be considered in relation to the exhaust discharge from the building, since this will typically be turned down against the potential ingress of rain, and may therefore 'drip' condensate on to any surface below. The exhaust pipe of a CHP unit will operate at a much higher positive pressure than is the case for boiler flues, perhaps of the order of 50 mbar (5 kPa) or more, and special provisions will be required for the trapping of condensate. This will typically employ a water-filled tank as a trap, from which the condensate can overflow into a suitable run of drainage pipe.

The exhaust pipework from CHP units may be susceptible to corrosion, and high-quality (and therefore expensive) stainless steel may achieve a longer life than mild steel. In any event the installation should allow the necessary access for periodic replacement during the life of the CHP unit. In particular, bends in the exhaust pipe may experience accelerated erosion by water droplets moving at high velocity in the exhaust gases, and provision for easy replacement will therefore be beneficial. Where condensing operation is employed, plastic exhaust pipework may in principle be used, although any leakage into the building from damaged pipework has the potential to carry with it poisonous carbon monoxide produced by the engine. An interlock is also recommended to shut down the unit in the event of a high-limit exhaust gas temperature that could damage the pipework. The exhaust system will typically incorporate one or more 'silencers' to reduce noise levels. Silencers may be of two types. 'Reactive' silencers rely on the *capacitance* of a hollow chamber and the *inductance* of the exhaust pipe to dampen out pressure pulses in the exhaust gases. This has its principal benefit at low frequencies, and is typically located close to the engine. The alternative, 'absorptive' type of silencer relies on a sound-absorbent material. This is the type most familiar to HVAC engineers, and has its principal benefit at higher frequencies. Where necessary, this type is located near the discharge termination of the exhaust pipe.

In order to achieve as many running hours as possible, the CHP unit will typically be operated as the 'lead' source of heat, with the boilers themselves being sequenced into operation as required. Sequence control of multiple boilers is typically carried out according to the common return water temperature in the boiler primary circuit (Chapter 10 considers this further, as well as the alternatives). This is generally satisfactory for boilers, since the boiler control thermostat is set to a value of approximately 10 K above the desired mixed flow temperature, and the last firing boiler (the lead boiler) has no difficulty in producing water at, say, 91°C when a low load on the heating circuits results in a common return water temperature at c.80°C. The same is not, however, true of a small-scale packaged CHP unit, which will typically favour a *constant return water temperature*, and will in any event have a protective cut-out function at a high-limit return water temperature. The control requirements and abilities of the CHP unit should be clearly established with the prospective manufacturer at an early stage in the design of an installation.

While the general electrical services engineering aspects of small-scale packaged CHP units are outside the scope of this book, it is important to note that specific technical requirements apply to the earthing of such units. This aspect will require early consideration, particularly where a CHP unit is being installed in an existing building or is intended to provide standby electrical power on mains failure.

4.3 Heat pumps

4.3.1 Principle of operation

Heat pump and refrigeration cycles are identical. In both cases a refrigerant evaporates to absorb heat (at the *evaporator*) from a source at relatively low temperature, and condenses to reject it (at the *condenser*) to a sink at a higher temperature. In the *vapour-compression cycle* the transfer (upgrading) of this heat between the evaporating and condensing temperatures is achieved by compression of the refrigerant vapour. This takes place in a compressor that is typically driven by an electric motor. The thermodynamic cycle is the same whether we are considering a heat pump or a refrigeration plant. The difference is in the application for which the equipment employing the cycle is optimised. Refrigeration plant is optimised to achieve useful cooling of some medium at the evaporator, the heat absorbed being simply rejected to waste. A heat pump is optimised to deliver the heat absorbed usefully for the purpose of providing space heating. In some circumstances it is possible to do useful cooling and heating simultaneously. This is termed *refrigeration with heat recovery*, and is common practice in supermarkets.

4.3.2 Heat sources

The heat pumps that will be considered in this chapter are of the *air-to-water type*, i.e. they absorb heat from ambient air and transfer it to an LPHW heating system. The schematic arrangement of such a heat pump is shown in Figure 4.5a. Other types of heat pump exist and may be used for space heating. Most common in this respect are the *air-to-air types*, although their use is typically linked to the provision of comfort cooling. Such heat pumps lie outside the scope of this book and will therefore not be considered.

In principle, heat pumps may draw heat from a variety of sources other than ambient air. These include surface water in lakes and streams, groundwater and even the ground itself. If available in continuous and adequate supply, sources of waste heat, such as process water or warm ventilation exhaust, may also provide a heat source of a usefully higher grade. Some of these sources are available at a more-or-less constant temperature, while the temperature of ambient air is, of course, highly variable according to the climate and pattern of weather experienced at a particular site.

(a) Vapour-compression air-to-water heat pump

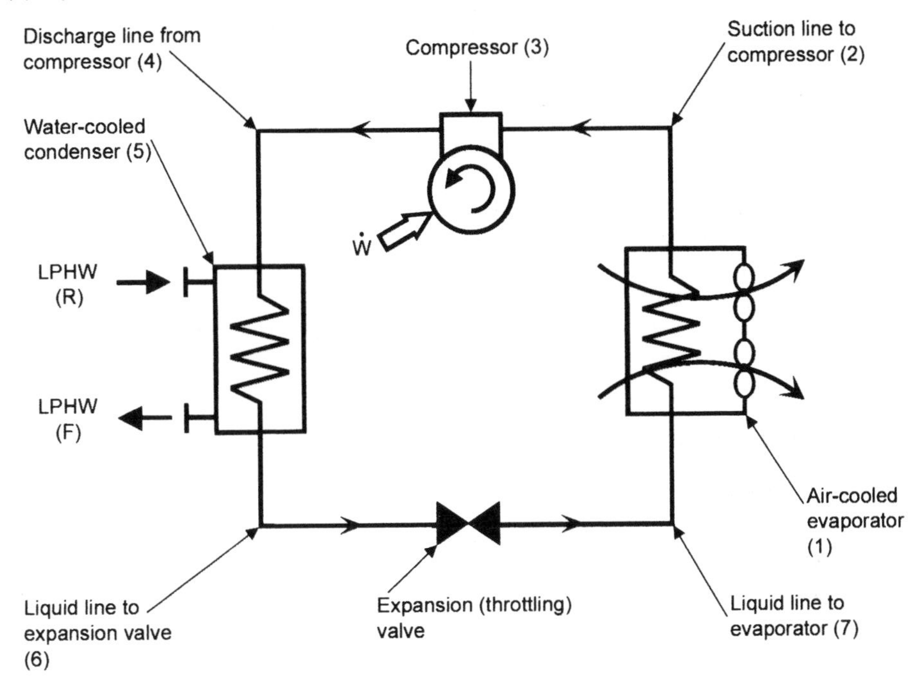

(1) $\dot{Q}e$ of low-grade heat absorbed from air.
(2) Cold (low-pressure) refrigerant gas.
(3) Low-grade heat upgraded to higher temperature by compressor work input $\dot{W}$.
(4) Hot (high-pressure) refrigerant gas.
(5) $\dot{Q}c$ of heat (= $\dot{Q}e$ + $\dot{W}$) rejected to LPHW heating system.
(6) Hot (high-pressure) refrigerant liquid.
(7) Cold (low -pressure) refrigerant liquid (plus some gas).

(b) Pressure~enthalpy (P~h) diagram for a 'practical' ideal heat pump cycle

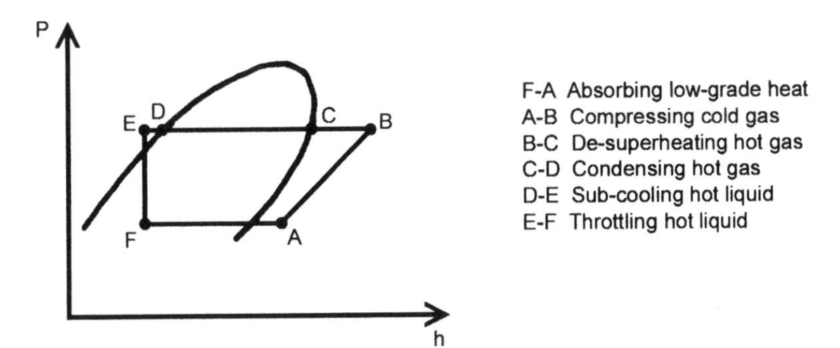

Figure 4.5 Schematic of a vapour-compression air-to-water heat pump.

4.3.3 Coefficient of performance

A 'practical' ideal heat pump cycle is shown in Figure 4.5b. When operating in equilibrium, the First Law of Thermodynamics requires that:

$$\dot{Q}_C = \dot{Q}_E + \dot{W} \quad (4.2)$$

where $\dot{Q}_C$ = the rate at which energy is rejected from the cycle at the condenser (kW); $\dot{Q}_E$ = the rate at which energy is absorbed into the cycle at the evaporator (kW); $\dot{W}$ = the rate at which work is input to the cycle at the compressor (kW).

Equation 4.2 is applicable whether we are considering a refrigeration plant or a heat pump. In the case of the former the measure of performance that is of interest is the ratio of the rate of cooling at the evaporator ($\dot{Q}_E$, the useful 'output') to the power input at the compressor ($\dot{W}$, which must be purchased). This ratio, $\dot{Q}_E/\dot{W}$, is termed the *coefficient of performance (COP)* of the refrigeration plant. COP is a pure ratio. For a heat pump the measure of performance that is of interest is the ratio of the rate of heat rejection at the condenser ($\dot{Q}_C$, now the useful output) to the power input at the compressor, and the ratio $\dot{Q}_C/\dot{W}$ is the COP of the heat pump. Dividing throughout equation 4.2 by $\dot{W}$ yields:

$$\frac{\dot{Q}_C}{\dot{W}} = \frac{\dot{Q}_E}{\dot{W}} + 1 \quad (4.3)$$

Substituting the above definitions of COP:

$$COP_{Heat\ pump} = COP_{Refrigeration} + 1 \quad (4.4)$$

Where confusion is possible, COP is typically qualified as $COP_{Cooling}$ or $COP_{Heating}$ as necessary. The value of COP that can be attained by a heat pump depends not only on its design but also on the operating temperatures. A heat pump operating between evaporating and condensing temperatures of 5°C and 35°C respectively might achieve a $COP_{Heating}$ of c.5. This means that each kWh of useful heat requires the purchase of 0.2 kWh of electricity. This compares very favourably with direct electric heating (which has an effective 'COP' of unity), and *may* be comparable with heat generated by a conventional boiler plant fired on natural gas – although peak-rate electricity typically costs between 2.5 and 5 times the cost of natural gas. However, such a potentially favourable $COP_{Heating}$ is unlikely to be achieved in practice in an air-to-water heat pump. There are several reasons for this. Firstly, practical heat exchangers require a temperature difference of some 5 K between the working fluid and the heat source or sink. Hence a condensing temperature of 35°C would only allow LPHW to be generated at a maximum of c.30°C, which is of limited use in most building services space heating applications. At the same time, for an evaporating temperature of 5°C the ambient air would need to be at a temperature of at least 10°C if the design rate of heat transfer was to be achieved at the evaporator. In winter, when space heating demands are greatest, UK ambient air temperatures fall well below this. (The implications of this for the sizing and operation of heat pumps are discussed further later in this chapter.) Reducing the evaporating temperature and/or increasing the condensing temperature (the latter being almost invariably essential if the heat pump is to generate LPHW at a temperature suitable for a conventional space heating system) results in a significant fall in the COP of the heat pump.

For practical application it is essential that the measure of performance of a heat pump takes into account the fan and pump power necessary to provide air flow across the evaporator and water flow through the condenser, both of which are essential to its operation. An *appliance coefficient of performance* is therefore defined as the ratio of heat output to the effective power input at stated conditions.[3] Under this definition the *effective power input* is taken to include the power input to the compressor and to all control and safety devices, the power input to the heat pump 'defrost' cycle, and the proportional power input to the fans and pumps that transport the heat source and sink fluids within the heat pump. The concept of the appliance COP of a heat pump is mirrored by the *energy efficiency ratio*, the ratio of total heat energy dissipated by the heat pump to the total energy supplied to it[4]. Ultimately, as in the case of boiler efficiency, the value of COP that is crucial to the energy use (and cost) and carbon emission of a heat pump is the *seasonal COP*, the value that is attained as an average over the whole heating season.

The *absorption cycle* may also be applied to a heat pump. In this, refrigerant gas leaving the evaporator is absorbed by a liquid, and the pressure of the resulting solution is increased by a pump. The significant input of electrical power required to the compressor in the vapour-compression cycle is therefore replaced by a much smaller power input to a refrigerant/absorbant solution pump. However, a significant input of heat is then required (at a relatively high temperature) to drive the refrigerant vapour back out of solution. The absorption cycle is therefore *heat-operated*, whereas the vapour-compression cycle is *work-operated*. At the time of writing absorption-cycle plant in UK building services applications is essentially limited to water chillers. While some units may provide condenser heat recovery, and certain direct-fired types may function as *chiller/heaters* (*not* heat pumps), discussion of such units does not lie within the scope of this book. We may note, however, that the maximum $COP_{Cooling}$ of such commercially-available chillers is low (typically in the range 0.6–1.2, depending on the design of the unit), although in principle absorption-cycle heat pumps may ultimately be attractive from the perspective of primary energy use.

For a more detailed explanation of heat pump/refrigeration cycles the interested reader is referred to any one of a number of standard texts on engineering thermodynamics or refrigeration.[5,6]

4.3.4 Refrigerants

Vapour compression heat pumps have typically employed the *hydrochlorofluorocarbon (HCFC)* refrigerant R22. While considered to have a very good performance in heat pumping applications, the *ozone depletion potential (ODP)* that results from the (albeit low) chlorine content of this refrigerant has led to a phase-out of its use in new refrigeration plant and heat pumps (and ultimately in existing plant). Ozone depletion potential expresses the capacity of a molecule to destroy ozone in the stratosphere region of the earth's atmosphere. R22 exhibits an ODP of 0.05 relative to that of the former *chlorofluorocarbon (CFC)* refrigerant R11, which is taken to have an ODP of unity. New applications for heat pumps should therefore employ suitable replacement refrigerants. The two most widely used examples of these are R134a and R407c. The former is a *hydrofluorocarbon (HFC)* refrigerant, while the latter is a *ternary blend* (i.e. blend of three) of the HFC refrigerants R32 (23%), R125 (25%) and R134a (52%). Both are non-toxic and have zero ODP. The low pressures and high volume flow rates characteristic of R134a make it more suited to vapour compression machines employing centrifugal or high-speed screw compressors, while the high pressures and low volume flow rates characteristic of R407c make it more suited to vapour compression machines employing reciprocating (positive-displacement) compressors. The non-azeotropic nature of R407c means that it experiences a temperature 'glide' during evaporation and condensation (since the pure components of the R407c mixture do not boil or condense at exactly the same temperature), and this effect must be taken account of in the design of heat exchangers for this particular working fluid. Different manufacturers have their own technical preferences regarding the use of R134a and R407c.

4.3.5 Application to LPHW heating systems

Several factors have significant implications for the application of heat pumps to LPHW heating systems. The first is that the temperature at which LPHW can be generated by a heat pump is typically limited to 50–55°C by the pressure at which the refrigerant system can operate. The higher the LPHW flow temperature required from the heat pump, the higher must be the condensing temperature *and pressure*. The use of very high pressures will result in an expensive heat pump.

Perhaps more significantly, however, as the ambient temperature falls in colder weather, both the rate of heat transfer to the evaporator and the compressor input power fall, thereby reducing the rate of heat output that the heat pump can supply. This is, of course, exactly the converse of the increase in space heating load that accompanies such a reduction in ambient temperature. It is typically not economic to size a heat pump to meet the heat output required by the heating system under UK winter design conditions that may lie in the range from, say, –1°C to –4°C. The concept of a *balance temperature* is therefore applied to heat pumps. Under this concept the object is to size the heat pump to supply *most* (only) of the total heat required during a typical heating season. The balance of the heat requirement is then supplied by a *supplementary* heat generator. The balance temperature is the outside air temperature at which the heat pump just satisfies the load requirement of the heating system, and in the UK typically lies in the range 1–4°C. On this basis, for a balance temperature of 4°C and internal air temperatures in the range 19–22°C, a heat pump would typically provide from approximately two-thirds to three-quarters of the design heat load, based on a range of external design conditions from –4°C to –1°C. The concept of balance temperature is illustrated in Figure 4.6. The actual value of the balance temperature is a complex function of the capital costs of the heat pump and the supplementary heat generator, the appliance COP of the heat pump, the efficiency of the supplementary heat generator, maintenance costs and the design of the heating system. The optimum balance temperature should therefore be determined by an appraisal of the life-cycle cost and environmental impact of various combinations of heat pump and supplementary heating capacity. LPHW heating systems employing heat pumps are designed as *bivalent heating systems*, in which heat is generated by two separate means; one is the primary means while the other is the secondary (or supplementary) means. The heat pump is naturally the primary means of generating heat.

What happens below the balance point? Under the balance-temperature concept two approaches are possible. Under the first or *bivalent alternative*

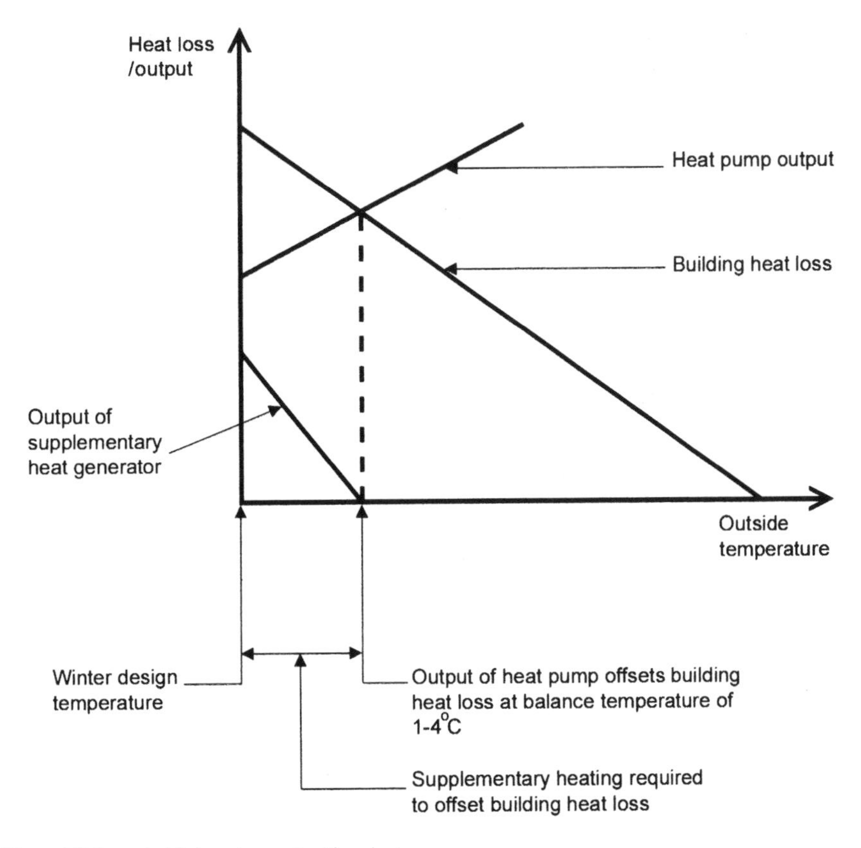

Figure 4.6 Concept of 'balance temperature' for a heat pump.

approach the heat pump is simply switched off at a predetermined ambient temperature at which it is known that the output of the heat pump is inadequate to satisfy the load on the heating system. The secondary means of generating heat then takes over the full heating load and must of course be sized to cope with the design load. It should of course be remembered that in the presence of substantial casual heat gains the heat pump will not simply cease to cope with the demand of the heating system as the ambient temperature drops below the balance temperature. Under the second or *bivalent supplementary (bivalent parallel)* approach, below a predetermined ambient temperature the output of the heat pump is simply progressively supplemented by the secondary means of generating heat. Both means of generating heat operate simultaneously to satisfy the load on the heating system. In the UK climate the balance temperature is inevitably below the mean temperature that exists throughout a typical heating season. Coupled to this, the number of 'design days' are relatively few and design heating loads do not, in any event, take any account of the casual heat gains that (although unpredictable from a 'worst-case' design perspective) inevitably exist in an occupied building. Hence the secondary means of generating heat will normally be required to operate for a relatively limited proportion of a typical heating season. While under design ambient conditions a heat pump may supply only 50–55% of the design heating load, leaving the balance of 45–50% to be made up by the secondary means of generating heat, over a typical heating season the total energy supplied by the latter may represent only one-half to one-third of this percentage, i.e. from 15–25% of the total, and the use of direct electric heating employing a simple 'flow boiler' may therefore be acceptable and certainly represents a low-cost option.

At low ambient temperature the heat output of an 'air-source' heat pump is also adversely affected by the need for a *defrost cycle*. Below ambient temperatures of c.5°C cooling of the air by the evaporator will result in freezing of its moisture content and the build-up of a layer of frost on the heat transfer surfaces. Since for an air-source heat pump the evaporator will typically take the form of a compact heat exchanger employing closely-spaced finned tubes, any build-up of frost will lead to an undesirable restriction of the air flow across the evaporator. The build-up of frost must therefore be periodically dispersed. This may be achieved by various means. One such method is to simply shut down the heat pump but allow the evaporator fans to continue to circulate ambient air across the evaporator to defrost it. Alternatively, hot refrigerant gas from the compressor may be bypassed to the evaporator, or the latter may be warmed externally by suitably-placed electrical resistance heating elements. Whichever of these means is employed to perform the defrost cycle, up to several percent of the seasonal energy input to the heat pump will be used for the purpose, the heat output of the heat pump will be reduced and the seasonal COP will be lowered. The secondary means of generating heat may also be required to make up for the loss of heat output from the heat pump during its defrost cycles. The range of ambient temperature which may result in 'frosting' of the evaporator is increased if the latter is undersized. Any increase in the need for the operation of the defrost cycle will penalise the seasonal operating costs of the heat pump.

Whether by design or 'accident', boilers typically provide substantial *overload heating capacity* that can be used for preheating or in the event of failure of one boiler in a multiple installation. This is not the case with heat pumps, where the provision of overload capacity is uneconomic. It may therefore be necessary to extend building preheat periods in cold weather. Since the secondary means of generating heat operates at a conversion efficiency of only unity, it may be preferable to run a heat pump continuously where this can avoid or minimise the use of the secondary heat generator for preheating, particularly where the heat pump can be operated on 'off-peak' electricity. In the event of failure of the heat pump the heating system must rely on the secondary heat generator, and under the bivalent supplementary approach the size of the latter may therefore be increased to provide an element of standby capacity that is perhaps sufficient to satisfy the heating load at the seasonal average ambient temperature. In any event the capacity of the secondary heat generator should be provided in suitable stages to allow a rapid and flexible response to a demand for heat.

Since the practical temperature at which LPHW can be generated by an air-source heat pump is rather less than is typically associated with conventional boiler installations, there are implications for the design of the heating system and the type and sizing of heat emitters. The performance of a heat pump is sensitive to the temperature of water returning from the heating system, while the heat output of emitters is sensitive to the flow temperature from the heat pump and also becomes more sensitive to the temperature of the space served, as flow temperature reduces. The power-law relationship that exists for the heat output from conventional radiators means that the heat output to a space will be substantially reduced (by a factor of 0.5 or more) at the lower flow and return temperatures that may typically be employed with an air-to-water heat pump, when compared with conventional boiler-generated flow and return temperatures of 82°C and 71°C. This subject is considered in more detail in Chapter 6. Heat pumps are therefore more suited to heating systems that can employ lower water temperatures, such as underfloor heating, or where the emitter area may be substantially increased to compensate for lower water temperatures. Where specific zones of a building cannot adequately be served at the flow temperature from the heat pump, it may be possible to heat these independently by alternative means or to provide local supplementary ('top-up') heating.

A heat pump will typically be installed in a primary heating circuit from which individually-pumped heating load circuits draw their flow needs, so that interaction between the circuits is avoided and the needs of both the heat pump and the load circuits may be optimised. For smaller heat pumps, capacity control may be achieved by on/off cycling, while larger machines have typically employed step control based on cycling or offloading of multiple compressors. The temperature associated with the high-pressure cut-out of the heat pump will typically form the reference point for defining the design flow temperature and control differential. It is recommended that the maximum flow temperature leaving the condenser should provide a margin of 8 K below the operating point of the high-pressure cut-out.[7] The design flow temperature to the heating system will be offset below this as necessary to allow control of the output of both the heat pump and the secondary means of heat generation, with a suitable 'dead' band between operation of the heat pump only and simultaneous operation of both means of heat generation.

Since a heat pump employs a compressor, and this will frequently be of reciprocating type, likely levels of noise and vibration need to be considered and taken account of in the design of an installation.

4.3.6 Cost and environmental performance

The mechanical complexity of a heat pump ensures that it carries a substantial capital or 'first' cost premium over a boiler of equivalent heat output to water. The extent of this premium naturally depends in part on the particular equipment compared, but may be of the order of two to three times the cost of the reference conventional (non-condensing) boiler plant. However, as far as energy costs are concerned, where a supply of natural gas is unavailable and the use of boilers fired on oil is either unacceptable or impractical, heat can be generated both more efficiently and cheaply by an electric heat pump than by direct electric heating. If the COP of the heat pump is better than approximately 2.5, the cost per kWh for 'daytime' heat output will be lower than the cost of off-peak electricity used at a conversion efficiency of unity. The exact breakeven point with the cost of off-peak electricity naturally depends on the actual purchase cost.

A comparison with the energy cost of a conventional LPHW boiler installation fired on natural gas can be made by considering the total requirement over a typical heating season for heat output from the boilers or from a heat pump and a secondary means of heat generation based on an electric flow boiler. For simplicity this will ignore any design differences in the way that the heat output from the boilers and heat pump is used, and any capital cost differences associated with this. Let A, B and C represent respectively the percentages of the seasonal requirement for heat output that are provided by operation of the heat pump on peak-rate electricity, its operation on off-peak electricity and operation of the electric flow boiler. Then a balance will exist between the energy costs of the boiler and heat pump installations when:

$$0.01\left[\left(\frac{(A\times E1)+(B\times E2)}{COP_{Seasonal}}\right)+(C\times E1)\right]=G\left[\frac{100}{\eta_{Boiler,\ Seasonal}}\right] \quad (4.5)$$

where E1 = purchase cost of peak-rate electricity (£ kWh^{-1}); E2 = purchase cost of off-peak electricity (£ kWh^{-1}); G = purchase cost of natural gas (£ kWh^{-1}); $COP_{Seasonal}$ = seasonal COP of heat pump; $\eta_{Boiler,Seasonal}$ = seasonal efficiency of boiler plant (%).

Substituting nominal values of 60%, 20% and 20% for the proportions of the seasonal heat output provided by the heat pump using peak-rate electricity (purchased at 5p kWh_e^{-1}), the heat pump using off-peak electricity (purchased at 2.5p kWh_e^{-1}) and the electric flow boiler (using electricity purchased at 5p kWh_e^{-1}), a boiler seasonal efficiency of 80% and a purchase cost of (1.2p kWh^{-1}) for natural gas, the seasonal COP for the heat pump that is necessary to achieve energy cost parity with the boiler plant may be determined as 7, which is unlikely to be achieved. This calculation assumes that the seasonal COP may be equally applied to operation of the heat pump on both peak-rate and off-peak electricity. The effect of electrical maximum demand charges are also ignored, although a heat pump will naturally be better-off in this respect than simple direct electric heating. If the purchase cost of natural gas is increased to (1.35p kWh^{-1}) and the seasonal efficiency of the boiler plant is reduced to 70%, the seasonal COP for energy cost parity reduces to slightly under 4 (c.3.77). Increasing the purchase costs of electricity, or the proportion of the season's heat output provided by the electric flow boiler, naturally requires an increase in the seasonal COP if energy cost parity with the gas-fired boiler plant is to be achieved. Conversely, reducing the proportion of the season's heat output provided by the flow boiler is beneficial. However, we have already seen that there are reasons fundamental to the sizing of the heat pump which may preclude this and whereby a reduction in the use of the flow boiler may adversely influence the seasonal COP that can be attained by the heat pump. The format of equation 4.5 is unable to reflect such effects. Where the comparison is made with a boiler plant fired on more-expensive LPG, the seasonal COP for energy cost parity will be substantially reduced. Actual energy costs may be analysed by considering either side of equation 4.5 in isolation, and multiplying it by the specified seasonal heat output.

A similar expression to equation 4.5 may be derived for parity of carbon emission between the heat pump and boiler installations for the same total heat output, and this is expressed by equation 4.6. In this instance it is only necessary to distinguish between the proportions of the total season's heat output provided by the heat pump and the flow boiler respectively.

$$0.01CEFe\left[\left(\frac{\text{Heat pump }\%}{COP_{Seasonal}}\right)+\text{Secondary }\%\right]=CEF_g\left(\frac{100}{\eta_{Boiler,\ Seasonal}}\right) \quad (4.6)$$

where CEF_e = carbon emission factor for electricity (0.12 kgC kWh^{-1})[8]; Heat pump % = percentage of season's heat output provided by the heat pump; $COP_{Seasonal}$ = seasonal COP of heat pump; Secondary % = percentage of season's heat output provided by the electric flow boiler; CEF_g = carbon emission factor for natural gas ($0.052\,kgC\,kWh^{-1}$)[8]; $\eta_{Boiler,Seasonal}$ = seasonal efficiency of boiler plant (%).

Substituting values of 80% and 20% respectively for the percentages of the season's heat output provided by the heat pump and the flow boiler, and a boiler seasonal efficiency of 80%, a seasonal COP of approximately 2.34 is required to achieve carbon emission parity with the gas-fired boiler plant. At boiler seasonal efficiencies of 70% and 60% this reduces to c.1.91 and c.1.53 respectively.

Finally, the common working fluids for vapour compression heat pumps, the refrigerants R134a and R407c, are both powerful greenhouse gases, with a global warming potential (GWP) respectively of 1300 and 1610 times that of carbon dioxide taken over a time-scale of 100 years. Strictly-speaking any comparison of the environmental impact of different equipment and refrigerants in a heat pumping application should compare the *total equivalent warming impact (TEWI)* of each option. This represents the overall global warming effect that may be expected to result from the operation of the heat pump over its entire life cycle, and is composed of both a direct contribution and an indirect contribution. The *direct contribution* to the TEWI of the heat pump installation arises from actual leakage of refrigerant from the machine throughout its life, which is also taken to include its disposal at the end of its operating life. The *indirect contribution* to TEWI of the heat pump installation arises as a result of the total carbon emission associated with the life-cycle input of electrical energy to the heat pump, and is naturally a function of the seasonal COP that is attained by the installation.

4.4 Energy crops and bio-diesel

So-called *'energy crops'* represent an interesting concept in renewable energy sources, giving a 'grow-burn-grow' cycle. Photosynthetic growth is principally based on the availability of carbon dioxide, water and sunlight. Carbon dioxide is therefore removed from the atmosphere during the 'growth phase', and is returned to it during the subsequent 'burn phase'. There is thus the potential to reduce carbon emissions when energy crops are substituted for fossil fuels. The latter represent carbon dioxide that was removed from the atmosphere in a growth phase that has been separated from the burn phase across a long geological time-scale, so that now only the burn phase remains. Energy crops are a 'renewables' technology that is considered to be ready for UK application. Currently the species most likely to be employed in any such application are *short-rotation willow coppice* and *miscanthus*.

As far as building services space and water heating applications are concerned, the handling and combustion of energy crops as a boiler fuel for individual building applications appears to be associated with significant practical disadvantages in comparison with natural gas and oil, and it is unclear as yet how (or if) these could be overcome in practice. If the 'harvest' of energy crops can be substantially affected by extreme weather conditions (drought, flooding, etc.) or the action of plant pests or diseases, there may also be an issue regarding continuity of supply, unless 'dual-fuel' installations are possible. It would seem likely, therefore, that the sphere of application of energy crops will be in electrical power generation or other applications requiring large-scale heating, rather than individual building services applications.

So-called *bio-diesel* may be considered as a particular form of energy crop in that it is fuel that is derived from a plant crop such as *rapeseed*. Since bio-diesel is a liquid fuel that may be burnt directly by a boiler and burner combination of suitable design, its potential for use in building services applications appears to be less problematic than that of other energy crops, assuming that considerations of fuel cost, combustion process and emissions can be satisfactorily addressed. At the time of writing it is not known whether any demonstrations of the use of this renewable source have been made.

4.5 Waste as a fuel

Ignoring questions of content, calorific value and emissions, the solid form of *community waste* (waste from households, commerce and business rather than industrial waste) puts it at a fundamental disadvantage in comparison with gaseous or liquid fuels in mainstream building services applications. Its potential sphere of application is therefore likely to be confined to electrical power generation or large-scale CHP and district heating. Use of community waste as a fuel in such applications may be influenced by government and European Union policies towards the 'landfilling' of such waste.

References

1. Building Research Energy Conservation and Support Unit (BRECSU) *The designer's guide to energy-efficient buildings for industry.* Energy Efficiency Office Good Practice Guide 303, p. 74. BRECSU, Garston.
2. Building Research Energy Conservation and Support Unit (BRECSU) (1989) *Guidance notes for the implementation of small scale packaged combined heat and power*. Energy Efficiency Office Good Practice Guide 1, p. 29. BRECSU, Garston.
3. Chartered Institution of Building Services Engineers (CIBSE) (1988) *Design Guidance for Heat Pump Systems.* Technical Memorandum TM15, p. 1. CIBSE, London.
4. Kew, J. (1985) *Heat Pumps for Building Services.* Technical note 8/85, p. 34. Building Services Research and Information Association, Bracknell.
5. Rogers, G.F.C. & Mayhew, Y.R. (1980) *Engineering Thermodynamics Work and Heat Transfer*, 3rd edn. Longman, London and New York.
6. Gosney, W.B. (1982) *Principles of Refrigeration.* Cambridge University Press, Cambridge.
7. Chartered Institution of Building Services Engineers (CIBSE) (1988) *Design Guidance for Heat Pump Systems.* Technical Memorandum TM15, p. 5. CIBSE, London.
8. Building Research Energy Conservation and Support Unit (BRECSU) *The designer's guide to energy-efficient buildings for industry.* Energy Efficiency Office Good Practice Guide 303, p. 29. BRECSU, Garston.

5 Flueing

5.1 Purpose of the flue

The basic products that result from the complete combustion of a hydrocarbon fuel in a boiler are carbon dioxide and water. Carbon dioxide is a natural, if minor, constituent of the atmosphere at c.0.03% by volume. A maximum concentration of c.0.1% by volume is typically considered acceptable for sedentary occupancy within a building, while concentrations between 2–3% by volume are deadly to human beings. Its concentration in the combustion gases discharged from a boiler typically vary from c.9–10% for natural gas boilers to c.12–13% for oil-fired boilers. Therefore the combustion gases cannot be allowed to collect and concentrate near ground level. Water vapour is another natural constituent of the atmosphere, in small and varying quantities. When condensation of water vapour in the combustion gases occurs on contact with ambient air it visibly identifies the presence of the gases, and may still be misinterpreted by building occupants and members of the public as 'smoke'. If combustion of the fuel is for any reason incomplete, carbon monoxide will also be produced, in lieu of some of the carbon dioxide, and this is toxic to human beings at very low concentrations. The production of carbon monoxide should of course be minimal from a burner that is both functioning correctly and set up with an appropriate proportion of excess combustion air. In some instances where oil-fired boilers are employed, particles of unburnt carbon may also be produced, and these may be discharged to atmosphere with the combustion gases as a 'smut'. Finally, depending on the chemical composition of the fuel, smaller quantities of other 'noxious' compounds are also produced as *pollutants*. Of principal concern among these are sulphur dioxide and the oxides of nitrogen that are loosely termed 'NO_x', although other substances may be present if the boiler combustion air is itself contaminated. The production of sulphur dioxide is typically only a factor in the design of flues for oil-fired boilers. In any event, if the pollutants that may result from the combustion of boiler fuel are allowed to 'concentrate' near ground level, their effects are harmful to human health and particularly to those who suffer with respiratory problems. The pollutants are also harmful to the environment.

A *boiler flue* provides the means for removing the 'spent' combustion gases (or *flue gases*) from a boiler to a position where it can safely and effectively *disperse them into the general atmosphere*. Dispersal of the gases is typically achieved by a combination of flue height above the building and *efflux velocity* from the flue. The efflux velocity is the average velocity of the gases at outlet from the flue. It is important to recognise that the flue typically cannot exert any influence on the *level of pollutants* discharged by a boiler. In the UK boilers flues are subject to both statutory legislation and local authority regulation.

5.2 Flue and chimney

Flue and *chimney* are both terms that are widely used. Where a flue may rise directly from a boiler to its termination above the building, there is no distinct chimney. Where a flue must run 'horizontally' (in practice there is typically a requirement for a minimum gradient) before rising vertically to its termination above the building, we will regard the vertical element of this *flue system* as the *chimney*.

5.3 Types of flue

Four principal types of flue system may be identified:

- Natural-draught flues
- Mechanical-draught flues
- Balanced flues
- Dilutant flues, or flue dilution systems as they are more commonly termed.

Mechanical-draught flues include *induced-draught*, *forced-draught* and *balanced-draught* systems, while balanced flues may be subdivided into *horizontal* and *vertical* (*balanced-compartment*) types.

5.4 Flue draught

The *draught* of a flue or chimney may be defined as the difference between atmospheric pressure and the internal pressure at some level in the flue. *Chimney draught* is the draught that exists at the base of the chimney. Draught may be derived naturally or mechanically. *Natural draught* is draught that derives from the buoyancy effect caused by the difference between the density of the hot gases inside the flue (lower) and the ambient air (higher). *Mechanical draught* is draught that is generated by a fan.

Natural draught is a function of the temperature of the combustion gases leaving the boiler (the *flue gas temperature*), the temperature of the ambient air, and the height of the flue or chimney. For any given flue gas temperature the natural draught will therefore be greater under winter operating conditions when the ambient air temperature is low than under summer operating conditions when it is high. The two extreme (maximum and minimum) draught conditions will be represented by the winter and summer design ambient temperatures respectively. Natural draught is a function of the difference in *height* between two points in a flue or chimney, and horizontal sections of flue do not contribute to it.

The value of natural draught, in Pa per metre of flue or chimney *height*, is given by the expression:

$$\text{Draught (Pa m}^{-1}) = T_{REF} \times g \times \left[\frac{\rho_{Air}}{T_{Air}} - \frac{\rho_{Flue\ gas}}{T_{Flue\ gas}}\right] \qquad (5.1)$$

where T_{REF} = a common reference (*absolute*) temperature for air and flue gas densities (K = °C 273); g = acceleration due to gravity at the earth's surface (9.81 m s^{-2}); ρ_{Air} = density of air at T_{REF} (kg m^{-3}); $\rho_{Flue\ gas}$ = mean density of flue gases at T_{REF} (kg m^{-3}); T_{Air} = air *absolute* temperature (K); $T_{Flue\ gas}$ = flue gas *absolute* temperature (K).

For approximate calculations it will typically be acceptable to regard the reference density of flue gases to be the same as that of ambient air. In this case equation 5.1 may be simplified to:

$$\text{Draught (Pa m}^{-1}) = T_{REF} \times g \times \rho_{Air}\left[\frac{\theta_{Flue\ gas} - \theta_{Air}}{T_{Air} \times T_{Flue\ gas}}\right] \qquad (5.2)$$

where θ_{Air} = air temperature (°C); $\theta_{Flue\ gas}$ = flue gas temperature (°C).

In equation 5.2, air and flue gas temperatures may be expressed in °C in the numerator since *for subtractions* the '273s' in the absolute temperatures cancel each other out. Corresponding temperatures in the denominator must, however, remain *absolute* (K = °C + 273). For conventional (non-condensing) boilers with fan-assisted or forced-draught natural gas burners, flue gas temperatures at the full design firing rate of the burner will typically lie in the range 190–250°C. For some high-efficiency modular boilers this may reduce to c.125°C, while for condensing boilers it may be as low as 45°C during condensing operation. Where a *draught diverter* is employed, as will typically be the case for boilers employing atmospheric natural gas burners, flue gas temperatures downstream of the draught diverter may be reduced to a level of approximately 130–160°C. Flue gas temperature will typically tend to fall when high/low/off or modulating burners are firing at a reduced rate. Flue gas temperature also depends on the mean water temperature in the boiler, taken as the average of the boiler flow and return water temperatures. Where the water temperature in the boiler falls at a constant heat output, flue gas temperature will also fall. A change in the water temperature in the boiler of 10K will typically result in a 5–8K reduction in flue gas temperature. Where 'baffles' are fitted in the combustion gas passages of the boiler block, their removal (or their inadvertent omission following removal for cleaning of a boiler) will result in an increase in flue gas temperature. Anything that reduces the transfer of heat from the combustion gases will tend to cause an increase in flue gas temperature and a reduction in boiler efficiency. Therefore, flue gas temperature will increase if the gas-side heat transfer surfaces of the boiler are dirty. Figure 5.1 plots the approximate natural draught obtained by solution of equation 5.2 for flue gas temperatures of 190°C, 125°C and 45°C over a range of ambient temperature from –4°C to 20°C.

The draught generated by the buoyancy of the flue gases is used to offset the pressure losses associated with the flow of gas through the flue system. As in the case of air flowing in ventilation or air-conditioning ducts, there are two components to these pressure losses. The first of these is due to 'pipe' friction, and may be expressed as:

$$\Delta P_{\text{friction}}\,(\text{Pa m}^{-1}) = \text{Friction factor} \times \text{VP} \times \left(\frac{\text{P}}{\text{A}}\right) \tag{5.3}$$

where Friction Factor = a non-dimensional parameter that is a complex function of the relative roughness of the flue 'pipe' and the flow regime, as represented by the Reynolds number for the flow in the flue pipe; VP = velocity pressure based on the mean velocity of the flue gases (Pa); P = perimeter of the flue cross-section that is in contact with ('wetted by') the flue gases (m); A = cross-sectional area of the flue (m^2).

For flue gas temperatures in the range 180–340°C values of the friction factor for various flue constructions and diameters are set out in the CIBSE *Guide*[1]. For flues of various cross-sections the ratios of perimeter to cross-sectional area are:

Circular	4/diameter
Square	4/length of side
Rectangular	2 × (width + height)/(width × height)

In all cases the relevant dimensions of the flue are expressed in m. The velocity pressure of the flue gases (i.e. their kinetic energy per unit volume) may be expressed as:

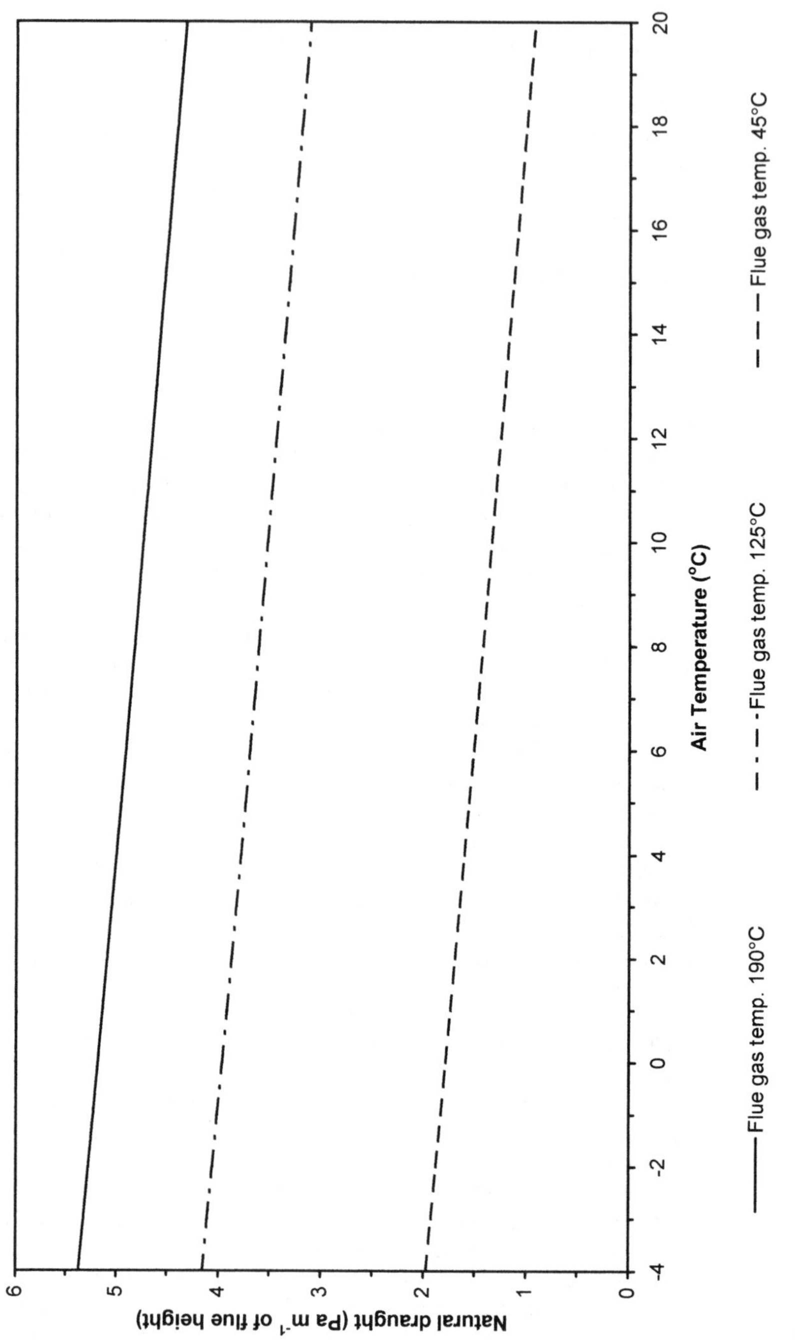

Figure 5.1 Approximate natural draught versus air temperature for flue gases at 190°C, 125°C and 45°C.

$$\text{Velocity pressure (Pa)} = 0.5 \times \rho_{\text{Flue gas}} \times v_{\text{Flue gas}}^{2}\left(\frac{T_{\text{REF}}}{T_{\text{Flue gas}}}\right) \tag{5.4}$$

where $\rho_{\text{Flue gas}}$ = mean flue gas density at T_{REF} (kg m^{-3}); $v_{\text{Flue gas}}$ = mean velocity of flue gases (m s^{-1}); T_{REF} = reference (*absolute*) temperature for flue gas density (K = °C + 273); $T_{\text{Flue gas}}$ = flue gas *absolute* temperature (K).

For approximate calculations it will again typically be acceptable to regard the reference density of flue gases as the same as that of ambient air, and if we assume this to have a nominal value of 1.2 kg m^{-3} at 20°C (293 K), equation 5.4 may be simplified to:

$$\text{Velocity pressure (Pa)} = 175.8\left(\frac{v_{\text{Flue gas}}^{2}}{T_{\text{Flue gas}}}\right) \tag{5.5}$$

For a flue gas temperature of 293K (20°C) this becomes the expression familiar to HVAC designers dealing with ducted air systems, (VP = 0.6 × velocity^2). At a flue gas temperature of 45°C, such as may be encountered during condensing operation, the velocity pressure of flue gases will not be too dissimilar to this, at approximately (0.55 × flue gas velocity^2). However, at a flue gas temperature of 190°C this will reduce to approximately (0.38 × flue gas velocity^2), a reduction of c.37%.

For typical flues of circular cross-section with a smooth internal ('gas-side') surface finish, such as welded steel, *approximate* values of frictional pressure loss rate at a flue gas temperature of 190°C and a flue gas velocity of 6 m s^{-1} may vary from 1.6 Pa m^{-1} for a flue of 150 mm nominal internal ('gas-side') diameter to 0.7 Pa m^{-1} at a nominal internal diameter of 300 mm and 0.3 Pa m^{-1} at a nominal internal diameter of 600 mm. At a flue gas temperature of 45°C the corresponding values are 2.3 Pa m^{-1}, 1.0 Pa m^{-1} and 0.4 Pa m^{-1} respectively. For the flue of 150 mm diameter this value is greater than the natural draught available, per vertical metre of flue or chimney, at –4°C ambient temperature. In such a case it would either be necessary to reduce the frictional pressure loss rate by increasing the diameter of the flue, or to employ mechanical draught (or a mixture of the two). Increasing the diameter of a flue reduces the flue gas velocity by a factor of (smaller diameter/larger diameter)^2, and hence reduces the velocity pressure of the flue gases by a factor of (smaller diameter/larger diameter)^4. For a given flue gas velocity a larger-diameter flue will have a lower friction factor due to its reduced relative roughness. However, for a given diameter the friction factor increases as flue gas velocity decreases below 6 m s^{-1} due to changes in the nature of the flow regime ('Reynolds number effects') – by c.8–10% between 6 m s^{-1} and 3 m s^{-1} and c.16–20% between 6 m s^{-1} and 1.5 m s^{-1}. Typically for an increase in flue diameter of up to twice the original diameter the combined influence of the change in relative roughness and Reynolds number effects is likely to be negligible. Friction factors and their relationships to flue diameter and gas velocity will be modified for flues of other cross-section, and where the assumption of 'smoothness' for the gas-side surface finish is affected by techniques of manufacture or installation.

The second component of the pressure losses associated with the flow of gas through the flue system arises from the turbulence generated at changes of direction and shape in the flue, at junctions in the flow, such as occur at the entry of a primary flue into a common flue header or at the entry of a flue into a chimney, and finally at the flue outlet. In common with building services practice generally where fluid conduits run full, the pressure losses in the flue 'fittings' that function as bends, tees, etc. are accounted for by allowing for each such fitting a loss of pressure equivalent to the product of a characteristic velocity (or 'dynamic') pressure at some point in the fitting and an empirically-determined velocity pressure loss factor that is characteristic of the type of fitting. Typical values of the velocity pressure loss factors for various flue fittings are given by the CIBSE and BSRIA[1,2]

It is of course the *cross-sectional area* of a flue that produces its *volume flow rate capacity* at any flue gas velocity. For natural gas at a gross calorific value of 38.6 MJ m^{-3}, each kW of rated heat input is equivalent to an input rate of $2.59 \times 10^{-5}\,m^3 s^{-1}$ of natural gas. If this volume flow rate of natural gas is burnt with sufficient excess air to produce flue gases with a nominal CO_2 content of 9% by volume, the corresponding volume flow rate of flue gases in $m^3 s^{-1}$ may be approximated by ($1.222 \times 10^{-6} \times$ flue gas absolute temperature). Strictly-speaking this is only applicable at or reasonably near to sea level, and a correction for the effects of altitude should be made above c.600 m above sea level. For a given rated heat input to the boiler an approximate flue diameter may be determined from the expression:

$$\text{Flue diameter} = 1.247 \times \left(\frac{\text{Rated heat input} \times T_{\text{Flue gas}}}{v_{\text{Flue gas}}} \right)^{0.5} \tag{5.6}$$

where Flue diameter is in *mm*; Rated heat input is in kW; $T_{\text{Flue gas}}$ = flue gas *absolute* temperature (K = °C + 273); $v_{\text{Flue gas}}$ = mean flue gas velocity ($m s^{-1}$).

For flue gas velocities of 6 and 7.5 $m s^{-1}$, the approximate flue diameter in mm then becomes (Rated heat input × flue gas absolute temperature)$^{0.5}$ multiplied by factors of 0.51 and 0.455 respectively. On the same basis the approximate maximum rated heat input for a given flue diameter would be:

$$\text{Rated heat input (kW)} = 0.643 \left(\frac{\text{Flue diameter}^2 \times v_{\text{Flue gas}}}{T_{\text{Flue gas}}} \right) \tag{5.7}$$

where Flue diameter is in mm.

For flue gas velocities of 6 and 7.5 $m s^{-1}$, equation 5.7 yields approximate solutions of [flue diameter (mm)^2 ÷ flue gas absolute temperature] multiplied by factors of 3.86 and 4.823 respectively.

Since in an economic installation a flue or chimney cannot be perfectly insulated, some heat loss will occur over its vertical height and this will result in the flue gases being discharged from the chimney at lower temperature than that at which they enter its base. The buoyancy forces of the flue gases are therefore typically based on the mean of their temperatures entering and leaving the chimney. The magnitude of the temperature drop experienced by the flue gases may be assessed from the energy balance:

$$UA(\theta_{\text{Flue gas, Mean}} - \theta_{\text{Air}}) = B \times (\theta_{\text{Flue gas, entering}} - \theta_{\text{Flue gas, leaving}}) \tag{5.8}$$

where U = overall *thermal transmittance* of the chimney ($W m^{-2} K^{-1}$); A = total surface area of the chimney to which U is referred (m^2); $\theta_{\text{Flue gas, mean}}$ = mean flue gas temperature, i.e. ($\theta_{\text{Flue gas, entering}} + \theta_{\text{Flue gas, leaving}}$) ÷ 2 (°C); θ_{Air} = ambient air temperature (°C); B = the product [mass flow rate ($kg s^{-1}$) × specific heat ($kJ kg^{-1} K^{-1}$)] for the flue gases, with the value of specific heat being that at constant pressure and at the mean flue gas temperature; $\theta_{\text{Flue gas, entering}}$ = temperature of flue gases entering the chimney (°C); $\theta_{\text{Flue gas, leaving}}$ = temperature of flue gases leaving the chimney (°C).

Equation 5.8 may be transposed to yield an expression for the leaving flue gas temperature:

$$\theta_{\text{Flue gas, leaving}} = \frac{[\theta_{\text{Flue gas, entering}} \times (2B - UA)] + (\theta_{\text{Air}} \times 2UA)}{2B + UA} \quad (5.9)$$

Once the leaving flue gas temperature has been established, the 'gas-side' surface temperature of the inner skin of the flue may be established by proportioning the temperature difference at the flue outlet across the total thermal resistance of the flue construction, so that:

$$\frac{U}{h_{si}} = \left(\frac{\theta_{\text{Flue gas, leaving}} - \theta_{\text{Flue inside surface}}}{\theta_{\text{Flue gas, leaving}} - \theta_{\text{Air}}} \right) \quad (5.10)$$

where U = overall thermal transmittance of the chimney ($W m^{-2} K^{-1}$); h_{si} = gas-side surface heat transfer coefficient of the flue ($W m^{-2} K^{-1}$); $\theta_{\text{Flue gas, leaving}}$ = temperature of flue gases leaving the chimney (°C); $\theta_{\text{Flue inside surface}}$ = gas-side surface temperature of the inner skin of the flue at a point immediately adjacent to the flue outlet (°C); θ_{Air} = ambient air temperature (°C).

Equation 5.10 may be transposed to yield an expression for the gas-side surface temperature at outlet from the flue:

$$\theta_{\text{Flue inside surface}} = \frac{[\theta_{\text{Flue gas, leving}} \times (h_{si} - U)] + (\theta_{\text{Air}} U)}{h_{si}} \quad (5.11)$$

In some instances it may be necessary to provide a means of *controlling* the draught within a flue. Variations in draught may result from variations in ambient air temperature, changes in the number of boilers firing in a multiple-boiler installation or changes in the firing rate of boilers employing high/low/off or modulating burners. Control or *stabilisation* of draught will typically be required where boilers employ atmospheric natural gas burners, and in this respect the function of draught diverters was described in Chapter 2. Where boilers employ fan-assisted or forced-draught natural gas burners or pressure-jet oil burners, the use of *draught stabilisers* is typically not favoured for several reasons. First, the draught stabiliser admits air from the boiler plant room, which is at a much lower temperature than the combustion gases, into the flue. In extreme circumstances the flue gases may reach the acid or water dew-point temperature before leaving the flue outlet, leading to corrosion of the flue and the possible formation of 'smut' in the case of oil firing, although the dilutant air will have some effect in actually lowering the dew-point temperatures of the flue gases. Avoidance of the acid dew-point is particularly important where grades of fuel oil with high sulphur content are burnt, albeit that this is typically rare in modern commercial building services applications. However, under such circumstances the use of modulating flue dampers is preferable, since this avoids the introduction of cool air, albeit that the damper works in a harsh environment. Second, if improperly set up, the action of the draught stabiliser may reduce the capacity of the flue or chimney to handle the products of combustion. Finally, it may lead to a positive-pressure condition existing at the base of the chimney. Conversely, an advantage that is claimed for the use of a draught stabiliser is that the flue or chimney is ventilated following the cessation of burner firing. While it is sometimes suggested that a draught stabiliser reduces the heat loss due to the flow of combustion air through a non-firing boiler, this is better dealt with by the use of a fully-closing combustion air damper (in the case of a fan-assisted or forced-draught natural gas or pressure-jet oil burner) or a *flue damper* (in the case of an atmospheric natural gas burner).

In most cases a flue damper is a basic two-position device to achieve shut-off of the flue pipe when a boiler is not firing. The desirability of this with regard to reducing heat loss from the boiler has already been noted. The use of flue dampers is not, of course, restricted to boilers with atmospheric natural gas burners. In the case of an installation of multiple boilers with forced-draught burners, cooler air drawn through a non-firing boiler will mix with the hot combustion gases from firing boilers and reduce their temperature and buoyancy. It must not, of course, be possible for a burner to commence firing against a closed flue damper, and if closure of the damper during firing is possible due to malfunction or failure, this must result in an immediate shut-down of the burner. In some types of atmospheric natural gas boiler a *two-stage* flue damper may be provided that is linked to high/low/off control of the burner, being fully open at full burner output and partially open at the low firing rate. In this situation the effect of the added resistance to the flow of flue gases that results from the partially-open flue damper is to reduce the induction of combustion air and more closely match it to the burner's low-firing requirements, to the benefit of boiler efficiency. The control provisions for the damper must naturally ensure that the burner may not operate at full output unless the flue damper is fully open. A flue damper operates in a harsh environment of high-temperature flue gases, and its construction must reflect this. Avoidance of this harsh environment is one reason why an approach may be favoured which moves shut-off of the combustion air path through the boiler to the burner side when forced-draught natural gas and atomizing oil burners are employed.

5.5 Natural-draught flues

Natural-draught flues will typically find building services application with cast-iron sectional LPHW boilers that employ atmospheric natural gas burners. In this case the buoyancy of the hot combustion gases leaving the boiler must be strong enough to draw combustion air into the combustion chamber of the boiler, overcome the pressure drop associated with the flow of the combustion gases through the boiler to the outlet connection to the flue, overcome the pressure drop associated with the flow of gases through the flue system and finally ensure that the latter are discharged from the flue outlet at an adequate velocity. The principle is illustrated schematically in Figure 5.2.

Where the combined pressure drop associated with the induction of combustion air and the flow of the combustion gases through the boiler block is low, variations in natural draught may have a significant effect on combustion. Provisions for draught stabilisation are therefore typically required where natural-draught flues are employed. Since natural draught is totally dependent on the buoyancy generated by the hot flue gases, it is essential to minimise the heat loss between the boiler and the outlet from the flue. This will particularly be the case where flue gas temperatures are low.

5.6 Mechanical-draught flues

In an *induced-draught flue* (Figure 5.3) an induced-draught fan located at or closely downstream of the outlet connection from the boiler to the flue deals

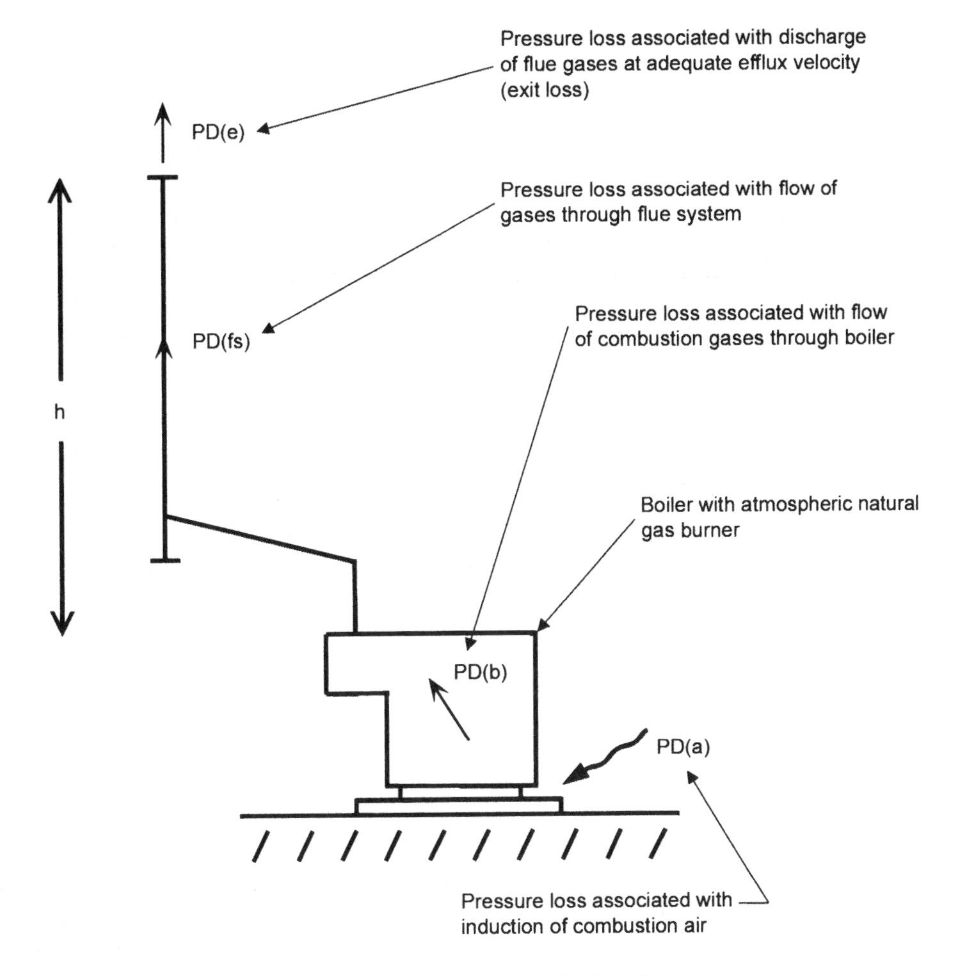

If ND(h) is the natural draught generated by the height h of the flue system,

ND(h) = PD(a) + PD(b) + PD(fs) + PD(e)

Figure 5.2 Principle of operation of a natural-draught flue.

with the pressure losses associated with the induction of combustion air into an atmospheric natural gas boiler, and with the flow of combustion gases through the boiler block. The pressure losses associated with flow through the flue may either be offset by the induced-draught fan itself or conventionally by natural draught. The impeller of the induced-draught fan operates in the harsh environment of the flue gases, and should therefore be made of a non-corrodible material, such as aluminium or stainless steel. It must be possible to inspect the fan periodically for any build-up of deposits on the blades of the impeller, which may lead not only to a drop in performance but, potentially more seriously, to the generation of an out-of-balance condition. The design of the fan casing should naturally not allow the accumulation of condensate. Even in the case of conventional (non-condensing)

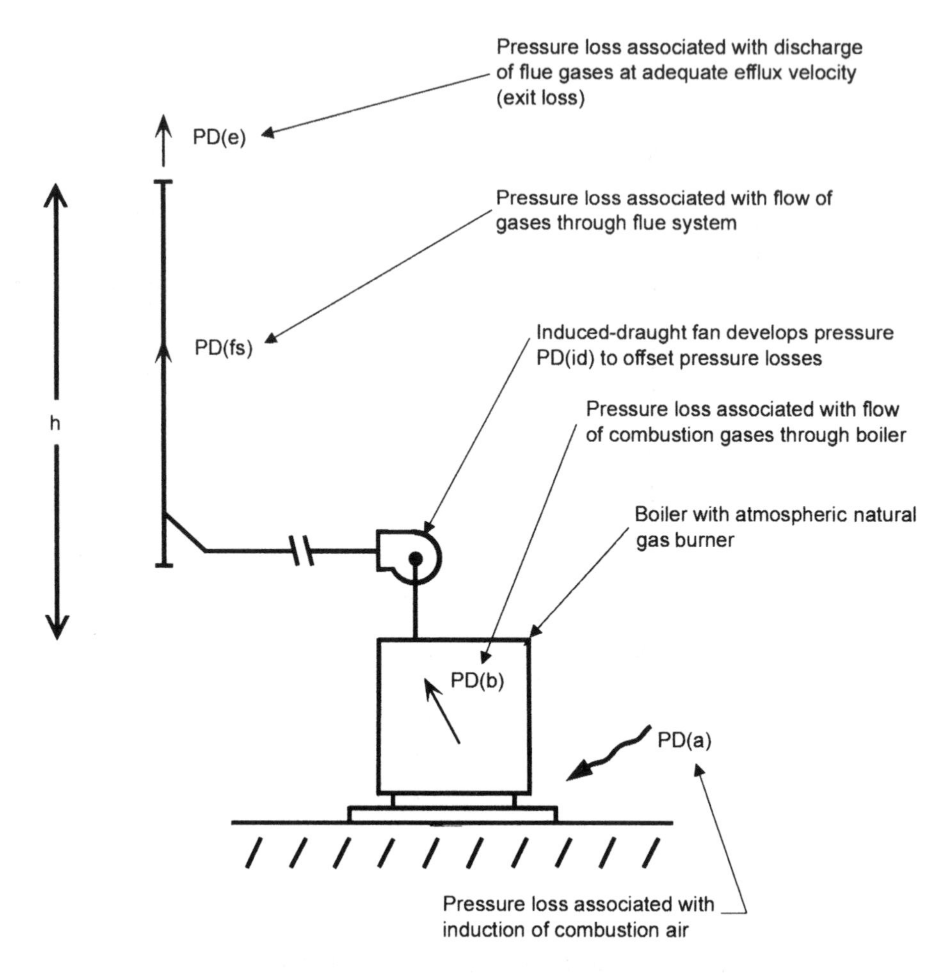

Figure 5.3 Principle of operation of an induced-draught flue.

boilers this will occur on start-up and may also occur under some low-load conditions. The induced-draught fan must be interlocked with the burner, so that firing is inhibited until operation of the fan is proven and terminated immediately in the event of fan failure. Since the resistance of the flue system to the flow of gases increases with reducing flue gas temperature, the motor of the induced-draught fan must be sized to cope with the increased load experienced under 'cold-start' conditions. A fan from a standard product range is unlikely to match the precise duty required for a particular combination of boiler and flue. An induced-draught fan will therefore typically be selected with a margin of excess duty that must be trimmed out during the commissioning process. Unless the fan speed is adjustable, provision must be made to achieve the required flow rate of combustion gases from the boiler

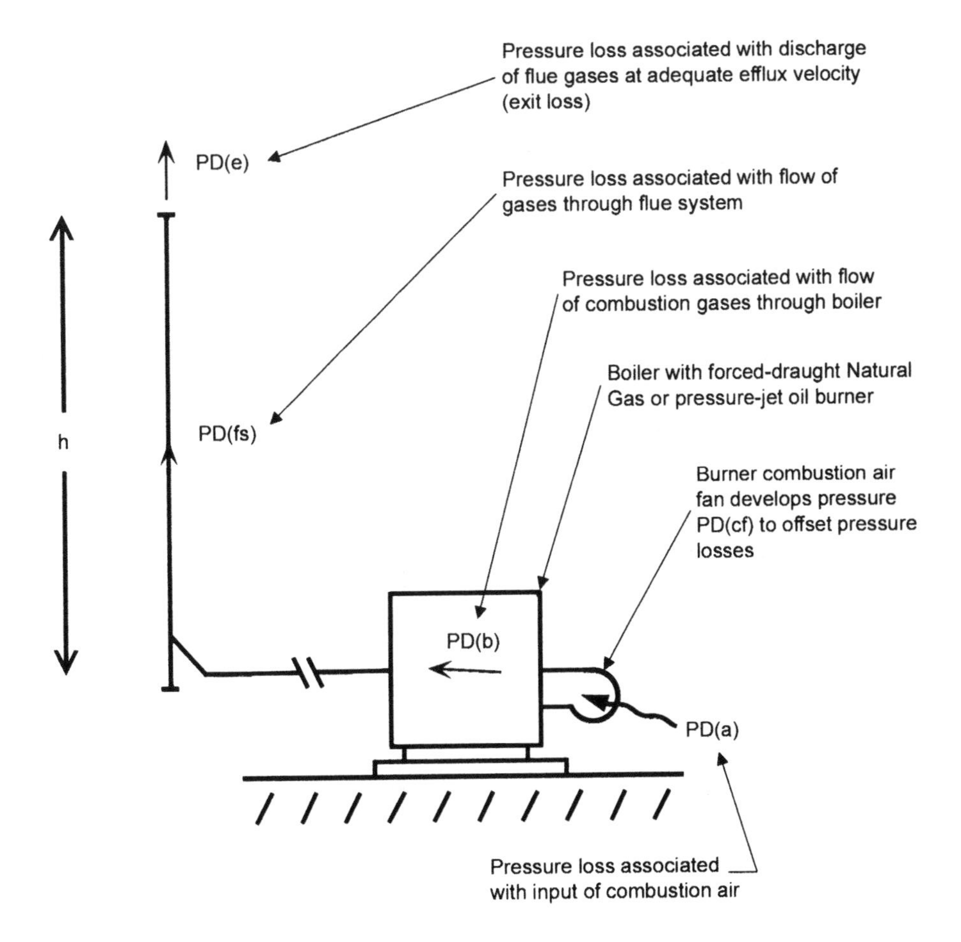

Figure 5.4 Principle of operation of a forced-draught flue.

via an adjustable bleed of air from the plant room into the flue at the fan inlet. Finally, if operation of the boiler on natural draught alone remains possible in the event of failure of the induced-draught fan, a bypass may be provided, with bypass dampers interlocked with the fan.

In a *forced-draught flue* (Figure 5.4) the burner fan offsets the pressure losses associated with the supply of combustion air and the flow of combustion gases through the boiler block. The burner fan may provide either a neutral (zero) or positive pressure at the boiler outlet connection to the flue. In the former case the pressure losses associated with flow through the flue are offset conventionally by natural draught. In the latter case the mechanical draught available at the boiler outlet to the flue may be utilised either to reduce the requirement for natural draught (and potentially, therefore, the height of the flue or chimney) or, where necessary, to increase the velocity of the flue gases (either generally or only the efflux velocity) for the same

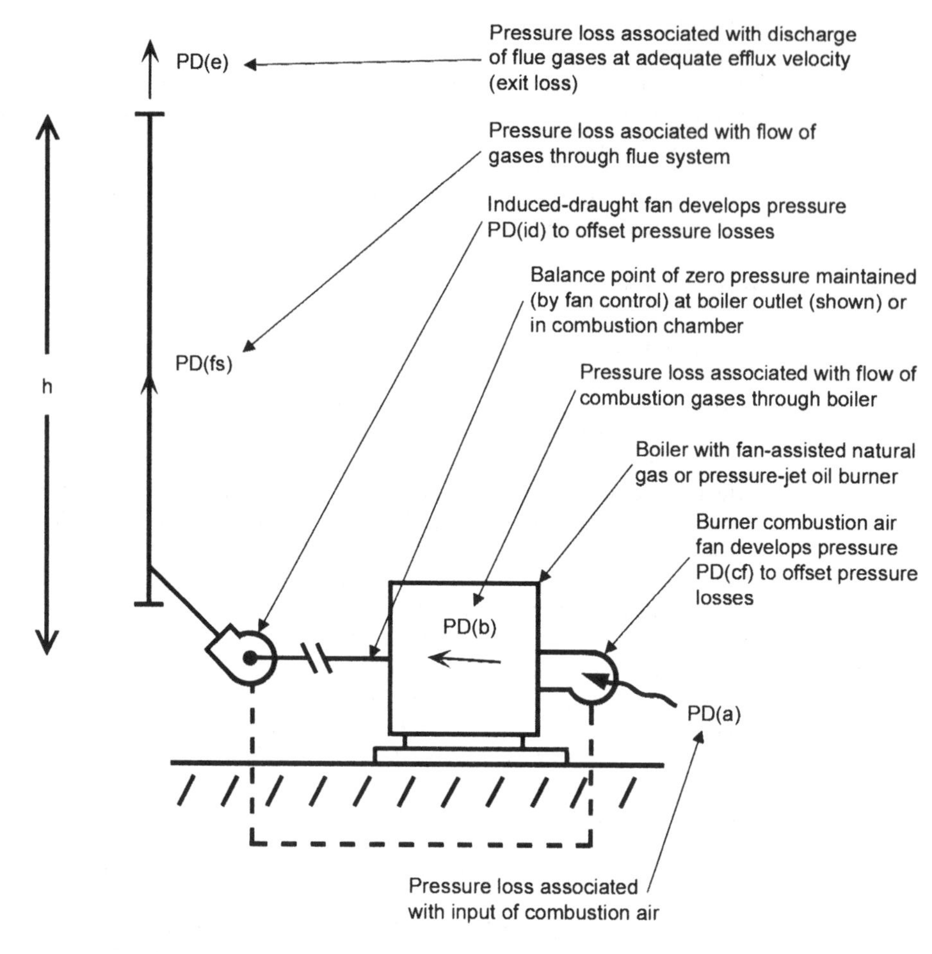

Figure 5.5 Principle of operation of a mechanical balanced-draught flue.

availability of natural draught. Forced-draught flues will typically be employed where cast-iron sectional boilers are fitted with fan-assisted or forced-draught natural gas burners or oil pressure-jet burners.

In a *balanced-draught flue* (Figure 5.5) the boiler (gas-side) and flue pressure losses are apportioned between the burner combustion air fan and an induced-draught fan. The two fans must be controlled to maintain a specified 'balance' point of zero pressure either in the boiler combustion chamber or at the outlet connection to the flue.

The effect of variations in natural draught will inherently be less for mechanical-draught flues than for natural-draught flues, and will be minimal where the flue is designed without significant reliance on buoyancy forces. The use of draught stabilisers is therefore typically unnecessary in such systems. Appropriately-designed induced-draught and forced-draught flue systems have the potential to ensure more-or-less optimum combustion performance under all load conditions. Since fans are essentially constant-volume-flow-rate devices, where mechanical-draught flues are employed at altitudes in excess of approximately 600 m above sea level, the volume flow rate of flue gas to be handled by the fan, for a given rated heat input, should be corrected for the effects of altitude. An increase of c.4% for every 300 m above sea level is typically considered appropriate.

5.7 Balanced flues

In a *balanced flue* the air inlet and flue gas outlet are located in the same pressure zone, so that wind pressure affects both more-or-less equally. This is typically achieved by combining both air inlet and flue outlet within the same *flue terminal*, the air inlet being concentric to (around) the flue outlet. The terminal of a balanced flue is therefore less sensitive to location and wind effects. The term 'concentric' has been used for convenience, since the air inlet is typically positioned around the perimeter of the flue gas outlet at its centre. However, neither are of necessity circular in cross-section.

Balanced flues may be of horizontal or vertical type. The former have traditionally been used for 'room-sealed' and 'wall-hung' domestic natural gas boilers. In this case the boiler is mounted directly on an external wall, with a short balanced flue extending through the adjacent wall to terminate outside the building (Figure 5.6). This type of balanced flue handles only the combustion air for the burner. Ventilation to limit the temperature of the boiler enclosure is dealt with by separate natural ventilation openings at low and high level. In many instances the boiler is in any event located within the dwelling itself, typically in a kitchen or utility room. In such applications the balanced flue provides a very compact, and hence economical, solution, the extent of the flue comprising the terminal itself and a short 'concentric' sleeve of length just sufficient to extend through a typical domestic cavity-wall construction. The balanced flue arrangement allows the incoming combustion air to be pre-heated by the outgoing flue gases. This improves boiler efficiency and the reduction in the discharge temperature of the flue gases is beneficial, since the flue terminal is typically located at more-or-less head height (the height of the wall-hung boiler is limited by the need for access to controls and for periodic inspection and servicing). Since the flue terminal is typically an angular metal device, there is a risk that occupants of the building may be hurt by inadvertently walking into it. This may be avoided by fitting a simple metallic mesh 'basket' around the terminal. The terminal must, of course, prevent the ingress of small birds, vermin and insects generally. Fan assistance may be applied to allow the siting of a boiler remote from an outside wall, and in this instance the balanced flue and its terminal may be of fairly small diameter.

Despite its primary sphere of application to domestic boilers, wall-mounted LPHW boilers with premix natural gas burners are available for room-sealed, balanced-flue use, offering a design heat output to water in the range 50–70 kW. Such units are therefore suitable for 'light commercial' heating applications. Naturally the necessary design heat output may be built up by operating several such units in parallel, each being provided with an individual balanced flue. In this instance the balanced flue connection will typically be made off the top of the boiler, and may either be turned through 90°

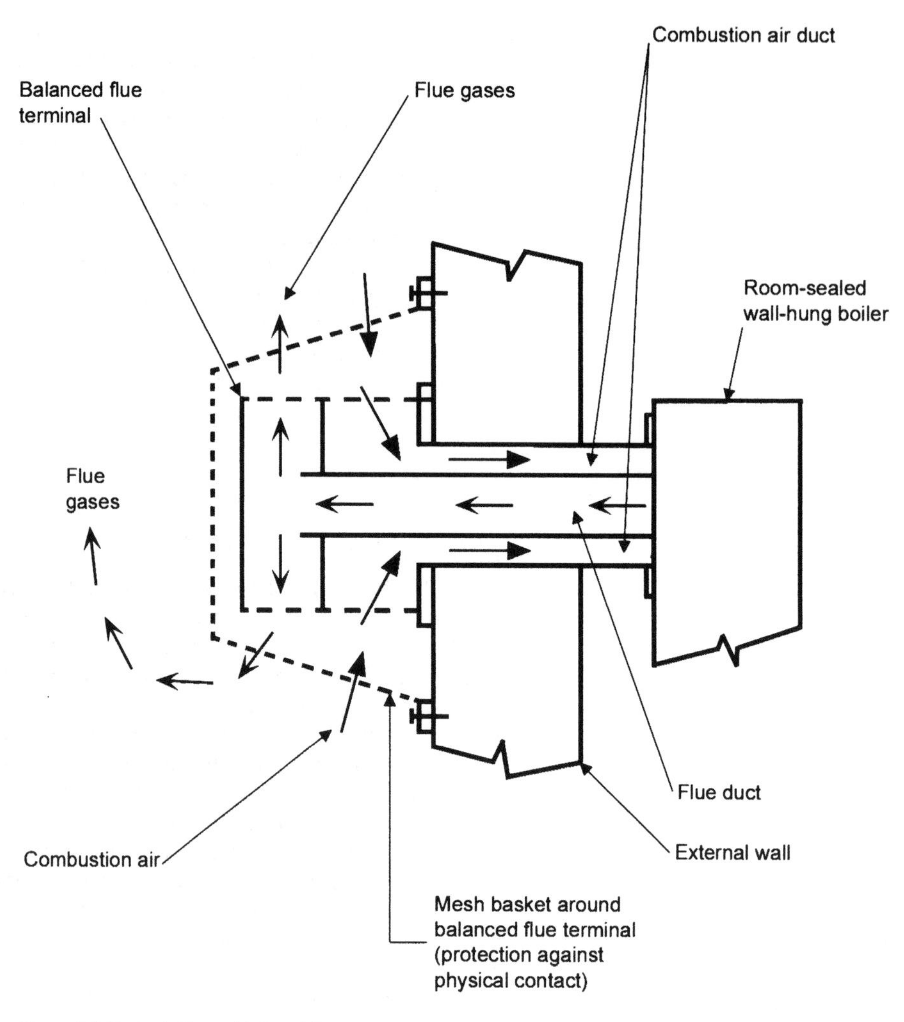

Figure 5.6 'Horizontal' type of balanced flue.

to terminate through an adjacent outside wall or run vertically to terminate above roof level. Such systems typically rely on the positive pressure made available by the premix fan at the boiler connection to the flue, and diameters of as little as 75/125 mm may be possible for the concentric flue gas and combustion air inlet ducts.

Figure 5.7 illustrates schematically a *vertical* balanced flue arrangement for the type of conventional installation of multiple floor-standing boilers that is commonly encountered in commercial space and water heating applications covering the whole size range. Following the balanced-flue principle the air inlet and flue outlet are in the same pressure zone at the top of the chimney, and are thus affected equally by wind pressure. However, in this case the flue terminal not only provides an air inlet for combustion air but also both an air inlet and exhaust for ventilation air to limit plant room temperature. Strictly-speaking, therefore, such a system combines the functions of both a

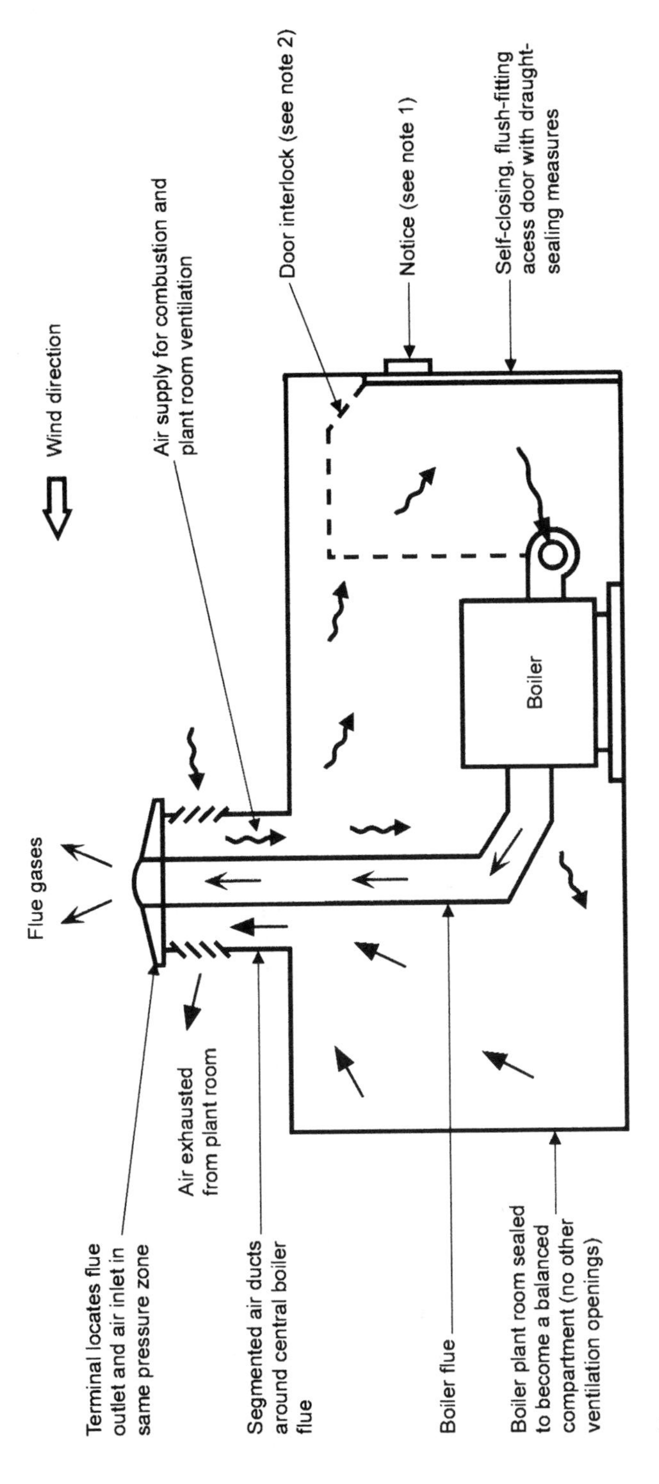

Figure 5.7 Principle of a balanced-compartment system for boiler flueing and ventilation of the boiler plant room.

boiler flue system and a boiler plant room ventilation system. This requires two fundamental changes from the simple 'domestic-type' of balanced flue. First, the handling of both air inlet and exhaust flows cannot be achieved in a simple concentric arrangement, and therefore requires internal division or 'segmentation' of the air duct around the flue. Second, and probably the most fundamental of the two, since both combustion and ventilation air is provided by the balanced-flue system, the plant room in which the boilers are installed is 'sealed', with no other ventilation openings permitted. The boiler plant room therefore becomes a *balanced compartment*, and this is the term by which such systems are now formally identified under British Standards. Doors to the plant room must be self-closing and flush-fitting, with effective draught sealing measures applied. Apart from during ingress or egress of personnel, doors to the plant room should remain shut and notices to this effect will typically be prominently displayed. If any door opens from the plant room on to an occupied space, a suitable control interlock will typically be required to terminate and inhibit further firing of any boilers in the event that the door is inadvertently left open for longer than is required for simple ingress to or egress from the plant room.

The flue terminal in such a system typically takes the form of a multi-sided 'penthouse' or similar arrangement of air inlet louvres disposed around a central vertical flue discharge. The terminal is internally divided or 'segmented' around the chimney. The 'concentric' arrangement of a flue within segmented air ducts is extended only as far as necessary to ensure that input air for combustion and ventilation reaches the plant room and ventilation exhaust air is received from it. The configuration of the flue terminal and segmented air ducts is flexible to suit the needs of the building and its structure, within the limits of what can be physically formed, fabricated and installed. The design of the flue terminal will typically be a considerable improvement visually over more conventional flue outlets and terminals, and can accommodate most architectural preferences (at a price, of course).

In operation the flow of the boiler combustion gases through the central flue may employ either natural-draught or mechanical-draught techniques. In calm ambient conditions the flue natural draught, or the action of the induced-draught or forced-draught fans, will tend to support an inflow of air through one or more of the segmented air ducts, while the temperature difference that will inevitably exist between the plant room and ambient air will act to both enhance the inflow of air and generate an exhaust of air from the plant room via the remainder of the segmented ducts. The action of wind will be to force outside air through the inlet louvres on the windward side of the terminal and through the associated duct segment to the plant room, with air being exhausted from the plant room to discharge through the louvres on the leeward side of the flue terminal. Again the action of natural or mechanical draught will support the inflow of air while the temperature difference between the boiler plant room and ambient air will support both the inflow and exhaust process. While the duct segments that carry air to and from the plant room may change with the prevailing direction of the wind, the principle is that there will always be sufficient segments carrying out each function to meet the needs of the boiler plant. The boiler plant room acts as a ventilation 'balance plenum' or 'balance chamber', and this can only be successful if the plant room is sealed from the action of any other ventilation systems or the effects of significant infiltration or exfiltration.

This type of balanced-flue system may be employed where the boiler plant room does not have an external wall, or where the provision of conventional natural ventilation openings at low and high level may be structurally difficult or inappropriate due to local sources of external pollution. A principal benefit of the system is that it allows a design solution to be employed that remains based on natural air movement in circumstances in which a mechanical system might otherwise be necessary. Since the system depends entirely on the use of high-level ventilation openings, the free area of the flue terminal is typically increased in comparison with that required for conventional high-level natural ventilation openings (by a factor of 2.5 or more, depending on the practice of the particular manufacturer).

5.8 Flue dilution systems

In a *flue dilution system* the combustion gases from one or more boilers are discharged via individual short primary flues into a common 'header' duct that is connected to atmosphere at both ends via louvred openings. A fan, which is located downstream of all primary flue connections from the boilers, draws a sufficient volume flow rate of ambient air through the duct to dilute the combustion gases to a degree that allows their discharge from the building at relatively low level on an external wall of the boiler plant room. The degree of dilution is measured by the volumetric content of carbon dioxide (CO_2) at the flue outlet, which must typically be equal to or less than 1%. The principle is illustrated schematically in Figure 5.8.

The degree to which the combustion gases from the boiler or boilers must be diluted to achieve a maximum CO_2 content of 1% by volume at the discharge from the header duct will naturally depend on the fuel used to fire the boilers. In the case of natural gas a dilution factor of approximately 10, relative to the *stoichiometric combustion volume*, will be necessary, which translates into a flow rate of *diluted flue gases* equivalent to approximately [2.7 + (0.01 × temperature of diluted flue gases in °C)] $l\,s^{-1}$ per kW of boiler *rated input*. This represents a factor of slightly in excess of 100 relative to the rate of gas input to the burner. In the case of LPG the dilution factor will be higher. The *temperature* of the combustion gases will naturally also be diluted. The mixed temperature of the diluted combustion gases at the outlet from the system will vary with the ambient temperature. However, it is unlikely to exceed 50°C even at ambient temperatures up to 20°C for *primary* flue gas temperatures up to 200°C. In a space heating application a single boiler will naturally be firing for less of the time as the ambient temperature increases. In a multiple-boiler installation fewer boilers will be firing, so that for a nominally constant volume flow rate of dilutant air the extent of both CO_2 and temperature dilution will increase. In the case of a condensing boiler the temperature of the diluted combustion gases at outlet from the system is unlikely to differ by more than c.10 K from that of the ambient air itself, even if the boiler is operating in non-condensing mode.

The ambient air intake and diluted flue gas discharge are both located on the same external wall of the boiler plant room, so that the effects of wind pressure are balanced. The intake of air for flue dilution is independent of the supply of air to the plant room for combustion or ventilation to limit temperature, and it is preferable if the conventional openings that are required at low and high level for these functions are located on different walls to the flue dilution openings. The motor of the flue dilution fan will typically require a supply of cooling air, which may be ducted from outside the building or

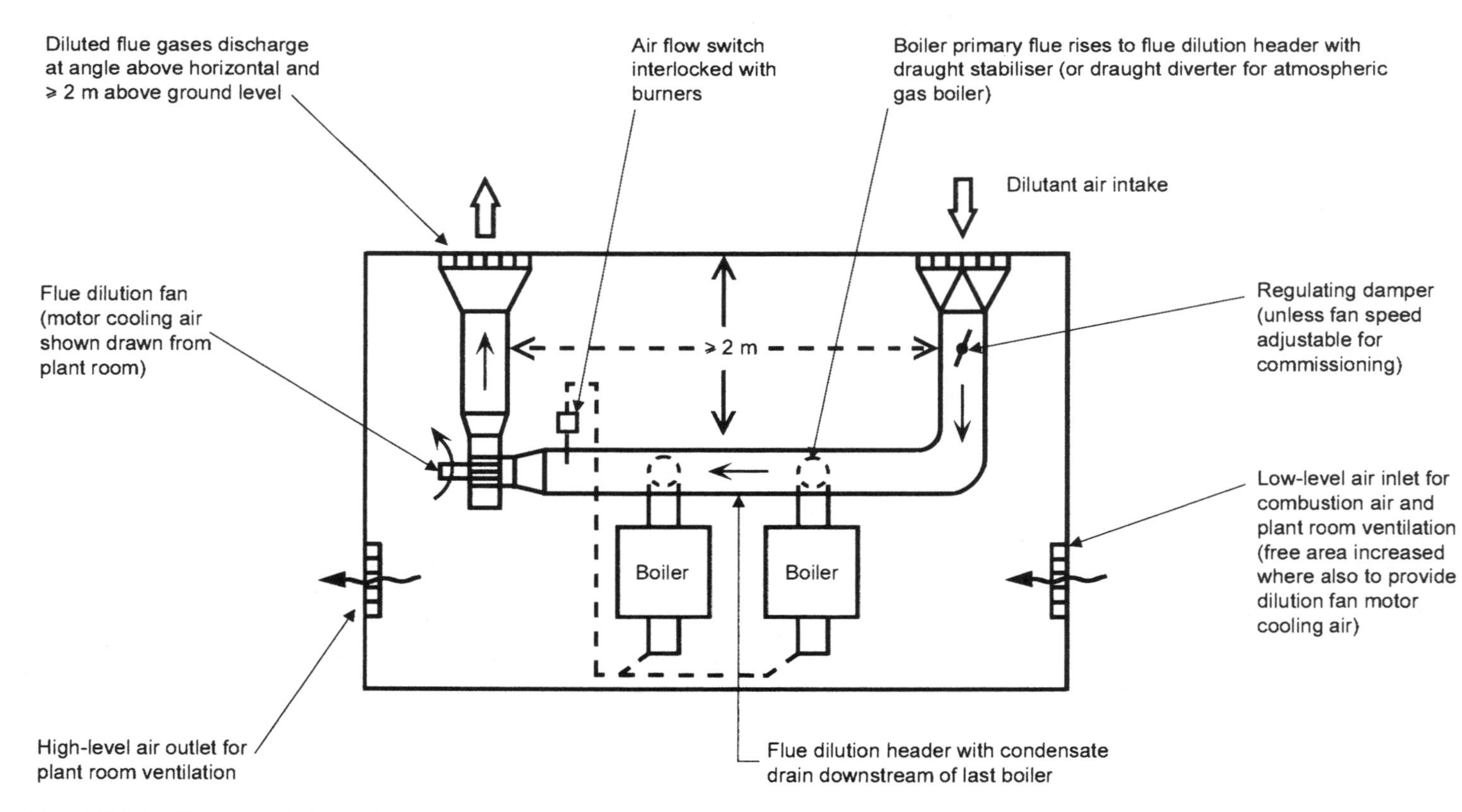

Figure 5.8 Schematic arrangement of a typical flue dilution system.

drawn directly from within the boiler plant room. In the latter case the free area of low-level ventilation openings in the plant room will need to be increased (one manufacturer recommends a twofold increase) to allow for the additional air flow required. Since it is essential to avoid the possibility of recirculation of the diluted flue gas discharge leading to a build-up of combustion gases in the system, the system intake and discharge openings should be horizontally separated. In the case of a single boiler, a minimum separation of at least 2 m is likely to be required, and this will naturally be considerably exceeded where multiple boilers are employed. Where the available horizontal separation between the openings is limited, a vertical separation may also be introduced. In this instance, since the warm diluted flue gases will naturally tend to rise once discharged into ambient air, it will be preferable for the diluted flue gas discharge to be positioned uppermost. It will in any event be preferable for the discharge to be directed at an angle of approximately 30° above the horizontal. Intake and discharge openings should be located so that any risk of inadvertent blockage or obstruction can be controlled, and it is of course preferable that any such risk should inherently be minimal. The discharge should not, of course, be directed into an enclosed area, such as an internal courtyard, or beneath a 'canopy' or overhang. The peak concentration of CO_2 in the diluted flue gas discharge depends on the thoroughness of the mixing that occurs between the flue gases and the dilutant air. This mixing commences in the header duct upstream of the flue dilution fan and continues through the fan itself and in the downstream discharge duct. It is therefore essential to ensure that an adequate mixing distance is available between the discharge from the flue dilution fan and the outlet from the system, and 2 m is again likely to be the minimum acceptable. If adequate mixing cannot be ensured by such means, it may be necessary to increase the volume flow rate of dilutant air. Use of mean diluted-gas velocities in excess of approximately $7.5\,\mathrm{m\,s^{-1}}$ in the flue dilution header may give rise to noise problems.

The use of flue dilution systems is favoured for boilers that are fired on natural gas or LPG, where the sulphur content of the fuel is minimal. Use may be particularly relevant to condensing operation with natural gas, since the high dilution of the flue gases inherently lowers their dew-point temperature so that it approaches that of air at the diluted temperature. However, since condensing boilers typically (although not, it must be said, universally) employ induced-draught fans, these must be balanced with the flue dilution system to avoid *fan interaction*. Since the static pressure in the header duct is increasingly negative from the ambient air intake to the fan suction, this will naturally tend to translate into a suction at the boiler outlet connection to the primary flue. It may therefore be necessary to stabilise the draught at the boiler outlet in order to avoid a condition of excessive suction arising. In the case of atmospheric natural gas boilers the situation is not too dissimilar from what exists in a conventional natural-draught flue (merely the source of the suction at the boiler outlet has changed), although the operating limits of the particular boiler will need to be considered, and its suitability for a flue dilution application discussed with the manufacturer at an early stage in the planning of the installation. However, for conventional boilers with fan-assisted or forced-draught natural gas or oil pressure-jet burners, it will typically be necessary for a draught stabiliser to be installed in the primary flue connection from each boiler.

Precautions will be necessary against corrosion of the flue dilution system.

Flue dilution fans will typically employ impellers manufactured from aluminium or mild steel coated with 'aluzinc'. Fan casings may employ stainless steel or, again, mild steel coated with 'aluzinc' and painted with a polyester paint suitable for a high-temperature working environment. It is especially important that condensate does not collect in the fan casing, which will result in accelerated corrosion. Where condensation may occur in the flue dilution header duct itself, this should be fabricated from stainless steel. Provisions for condensate drainage (via a suitable trap) should in any event be made to drain the header downstream of the boiler primary flue connections. If construction of the flue dilution header employs 'mixed' materials, it is essential to avoid direct (surface-to-surface) contact between dissimilar metals, which will again produce accelerated corrosion.

The flue dilution system must include provision for regulating the volume flow rate of dilutant air. At its simplest this may take the form of a conventional throttling damper positioned downstream of the ambient air intake, or alternatively a fan may be employed in which the speed is adjustable during commissioning of the system. Burner controls must be interlocked with the flue dilution fan to inhibit burner firing unless operation of the flue dilution fan is proven, and to terminate firing on fan failure or in the event of a restriction in the dilution air flow. This is typically achieved via an *air flow switch*.

For boiler plants with rated heat input below approximately 1 MW (1000 kW, which therefore covers a wide range of typical commercial space heating applications) the diluted flue gases may be discharged as low as 2 m above ground level if certain criteria are met regarding efflux velocity of the diluted flue gases and separation from other buildings, opening windows and other air intakes not associated with the boiler plant.

5.9 Sizing of the flue

In the UK, constraints on the location of flue outlets and the sizing of boiler flues typically focus on the requirements of the Clean Air Acts and their associated Memoranda, on Building Regulations and on local authority requirements, although environmental protection legislation may also be relevant. Future legislation from the European Union is also anticipated. General design requirements and good practice recommendations are also addressed by various British Standards and the publications of the former British Gas. In the UK the *Guide* published by the Chartered Institution of Building Services Engineers encapsulates the procedures and practices recommended by the building services profession. The *Guide* also provides a source of references to much of the relevant documentation on standards and requirements[1]. All quantitative information given in this section is for the purposes of discussing the factors that are typically involved in the sizing of a boiler flue or chimney. *It is not intended for use in the design of any specific installation*, which should always be based on reference to the latest editions or versions of the relevant publications. Texts age, and the material content or interpretation of published standards change.

Satisfactory performance of the flue or chimney depends on *appropriate sizing and dimensioning*. Constraints on this typically take the form of minimum requirements for height of a flue or chimney and for the efflux velocity from its outlet, and are more stringent when boilers are fired on fuel oil than when fired on natural gas or LPG, since these latter fuels have negligible sulphur content. In situations where an 'interruptible' natural gas supply is backed-up by oil as a standby fuel for 'dual-fuel' burners, the oil-firing case may provide the limiting constraints. The location of the flue or chimney outlet will also typically have to maintain a minimum vertical and horizontal separation respectively above the roof of a building and from any opening window or air intake.

If the efflux velocity from the flue outlet is weak, there is a risk that conditions of *down-draught* or *inversion* may occur. In the former a 'plume' of cool flue gases 'spills' downwards on the downwind side of the flue or chimney. In the latter a cold down-draught occurs internally within the flue on its upwind side. A down-draught condition inevitably compromises the purpose of the flue by generating a concentration of flue gases at ground level. The influx of cold air into the flue that is associated with a condition of inversion leads to corrosion of the flue. Efflux velocity may be increased by the use of *outlet cones or nozzles*, the increase in the efflux velocity, relative to that in the flue itself, being inversely proportional to the reduction in cross-sectional area effected by the outlet cone. The use of a 'cone', however, results in the introduction of an additional pressure loss associated with the flow of flue gases, and one that is proportional to the *fourth power* of the (inlet/outlet) diameter ratio of the 'cone'. If the mean flue gas velocity in the flue is increased by 50% by the introduction of an outlet cone, the pressure drop associated with this will be slightly in excess of 2.25 times that for a simple untapered flue outlet. A maximum included angle of 30° is recommended for an outlet cone.

The height of a flue or chimney that is required to provide a given natural draught may be determined from the ratio of the sum of the pressure losses associated with gas flow through the flue at the design firing rate of the boiler (in Pa) to the available draught in Pa per vertical metre of flue. The HVAC designer has, at his or her disposal, the ability to vary the configuration and size of the flue to match the pressure loss through it to the available natural draught. In appropriate circumstances mechanical draught may be introduced and its contribution varied (within practical limits). Mean flue gas velocities will typically lie in the range 3.5–12 $m\,s^{-1}$, with values at the lower end of the range being appropriate for natural-draught flues in the height range up to c.12 m and values in the middle of the range appropriate for forced-draught flues in the same range of flue height. At substantially greater flue heights values of flue gas velocity in the mid-to-upper part of the range will be more appropriate[1]. With well-insulated flues or chimneys, a temperature drop of c.1 $K\,m^{-1}$ may be experienced, so that the entry velocity to the chimney will more or less approximate to that at its outlet.

For our present purposes two cases of flue or chimney sizing are worthy of note. The first is the case of a natural gas boiler served by a single chimney that rises through or adjacent to a building, which is a situation commonly encountered in commercial building services applications. As far as the implications of the Clean Air Acts and their associated Memoranda are concerned in the UK, the height of such a chimney will typically be determined by the height of the building, with an *additional height correction* that is related to the rated heat input to the boiler *in Megawatts* (i.e. kW ÷ 1000). The height correction therefore inherently accounts for boiler efficiency. This height correction, in metres, has a value of [$0.816 \times (\text{rated heat input in MW})^{0.6}$], and is always positive. However, the height of the chimney should not in any event be less than the height of a 'free-standing' chimney (i.e. one not rising through or adjacent to the building, but separated from it by a horizontal distance), which is defined by [$1.36 \times$ (rated heat input in MW)$^{0.6}$] or 3 m, whichever is the greater (3 m will be the greater for rated heat inputs up to approximately 3.74 MW). The second case of particular interest in typical building services applications is that in which closely-spaced multiple boilers employ individual flues. While we will not analyse the rules that apply to multiple chimneys, it should be noted that in this situation the height of each flue must be increased above the corrected height that would apply if treating each of the flues in isolation. In either case the final chimney height may be overridden by the need either to generate natural draught or to comply with Building Regulations or local authority requirements.

Where boilers are fired on fuel oil and the total rate of emission of sulphur dioxide (SO_2) is less than $0.38\,g\,s^{-1}$, the height of a single flue or chimney will typically be determined by reference to the requirement for generation of natural draught, but subject to the proviso that the flue or chimney should not terminate less than 3 m above the building to which it is attached or adjacent. The *Guide* published by the Chartered Institution of Building Services Engineers details a procedure for the sizing of flues or chimneys for boilers fired on high-sulphur fuels[1]. These include fuel oil grades 'E' and above, which have sulphur content in excess of 2%, although the popularity of natural gas as a boiler fuel has made the use of such fuels comparatively rare in modern building services applications.

As far as efflux velocities from the flue or chimney outlet are concerned, the requirements of the Clean Air Acts are typically interpreted to mean that a minimum efflux velocity of $6\,m\,s^{-1}$ should be provided from a natural-draught flue or chimney, where this serves a boiler plant with rated heat input up to 2.2 MW (i.e. 2200 kW). This level of heat input naturally covers a wide range of typical commercial building services space and water heating applications. However, it may not be possible to achieve this level of gas velocity throughout the flue or chimney in an installation that relies on natural draught, especially where flue gas temperatures are low, as is the case both for condensing boilers and for some high-efficiency modular boilers. Hence the use of a flue outlet cone or nozzle may be necessary, with the flue system being generally designed on a lower mean gas velocity. In the case of induced-draught or forced-draught flues the recommended minimum efflux velocity is increased to $7.5\,m\,s^{-1}$ for plant with a rated heat input up to 9 MW (i.e. 9000 kW), with a progressive increase up to a value of $15\,m\,s^{-1}$ at a rated heat input of 135 MW – although the lower range of heat input should be more than enough to cover virtually all building services applications for space and water heating that are likely to be encountered in practice. Where boilers employ burners with high/low/off control, the minimum recommended efflux velocity should be achieved with burners operating at their low firing rate.

Again where boilers are oil-fired, the temperature throughout the flue or chimney should not fall below the *acid dew-point* during normal firing. Should this occur, an acid film may form on the inner surface of the flue, to which unburnt carbon and other solid deposits and corrosion products will tend to adhere as a loose-layered agglomerate that may become dislodged on changes in boiler firing rate and carried from the flue (as 'smut') by the flow of gases. For LPHW heating boilers the acid dew-point typically lies in the temperature range 115–140°C, with 'acid corrosion' reaching a maximum level at approximately 30–40 K below this. The effect is typically negligible where 'sulphur-free' natural gas or LPG are used as the boiler fuel.

5.10 General design and construction of the flue

While boiler flues may be of single-skin construction, in modern commercial space and water heating applications they are typically of 'twin-wall' construction, comprising an inner and outer skin separated by a layer of thermal insulation (or in some instances simply an air gap). The layer of thermal insulation is typically a minimum of 25 mm thick, with additional increments of 25 mm possible up to a maximum of 100 mm. Modern practice typically considers a minimum 50 mm thickness of insulation to be preferable, while use of a 100 mm thickness is rare. Levels of thermal insulation may be more critical where high-efficiency modular boilers transfer more heat from the combustion gases, resulting in a lower flue gas temperature and a consequent reduction in the available buoyancy forces. The material of the flue should be able to withstand rapid changes of temperature, and should have a low thermal capacity to minimise the heat-up time during which condensation may occur within the flue. The system of construction and installation must be able to accommodate repeated cycles of thermal expansion and contraction, and should be economic to manufacture and install and simple to inspect and maintain. In induced-draught and forced-draught flues, where part or all of the flue may be under *positive pressure*, it is essential that the construction and jointing of the flue are effectively 'gas-tight'. Not only is containment of the products of combustion of paramount importance because of the associated health and safety considerations, but also due to the risk of damage being occasioned by the egress of water vapour held in the flue gases. This may wet the flue insulation and permeate into adjacent areas of the building fabric where condensation (slightly acidic) subsequently occurs. In such circumstances the sections of the flue will typically employ continuously-welded longitudinal seams, leaving just the transverse or 'cross-joints' between individual components to be site-made using one or another proprietary form of 'socket-and-spigot' jointing system. However, while leak-tightness may be most significant in positive-pressure flues, it is nevertheless also a significant issue in natural-draught flues, since the ingress into the flue of dilutant external air will cool the flue gases and reduce the available draught.

The inner skin of a twin-wall flue is in contact with the combustion gases, and therefore faces a harsh working environment. Smooth-wall, thin-gauge aluminium sheet may be used with boilers employing atmospheric natural gas burners and flue gas temperatures up to c.260°C, while stainless steels of varying grades are suitable for all the common boiler fuels and burner types. However, some grades of stainless steel may not be suitable for applications where either a saline atmosphere or a high level of atmospheric pollution exists. The former is naturally likely to be encountered in coastal locations. The highest (and therefore most expensive) grade of stainless steel is the *austentitic* grade '316'. The outer skin of a twin-wall flue may employ aluminium, stainless steel or 'aluminised' (aluminium/zinc-coated or 'aluzinc') steel. Where boilers are fired on an 'alternative' fuel, such as waste, a specific flue construction may be required. In current practice brick, block or masonry chimneys are unlikely to be used in typical commercial building services applications. Flexible flue liners may be employed for the refurbishment of existing chimneys, or to suit the needs either of new boilers or of a change of fuel.

Where multiple boilers are employed, it is preferable, wherever possible, to *provide each boiler with an individual flue*. This avoids the difficulties typically associated with maintaining adequate efflux velocity from a common chimney under part-load operation, and the risk of non-firing boilers experiencing 'back pressure' where induced-draught or forced-draught flue systems are employed. Failing this, multiple boilers may connect to a single chimney via a common flue 'header'.

In order to limit the pressure loss associated with the flow of flues gases, boilers should be located as close to the chimney as possible. Flues of circular or square cross-section are preferable, but where rectangular cross-sections are unavoidable, aspect ratios (width of the cross-section divided by its height) should be less than approximately 1.5. Short-radius bends and abrupt changes of cross-section are undesirable – the pressure loss associated with a 30° bend is approximately one-third of that associated with a 90° bend, and a similar relationship exists between a sharp bend and an 'easy-sweep' version. Since 'straight' (90°-to-the-vertical) connections of one section of flue to another have the highest pressure drops, connections of boiler primary flues into a common header, or of a flue into a chimney, should be swept at an included angle to the direction of gas flow of 45° or less. Such connections should in any event not protrude into the header or chimney. Where atmospheric natural gas boilers incorporate a draught diverter, a minimum height will typically be required above this before any change of direction is permitted in the flue or a connection made to a common flue header. This is typically of the order of 0.5 m or so, but boiler manufacturers will specify precise requirements for their products. Where a flue must run 'horizontally' to connect to a chimney, it is preferable if it can be arranged to slope slightly upwards toward the chimney. Since condensation is to be expected on start-up of the plant, and potentially under low-load conditions, a suitable 'fall' will in any event be required (a minimum slope of 2.5° is typically recommended) to a suitable drain point, which should be trapped. Joints in the flue should be installed in such a manner as to direct condensate to the drain point without local accumulation or leakage, and long individual lengths of flue will justify separate drainage provisions. Resealable gas-tight access and cleaning doors will typically be required at locations which may include bends, the ends of horizontal runs of flue and at the base of a chimney. In some instances *pressure-relief or 'blow-out' doors* may be required. The base of the chimney should not be left open since this will increase the pressure drop associated with the flue connection by at least a factor of two. The depth of the chimney below the lowest flue connection, or the connection of the common header, should be just sufficient to allow the necessary provisions for drainage, inspection and cleaning. Should any additional depth be necessary, it should be 'infilled' to minimise both gas turbulence and heat loss, and to avoid any risk of such 'dead' space generating *harmonic pulsations*. The routes of flues and chimneys will naturally have to maintain appropriate separation from combustible materials. Last, but by no means least important, careful attention must be given to the support of all sections of the flue system.

The location of the flue outlet should avoid any zones of wind pressure that are likely to cause conditions of down-draught, inversion or 'wind suction'. In this respect the effects of the building itself, local land contours, adjacent buildings and even trees should be considered. Attention to detail will be required to ensure weatherproofing where a flue or chimney penetrates the roof structure of a building. Internal flues and chimneys suffer reduced heat loss. However, where the chimney must be external to the building, it is preferable for it to be located on the leeward (downwind) side of the building, where it will be sheltered from the effects of the prevailing wind. External chimneys should be as slender as structural considerations permit.

5.11 Flues for condensing boilers

Condensing operation with natural gas results in certain characteristics that need to be taken into account in the design of the flue systems that serve this type of boiler plant. Fundamentally, the temperature of the combustion gases leaving the secondary heat exchanger is low (typically of the order of 45°C) during condensing operation, and will not be substantially higher (perhaps only by some 20 K) when not condensing. The natural buoyancy of the flue gases is therefore low. Unless an induced-draught fan is employed, which is typically but not exclusively the case, any heat loss from the flue gases is undesirable. Where an induced-draught fan is employed, single-skin construction may be employed for the flue.

The combustion gases typically leave the secondary heat exchanger of a condensing boiler in a saturated or near-saturated condition. Hence the occurrence of a condensate 'plume' is to be expected at the flue outlet, and the nuisance potential of this should be taken into account when siting the outlet. Flue terminals are considered to accentuate the formation of a condensate plume and should not be necessary where the flue is well-drained. The flue outlet should, however, be protected against the ingress of birds and debris, to which end a suitable wire mesh is typically a functional solution. A high efflux velocity may be required from the flue outlet to reduce the risk of conditions of down-draught or inversion occurring. Condensation *will occur* within the flue system, and it is therefore particularly important to apply throughout the system the general good-practice provisions for drainage and trapping of condensate described under section 5.10. The internal bore of the flue should be smooth and should not permit the accumulation of condensate at joints or in other local crevices or traps, where re-evaporation could lead to locally-enhanced levels of acidity and an 'aggressive' condensate. Joints need to be watertight to avoid leakage of condensate, and gas-tight to avoid leakage of flue gases (and the water vapour they still carry) where the flue operates under a positive pressure.

Condensate formed within the flue will in any event be *slightly acidic* (pH value 3–5). In industrial applications it will be advisable to check on the likely presence of contaminants in the combustion air, as may lead to the formation of an 'aggressive' condensate. The material of the flue must be not only impervious to the passage of condensate through the material, but also resistant to attack by its acidic nature. Austentitic stainless steels, such as grades '304' and '316', and aluminium in commercial-purity (99%) sheet are generally considered appropriate materials, with some types of aluminium exhibiting better resistance to 'pitting' corrosion than stainless steel[2]. In some instances the use of *polymeric materials*, such as 'ABS' plastic, may be acceptable for the flue where the maximum flue gas temperature that can be encountered in non-condensing operation is within the capability of the material, and its use is permitted. Hard plastic materials, such as unplasticised polyvinyl chloride (uPVC) and polypropylene (ABS) are in any event favoured for use in condensate drains. Conventional site drains using vitrified clay will also be impervious to the condensate. In the UK, disposal of the condensate from natural gas condensing boilers to a public sewer is typically acceptable, unless its temperature is excessive. In any event it will typically be good practice to dilute the condensate with other drainage flows within the building.

References

1. Chartered Institution of Building Services Engineers (CIBSE) (1986) Combustion Systems, Section B13 of the CIBSE *Guide*. CIBSE, London.
2. Walker, D.C. (1989) *Factors in the design of flues for condensing boilers*. Technical Note 2/89. Building Services Research and Information Association (BSRIA), Bracknell.

PART B
SYSTEMS AND CONTROL

6 Room Heat Emitters

6.1 Introduction

Room heat emitters are the interface between the heating system and the building, including its occupants. More than any other aspect in the design of the heating system, the selection, location and control of the heat emitters will determine the occupants' thermal comfort. In addition, there will be an impact on running, maintenance and capital cost; space requirements; appearance; and safety. It is important, therefore, that the designer of heating systems has a clear understanding of these issues so that the most appropriate emitter will be chosen and properly utilised.

This chapter looks at all these issues, starting with a brief physical description of the various types of room heat emitter applicable to centralised hot water heating systems. These have been collected into five groups based on physical construction:

- Natural convectors
- Fan convectors
- Radiators
- Radiant panels
- Underfloor

In order to understand the performance and methods of control of heat emitters, it is necessary to apply the principles of heat transfer theory. It is presumed that the reader has some prior knowledge of this. Readers without prior knowledge are advised to read one of the many standard texts[1]. To assist in the fluency of this chapter, much of the theory (and the inevitable mathematics) have been placed in appendices at the end of the chapter which can be treated as supplementary reading. Information on thermal comfort is less available. As an understanding is essential in the design of heating systems, an appendix on thermal comfort has also been included.

6.2 General description of heat emitter types

6.2.1 Natural convectors

Convectors give out heat by convection only (or virtually so). This is achieved by locating the heating element within a case through which air flows to the room. As the case surface temperature is very close to the room temperature, radiant heat output is negligible. The air flow is achieved entirely by buoyancy (natural convection) and convectors are sometimes referred to as 'natural convectors'. The heating element consist of a finned tube through which hot water flows. The fins increase the effective heating surface area in contact with the room air but also increase resistance to air flow. The size and spacing of the fins is designed by the manufacturer to optimise heat transfer rate and results in large, thick fins spaced well apart. Typical installations are shown in Figure 6.1.

Increasing water flow temperature significantly increases heat output rate. In the past, medium pressure hot water and steam have been used but with

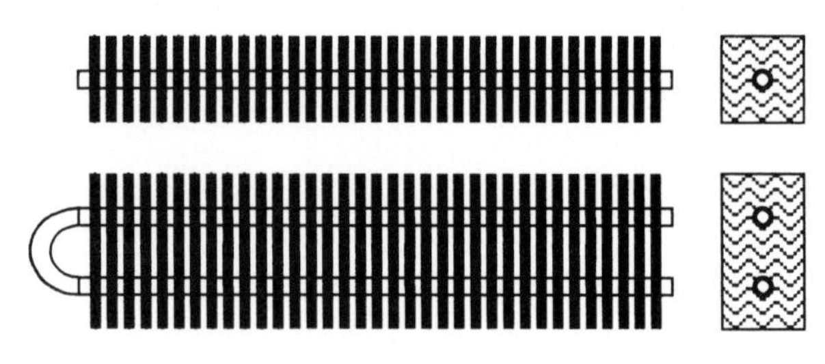

Figure 6.1 Natural convector.

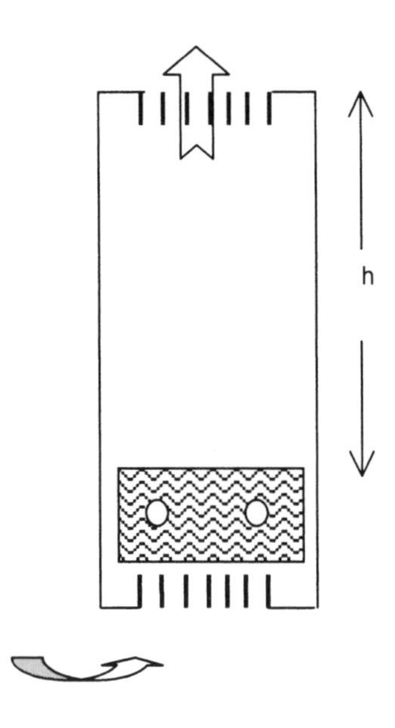

Figure 6.2 Perimeter convector.

modern buildings requiring far less heating, simpler (and safer) LPHW systems are fast becoming the norm.

Perimeter convectors

Perimeter convectors consist of a finned tube contained within a case. The case is not merely for decorative or protective purposes; it plays an important role in improving heat output rate. Room air enters through the bottom of the case and leaves by the top. The finned tube is located just above the air inlet (see Figure 6.2). This results in a column of warm air (a 'stack') within the height of the case. The taller the case, the greater the stack height, air flow rate and heat output rate.

Such convectors are normally run along external walls in a continuous length. A common approach used in the past in office buildings with extensive glazing involved a case height of about 500 mm so that it could be accommodated beneath the window cill but provided high heat output rates. Manually adjustable vents are often incorporated in the air outlet to provide local but crude occupant control. Only the case is continuous: the finned tube has a maximum practical length. As the length is increased, so the water temperature reduces and heat output rate reduces. For this reason connections are made every 2 m or so.

Skirting heating is so called because the case is only about 100 mm high and so can be run along the top of the skirting. This is less obtrusive but has a relatively low heat output rate. However, significant reductions in heating requirements brought about by the regularly updated Building Regulations[2] have made this form of convector increasingly applicable. Finned tubing can also be incorporated within architect designed cases providing great freedom in use.

Automatic zone control is usually by means of two- or three-port control valves and room temperature sensors. This is frequently used in combination with outdoor temperature compensation of the flow water temperature (see section 6.7).

Trench heating

Where full height glazing makes conventional wall-mounted perimeter or skirting heating undesirable (for example in the foyers of modern commercial buildings), the finned pipe may be located within a floor trench. Air enters and leaves by a single grille, flush with the floor and strengthened to allow occupants to walk on it. Heat output rate per metre is less than for wall-mounted convectors due to the reduced stack height and increased resistance to air flow. The grille is often designed to be rolled up to give easy access to the heating element. Careful consideration needs to be given to air venting and draining.

Trench heating is also commonly used in public greenhouses (for example the Princess of Wales conservatory at Kew Gardens where the floor grilles double up as public walkways). Large diameter unfinned pipes are used to avoid dirt build-up. Similar systems have also been installed in churches, though this is not usually the best choice of heat emitter for this type of building, as will be seen later.

6.2.2 Fan convectors

Fan, or ‘forced’, convectors may be floor, wall or ceiling mounted and consist essentially of a fan and a heating coil within a case. (Fan convectors provide heating only, while the term ‘fan coil unit’ is normally reserved for units which can provide both heating and cooling.) Large fan convectors used in industrial buildings are often referred to as ‘unit heaters’. In addition, heating coils (‘heater batteries’) may be duct-mounted with the air flow provided by a central fan. Fan convectors are often used above shop doorways, discharging downwards to offset the cold draught through the frequently opened door. These should not be confused with ‘air curtains’ used in industry to create a wall of air over open doorways to reduce ingress of cold outdoor air.

In most cases, the coil is constructed from small gauge copper pipe with closely spaced thin aluminium fins. The pipe may be serpentine to create a number of passes or may consist of a number of parallel tubes connected to manifolds (see Figure 6.3). This provides a very compact emitter in comparison to the natural convectors as the need to design for low airflow

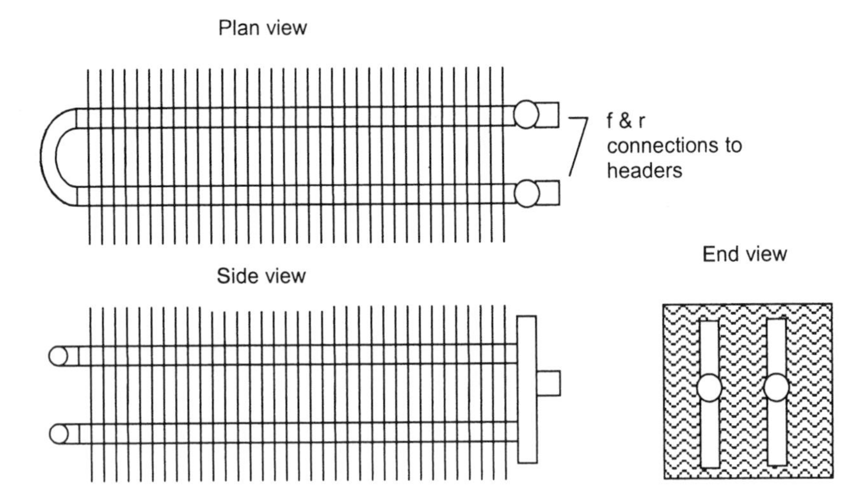

Figure 6.3 Fan convector coil.

resistance is removed. (An exception is fan-assisted trench heating; some manufacturers produce a fan assisted version to greatly increase heat output rate whilst using the same finned pipe as the natural convection model.)

Heat output is normally controlled by throttling the flow rate of hot water using two- or three-port valves. Provided the convector merely recirculates room air, the water flow temperature can be outdoor compensated. Occasionally, control is achieved by modulating the air flow rate, though this can give rise to complaints of noise and poor air distribution by occupants.

6.2.3 Radiators

These emitters give out heat to the room both by convection and radiation, despite their name. In fact, most 'radiators' produce far more convection than radiation, and confusingly those which produce substantially more convection than radiation are sometimes referred to as 'convectors'. The surface emissivity affects radiant heat output rate. Most paint finishes, regardless of colour, have high emissivities (typically 0.9 to 0.95) but some metallic finishes can have much lower values and should be avoided unless used in conjunction with certain varnishes[3].

They are easily the most common form of heat emitter used in virtually every type of building. An indication of their popularity is the large number of manufacturers in Europe. They may be constructed from modular cast-iron or aluminium sections, though the most popular, due to low capital cost, is the pressed steel panel. An example is shown in Figure 6.4.

Heat output rate is controlled by water flow rate throttling, nearly always by means of inexpensive thermostatic radiator valves (TRVs) fixed to individual radiators. This is best used in conjunction with flow water temperature compensation as will be shown in section 6.7. As the surface of the radiator is normally accessible to occupants, it is important that excessive water flow temperatures are not used so that these are normally limited to

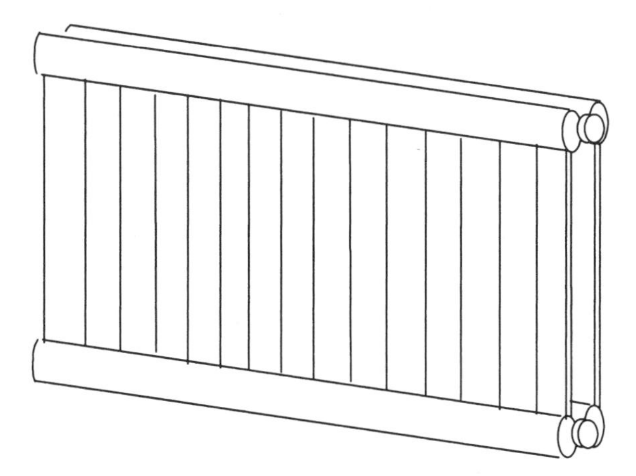

Figure 6.4 Typical radiators.

LPHW. Occasionally, the radiator may be installed within a purpose designed grille for aesthetic or safety reasons, though this greatly reduces overall heat output rate and almost entirely eliminates radiant heat output.

Radiators are conventionally located below windows to improve thermal comfort. This has the combined effect of offsetting radiation losses to the relatively low surface temperature of the glass and the cold down-draught that would otherwise occur, though this does result in slightly higher energy costs. It is preferable that radiator shelves are used which deflect the rising warm air convection current away from the surface of the window as it is substantially above room temperature and would increase heat loss rate. In domestic premises, it is particularly important that curtains do not trap this current of warm air against the window.

Pressed steel panel radiators

The pressed steel panel radiator consists of two thin corrugated sheets of steel (formed by pressing) which have then been welded together to form a series of vertical, parallel channels connected by a header at top and bottom through which water can flow. Such radiators are available as a single, double or triple panel construction set face to face. The corrugations form the waterways, strengthen the panel and increase the total surface area, so improving heat output. Fins may also be fixed to the rear surface (and between panels). This significantly increases convection output but has no effect on radiant heat output. So, although the total heat output is increased, the proportion of radiant heat to total falls quickly. Such radiators are often confusingly referred to as 'convectors' by manufacturers. The panel is fixed to the wall leaving a clearance of about 50 mm to allow convection from the rear surface.

Column radiators

Column radiators were traditionally fabricated in cast-iron (very similar to the design of the heat exchangers for cast-iron sectional boilers), but are now also available in aluminium and steel in a wide range of designs. Essentially, they consist of a number of identical hollow vertical sections which are joined together side by side to form a radiator of whatever length is required. These radiators are usually much heavier than the panel type and are floor standing. The design of the traditional column radiator results in most of the heat output being convective.

Pipework connections

Four pipework connections are provided to both types of radiator: top and bottom at each end. Ideally, the flow connection should be made at the top and the outflow at the bottom on the opposite side (known as top bottom opposite end (TBOE)). Hot water will then flow across the top header and be distributed reasonably evenly through all the vertical channels. As the water cools down so its loss in buoyancy will assist with the flow downwards towards the lower header. A top connection is not as convenient as a bottom connection, so it is common practice to connect flow and return to opposite bottom ends (BBOE). This reduces the heat output rate of the radiator (see section 6.3.3).

Connections to the radiator are normally made using angle valves. It is common practice to install a tamper-free angle valve on the flow side, which has been adjusted to give roughly the desired flow rate. As will be discussed later (section 6.7.1), the water flow rate has little effect on heat output rate and only approximate balancing is required. The return connection is made using either a manual throttling angle valve to provide some control to the occupant over the heat output rate, or preferably a thermostatic radiator valve. Both also double up as isolating valves.

Radiators are a natural collecting point for air in the system. Iron (steel) will also corrode in oxygenated water to form iron oxide and hydrogen gas. The air and hydrogen will displace the water and reduce heat output rate. A vent is provided so that these gases can be removed ('bled').

6.2.4 Radiant panels

A typical construction of radiant panel is shown in Figure 6.5. This consists of a metal plate, to the back of which is bonded a copper or steel pipe through which the hot water flows. This is then insulated so that the heat output is concentrated through the front face. Thus, a given size of panel will produce less heat output than a radiator as the effective area is reduced but the proportion of radiant heat to total heat output will be greater. The panel may be wall or ceiling mounted, often flush with the wall or ceiling finish to provide an unobtrusive heating system, or where ceiling height permits, suspended from the ceiling. The same constraints on surface temperature apply as for radiators where these are accessible to occupants. For conventional ceiling heights, although out of reach, relatively low surface temperatures are still required to prevent thermal discomfort (see section 6.4) at head height from the radiant heat. The higher the water flow temperature, the greater the heat output rate for a given panel so that in industrial buildings and sports halls etc. with high ceiling heights, high water flow temperatures are preferred. This also offers reduced running costs as will be discussed in section 6.5. As the construction is very simple, curved or flat-faced panels are possible. These are usually painted in the same colour as the wall or ceiling. The

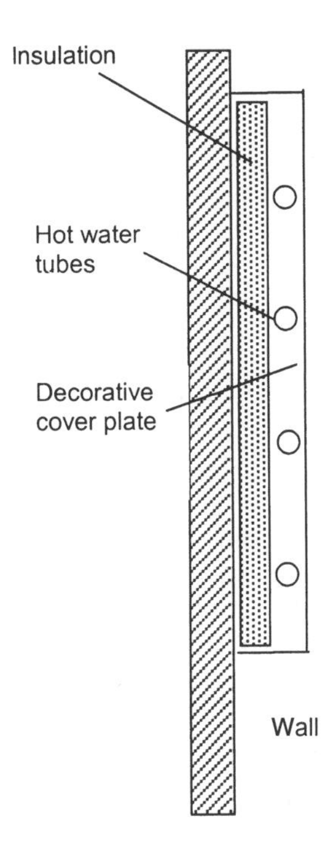

Figure 6.5 Radiator panel.

same consideration as to surface emittance applies as to radiators, but more so given the higher proportion of radiant heat. Horizontally mounted ceiling panels will have a lower heat output rate than vertical wall mounted panels due to natural convection being reduced in the former.

For wall-mounted panels, the water pipe is usually in the form of a serpentine to give good distribution of heat with a simple construction. Ceiling mounted panels are often in the form of narrow strips running the length of the room or around the perimeter where it is more convenient to use a number of parallel pipes connected to flow and return headers. Heat output from wall panels can be controlled in the same way as radiators.

6.2.5 Heated floors

A heated floor usually comprises a number of plastic hot water pipes laid on top of a concrete slab onto which the screed has been poured. The slab is insulated on the underside. The screed is covered with a suitable floor covering. It is also possible to incorporate heating into suspended wooden floor structures. Typical construction is shown in Figure 6.6. For ground floors, the installation of adequate insulation over the whole of the floor is essential, with particular attention being paid to the joint between floor and walls.

With solid floors the pipework must be safe against leakage as repair is prohibitively expensive. A plastic pipe is used incorporating an oxygen barrier. This is laid out and fixed to the slab before the screed is poured. To provide even heating, a number of parallel circuits must be used which are terminated at flow and return distribution headers. The screed thickness may

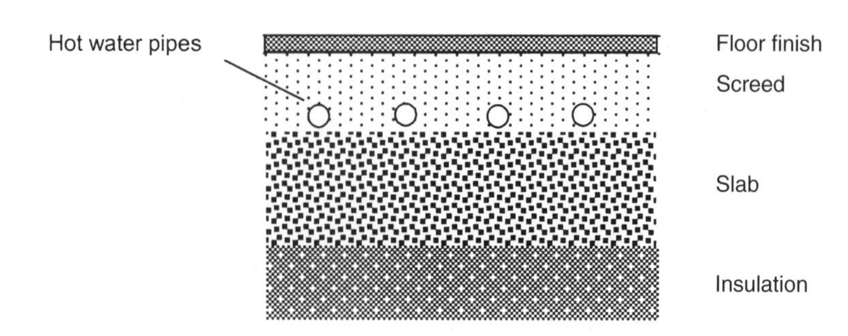

Figure 6.6 Underfloor heating.

need to be increased so as to avoid cracking. The circuits must take into account any expansion joints in the slab.

There is a strict limit on the surface temperature of the floor covering to prevent occupant discomfort. It is recommended that upper limits of 25°C be used where the occupants are often seated, due to prolonged contact with the floor. For transient areas, 28°C may be used. To achieve this, low water flow temperatures are required, making underfloor heating particularly suited to condensing boilers. However, such low surface temperatures reduce the heat output rate that may be achieved to between 50 and 80 W/m^2 (see section 6.3.4). This may not be sufficient to meet the heat loads of some rooms. In particular, heated floors are popular in office entrance foyers, especially where full-height glazing has been used. The heated floor removes the need for trench heating (which itself has a low heat output rate) or radiators in front of the glass, but will not necessarily provide sufficient heat due to the high heat loss rate through the extensive glazing. Additional heating of some form is required. In addition, with concrete floor constructions, the thermal weight of the floor leads to a very slow response. As a consequence, underfloor heating is sometimes used as background heating with a fast-acting top-up system, usually a fan convector. The underfloor heating is run continuously and the fan convector operated as required.

Heat output of heated floors is by both convection and radiation. Excellent levels of thermal comfort can be achieved (as will be discussed in section 6.4), and in high ceilinged rooms running costs can be reduced. The main advantage of the heated floor is probably its unobtrusiveness. Provided the water flow temperature is not excessive, there is no danger in siting furniture etc., although this will reduce effective area and heat output rate.

Control is by means of a proprietary mixing station which contains a mixing valve and circulation pump for each zone. Due to the thermal mass, floor heating systems have special considerations compared to other heat emitters and will be dealt with in detail in section 6.8.

6.3 Typical heat output rates

Manufacturers quote heat output rates for a standard set of conditions under which the emitter has been tested in accordance with British and European Standards. These conditions are unlikely to be those under which the emitter will operate. It is useful therefore to be able to make corrections to the manufacturers' data. Heat output rate of a particular emitter type and size depends on the emitter mean surface temperature and the room temperature. The mean surface temperature of the emitter in turn depends on the flow and return water temperatures. Tests are therefore carried out at standard flow and return water temperatures of 90° and 70°C. These are higher than would be used normally, so manufacturers' data overestimates heat output in most cases. Some manufacturers obligingly quote heat output for lower flow and return temperatures. Convective heat output depends on the room air temperature flowing onto the emitter; radiant heat output will depend on the surface temperatures in the room. To overcome this problem, tests involving radiators and radiant panels should be carried out in rooms with uniform air temperature and with all room surfaces at the same temperature as the air. The room temperature can then be quoted as a single value without ambiguity. The following sections discuss how manufacturers' data can be adjusted for actual operating conditions, and typical heat output rates are presented below in graphical form for various types of emitter under a range of conditions.

6.3.1 Natural convectors

Heat output rate follows the empirical relationship:

$$Q = kL(\bar{\theta}_w - \theta_{ai})^n \tag{6.1}$$

where Q is in W or kW; k is a constant for the particular emitter; $\bar{\theta}_w$ (°C) is the mean water temperature; (this is given by the arithmetic mean of the flow and return temperature. It is customary to assume that the mean surface temperature of the emitter is approximately equal to the arithmetic mean of the flow and return water temperatures. This can lead to large errors in some cases and is discussed in Appendix 6.2); θ_{ai} (°C) is the indoor air temperature; L (m) is the length of the emitter and n is a constant. The value of n is typically about 1.33. These relationships are found from studying test data and are discussed further in Appendix 6.2.

Equation 6.1 can be written in terms of both the standard conditions (on which manufacturers' data is based) and the actual conditions at which the emitter is to operate:

$$Q_{standard} = kL(\bar{\theta}_w - \theta_{ai})^n_{standard}$$

$$Q_{actual} = kL(\bar{\theta}_w - \theta_{ai})^n_{actual} \tag{6.1a}$$

$$\frac{Q_{actual}}{Q_{standard}} = \frac{(\bar{\theta}_w - \theta_{ai})^n_{actual}}{(\bar{\theta}_w - \theta_{ai})^n_{standard}} = \left\{\frac{(\bar{\theta}_w - \theta_{ai})_{actual}}{(\bar{\theta}_w - \theta_{ai})_{standard}}\right\}^n$$

So that the heat output rate at the actual operating conditions is given by:

$$Q_{actual} = Q_{standard}\left\{\frac{(\bar{\theta}_w - \theta_{ai})_{actual}}{(\bar{\theta}_w - \theta_{ai})_{standard}}\right\}^n \tag{6.1b}$$

Figure 6.7 shows typical heat output per linear metre for a 100 mm deep trench heater, a skirting perimeter heater (180 mm high) and a typical wall convector (480 mm high). All three are based on the same size heating element (50 mm wide by 70 mm high fins). It is clear that the heat output rate of the

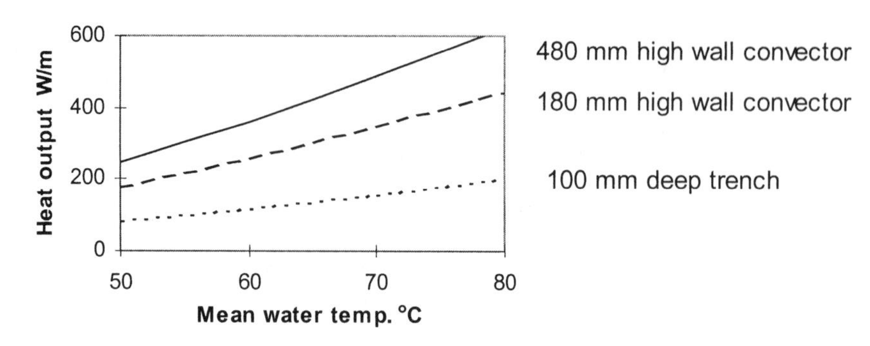

Figure 6.7 Typical heat output rates for natural convectors.

trench is very much less than the wall convector and for this reason is sometimes used in conjunction with other heat emitters where trench heating alone cannot meet all the requirements but some local heating is required at the base of full-height glazing. Also clear is the importance of the case height used with wall mounted convectors on heat output rate as it is this which determines the stack height.

6.3.2 Fan convectors

These emitters obey an equation of the form:

$$Q = \dot{m}c_pE(\theta_{w,in} - \theta_{a,in}) \qquad (6.2)$$

where Q is the heat output rate (in W or kW); $\theta_{w,in}$ and $\theta_{a,in}$ are the entering water and air temperatures respectively (°C); mc_p represents the product of mass flow rate and specific heat for either the water or the air, whichever is the smallest (W/K or kW/K); E is a property known as the 'effectiveness' of the convector (dimensionless). Provided the flow rates of the air and water remain constant, the effectiveness is constant and the heat output directly proportional to the temperature difference $(\theta_{w,in} - \theta_{a,in})$.

This is a similar equation to that for a natural convector although there are important differences. Firstly, it is not an empirical equation but can be derived from a study of heat transfer theory[4] with expressions for effectiveness given as functions of the air and water flow rates and the heat transfer properties of the emitter. Values of effectiveness range from 0 to 1 with typical values being about 0.6. High values of effectiveness are desirable for heat transfer but result in higher fan power, fan running costs and noise generation. The optimisation of fan convector design is a complex engineering problem. Secondly, the heat output rate depends not on the mean water temperature but the entering (flow) temperature. This is more likely to be known than the mean water temperature, making equation 6.2 particularly useful. Thirdly, exponent (n = 1.33) used in the natural convection equation has disappeared (or we could say is equal to 1).

The heat output rate at actual operating conditions can be found from that at standard conditions quoted by the manufacturer in the same way as for the natural convector:

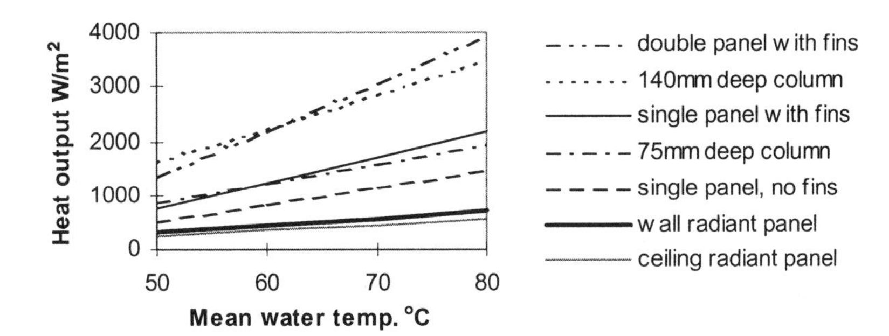

Figure 6.8 Typical heat output rates for radiators and radiant panels.

$$Q_{actual} = Q_{s\,tan\,dard}\left\{\frac{(\theta_{w,in} - \theta_{a,in})_{actual}}{(\theta_{w,in} - \theta_{a,in})_{s\,tan\,dard}}\right\} \qquad (6.2a)$$

Heat output rates vary considerably from manufacturer to manufacturer and model to model and are strongly dependent on air flow rate, so no typical values are presented for this type of emitter.

6.3.3 Radiators and radiant panels

The heat output properties of radiators and radiant panels are similar and will be discussed together. The heat output rate is found from test data to follow an empirical power law similar to natural convectors:

$$Q = kA_f(\bar{\theta}_w - \theta_{room})^n \qquad (6.3)$$

where k is a constant for a particular emitter; A_f (m^2) is the face area of the radiator or radiant panel seen from the front; θ_{room} is the 'room temperature' which assumes that the room surface temperatures and room air temperature are the same. The value of n is found to be about 1.3. Again, heat output rate at non-standard temperatures can be determined using the same method as for convectors given by equation 6.1b. As mentioned in section 6.2.3, the heat output rate depends on pipework connection arrangements. TBOE connections are the normal test method arrangement and can lead to heat output rates as much as 20% greater than BBOE.

Typical heat output rates (taken from manufacturers' literature) per square metre of face area are given in Figure 6.8. Strictly, the heat output rate varies with the height of the emitter when mounted vertically as this affects the natural convection. To overcome this, a typical height of between 450 and 600 mm has been used unless otherwise stated. As can be seen, the heat output rate for a radiator is significantly increased by adding fins or further panels. A column radiator 75 mm deep has a similar heat output rate to a single panel with fins radiator and a 140 mm deep column similar to a double panel with fins. Radiant panels produce only about half the heat output rate of a single panel (no fins) radiator. This is to be expected as the radiant panel has only half the surface area exposed to the room of a single panel radiator.

Radiant heat output

As we shall see later, the percentage radiant heat is important. Manufacturers do not normally quote the radiant component but it can be estimated using the standard equation for radiation:

$$Q_R = A_f \varepsilon \sigma \{\overline{T}_w^4 - T_r^4\} \tag{6.4}$$

where Q_R is the radiant heat output rate (W); ε is the emissivity of the emitter surface; $\overline{T}_w$ is the mean water temperature (K) (strictly, the mean emitter surface temperature should be used, but, as previously mentioned, it is commonly assumed that this is approximately equal to the mean water temperature); T_r is the room mean radiant temperature (K); σ is the Stefan-Boltzmann constant (equal to $6.67 \times 10^{-8}\,\mathrm{Wm^{-2}K^{-4}}$). Note the use of absolute temperature.

However, for the range of temperatures we are interested in, the above equation can be approximated to an acceptable accuracy (less than 2% error) by the simplified equation:

$$Q_R = 3.0 A_f (\overline{\theta}_w - \theta_r)^{1.21} \tag{6.4a}$$

Again, notice the similarity to the convection equation. The exponent (1.21) is less than that for pure convection (1.33), so that as the mean water temperature is increased, the convective heat output increases at a faster rate than radiant, at least over the range of temperatures normally encountered with LPHW heating systems.

It is important in estimating radiant heat output that the correct area is used. For a double panel radiator, for example, only the front surface can 'see' the room. There will be no net radiation between the panels as they will be at the same temperature. There will be some radiant output from the rear surface to the wall, but this will be greatly reduced due to the wall surface immediately behind the radiator being close to the panel temperature (and should be provided with a reflective film).

Radiant heat output has been calculated for each of the emitters shown in Figure 6.8 at various mean water temperatures and divided by the total heat output rate to determine the percentage radiation output. The results are presented in Figure 6.9. Clearly, the radiant panel is the only emitter in the group

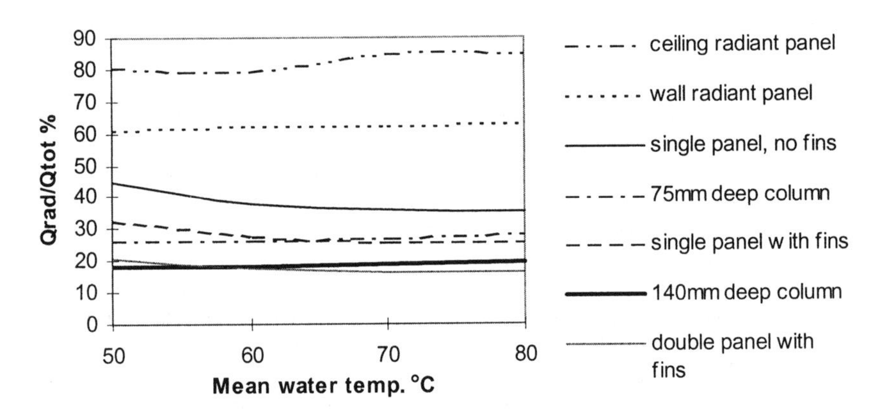

Figure 6.9 Fraction of radiant heat output from radiators and radiant panels.

that produces most of the heat radiantly, with the ceiling panel producing the highest proportion. However, this is at the expense of a reduced overall heat output rate. Ceiling mounted panels have low convective output as they are installed in the horizontal plane rather than the vertical plane with the heated surface facing downwards. Conversely, the increased total heat output of the radiators achieved by adding fins has increased the convective heat output, not the radiant heat output and it is for this reason that some manufacturers refer to their finned panel radiators as 'convectors'.

6.3.4 Heated floors

The heat output from a heated floor is best represented by the equation:

$$Q = Ah(\theta_f - \theta_{room}) \qquad (6.5)$$

where A is the floor area (m^2); h is a combined (i.e. radiant and convective) surface heat transfer coefficient ($Wm^{-2}K^{-1}$); θ_f is the temperature of the floor finished surface (°C). In practice, h is found to have a value of typically $10\,W/m^2K$, so that for a finished surface temperature of 28°C and a room temperature of 20°C, the heat output rate is $80\,W/m^2$ as mentioned in section 6.2.5. Where the floor is partly covered by furniture and fittings, the surface area (A) used in equation 6.5 will need to be reduced accordingly.

6.4 Effect on thermal comfort

The room occupant will lose heat to the room and room air by a mixture of convection, radiation and evaporation of water (see Appendix 6.1). The latter is not usually an issue with heating systems, so a simplified comfort temperature index can be defined in terms of the room air and radiant temperatures only:

$$\theta_c = \frac{\theta_{ai} + \theta_r}{2} \qquad (6.6)$$

where θ_{ai} and θ_r are the room air temperature and the room mean radiant temperature respectively (°C). This is the dry resultant temperature index used by CIBSE, often referred to as 'comfort temperature'. Values of around 20°C are recommended by CIBSE for sedentary occupants[5]; for other activities higher or lower values of comfort temperature apply.

Similar temperature indices are used in Europe and North America. The current International Standard on thermal comfort[6] uses operative temperature, which is almost identical to dry resultant temperature in the specification of the predicted mean vote method of assessing thermal comfort. This latter takes into account metabolic rate and clothing levels.

However, in order to feel thermally comfortable, not only must the occupant's total heat loss rate be acceptable, i.e. correct θ_c, but both the room air temperature and mean radiant temperature must be within acceptable limits (as must air velocity). Most importantly, both temperatures will depend on the room construction and the type of heat emitter used.

Although we talk of the room air temperature and the room mean radiant temperature, in reality these are not single values but will vary from point to point within the room. Hot and cold surfaces (such as heat emitters and windows) will cause room air currents, including draughts, to form. Air temperatures will tend to be above room average near ceiling level and lower at the floor due to stratification. Ideally, the heat emitter will reduce cold draughts while preventing excessive air temperature difference between floor and ceiling. CIBSE recommend a maximum air temperature gradient of about 1.5 K/m.

Radiant temperature is directional and depends on the shape factor between occupant and room surface. Discomfort arises when the radiant field is highly asymmetric. In poorly insulated buildings, wall surface temperatures in winter can be very low (see Appendix 6.2 for a method of determining room surface and radiant temperatures) and lead to both low mean radiant temperature and highly asymmetric radiant temperature in the horizontal direction. Large areas of single glazing are a particular problem, giving rise to a false sensation of draught and hence termed 'radiation draught'. Modern construction standards have mostly eliminated this problem with insulated wall surface temperatures being only one or two degrees below room air temperature, although large areas of standard double glazing can still cause problems. In winter, the inside surface temperature of standard double glazing will fall to about 10°C. CIBSE recommend a maximum asymmetry of radiant temperature of 10 K in the horizontal direction. Smaller values can be tolerated in the vertical direction, with CIBSE recommending a maximum of 5 K.

Purely convective heating will lead to a higher room air temperature than mean radiant temperature and increase stratification. Room air movement in the lower part of the room can be reduced so that the room feels stuffy, which can be exacerbated by a reduction in room air humidity, particularly with ducted systems. Mixed radiant and convective systems overcome these problems to some extent, depending on the mix. However, emitters with high surface temperatures will lead to localised hot spots of radiant temperature. Heated ceilings, unless above the typical height of 2.4 m of most buildings, should not be more than about 5 K above floor temperature. Desks should not be placed too near radiators or radiant panels.

Asymmetric radiant temperature can be reduced by appropriate location of the heat emitter, whether convective or radiant/convective. For this reason, emitters are best placed on external walls and in particular below windows. Convective systems will cause a relatively strong upward plume of warm air, helping to raise the wall or window surface temperature and countering the down draught. Radiant/convective systems will produce a weaker plume but have the advantage of raising the radiant temperature in that direction.

Heated floors can give particularly high standards of thermal comfort. Not only do they provide both convective and radiant heat, but they help overcome the problem of stratification and elevate foot temperature above head temperature. Figure 6.10 shows vertical temperature profiles for various heat emitter types.

More stringent building regulations have resulted in reduced heat loss rates overall and have reduced the difference between air and radiant temperatures so that with modern buildings the dry resultant, air and mean radiant temperatures will be almost equal.

6.5 Heat loss rate and energy consumption

The steady-state heat loss from a room is given by the equation:

$$Q = [\sum(UA)](\theta_{ei} - \theta_{ao}) + 0.33NV(\theta_{ai} - \theta_{ao}) \qquad (6.7)$$

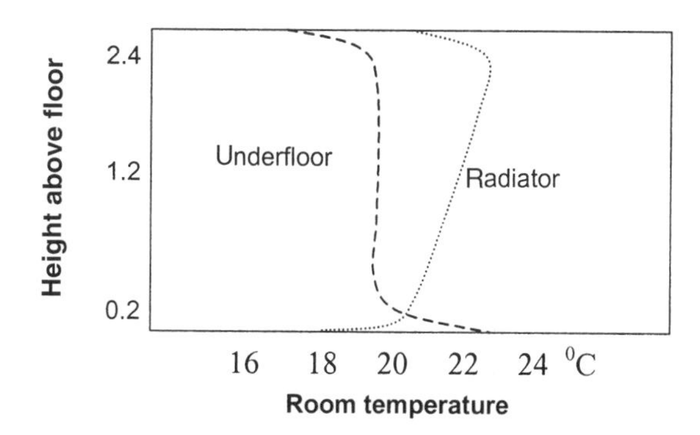

Figure 6.10 Room air temperature profile for heated floor.

where Σ(UA) is the sum of the products of U value and area for each of the room's external surfaces (W/K) and is often referred to as the fabric heat loss coefficient; N is the infiltration rate (hr^{-1}) and V is the volume of the room (m^3) giving Q in Watts. The first term represents heat loss to outside and is therefore based on the indoor environmental temperature, θ_{ei}. The environmental temperature is related to the room air and room mean radiant temperature by:

$$\theta_{ei} = \frac{2}{3}\theta_r + \frac{1}{3}\theta_{ai} \tag{6.8}$$

Strictly speaking, the outdoor environmental temperature should also be used, but in the design of heating systems it is assumed that the environmental and air temperatures outdoors are the same. The second term represents heat loss due to infiltration and so is based on the indoor air temperature θ_{ai}. The term 0.33 NV is known as the infiltration heat loss coefficient. However, as heating systems are designed to provide an acceptable dry resultant temperature, the above equation can be modified to give:

$$Q = [F_{1cu}\sum(UA) + 0.33F_{2cu}NV](\theta_c - \theta_{ao}) \tag{6.9}$$

where the factors F_{1cu} and F_{2cu} are simply temperature ratios used to eliminate the indoor air and environmental temperatures:

$$F_{1cu} = \frac{(\theta_{ei} - \theta_{ao})}{(\theta_c - \theta_{ao})} \quad \text{and} \quad F_{2cu} = \frac{(\theta_{ai} - \theta_{ao})}{(\theta_c - \theta_{ao})} \tag{6.10}$$

Notice that when $\theta_{ei} = \theta_{ai} = \theta_c$ then $F_{1cu} = F_{2cu}$.

For a given dry resultant temperature, the environmental temperature and air temperature in the room will depend on the thermal properties of the structure and infiltration rate and also on the proportion of radiant/convective heat output of the installed heat emitter. Therefore, F_{1cu} and F_{2cu} will depend on these structural and heat emitter properties. Equations for determining F_{1cu} and F_{2cu} are given in CIBSE[7]. Thus, the steady state heat loss will not only depend on the thermal properties of the room but also the type of heat emitter employed. Figure 6.11 shows the heat loss from two typical buildings, a factory and an office, together with the mean radiant temperature and air temperatures as a function of emitter radiant output based on achieving a dry resultant temperature of 19°C.

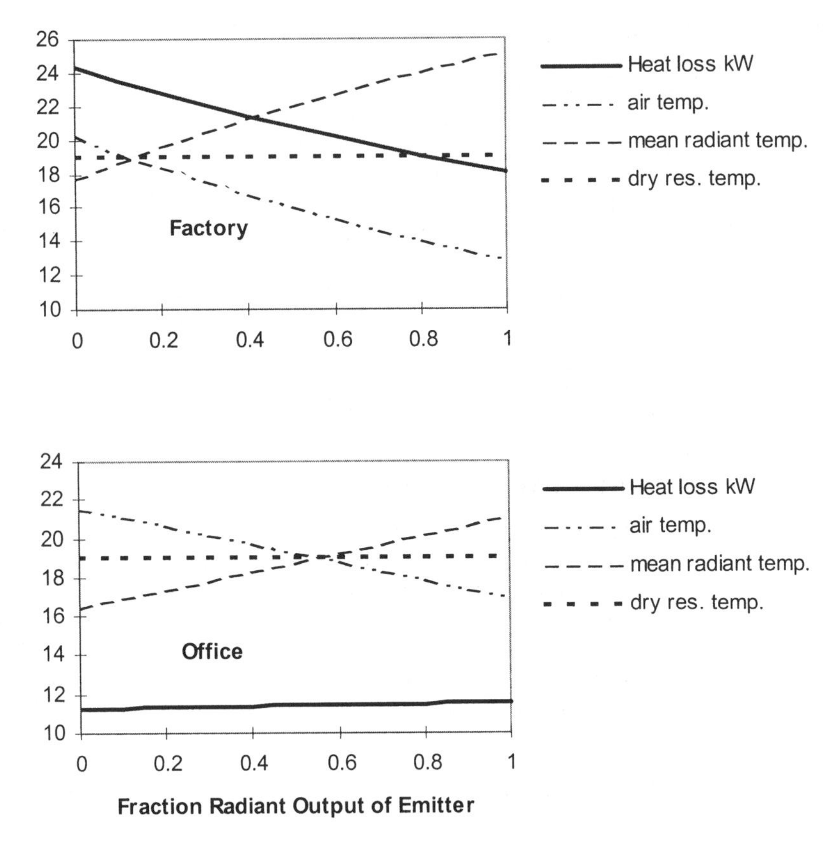

Figure 6.11 Heat loss rate from a factory and an office.

In the case of the factory, which has a relatively high infiltration heat loss coefficient (0.33 NV) compared to fabric heat loss coefficient ($\Sigma(AU)$), an emitter with a high radiant output results in a low air temperature (low value of F_{2cu}) and a high radiant temperature (high value of F_{1cu}) leading to a lower total heat loss rate. A lower heat loss rate will result in lower annual running costs whilst maintaining the same comfort temperature. For this reason, ceiling radiant panels are frequently used in factories and other buildings where infiltration dominates. Also, of course, ceiling panels take up no floor space and are safe from mechanical damage.

In comparison, the typical office has similar values of infiltration and fabric heat loss coefficients and so varying the proportion of radiant heat output has little effect on either air or radiant temperatures and virtually no effect on heat loss rate.

The above analysis assumes a uniform air temperature throughout the room. In reality, stratification will lead to increased temperature at high level in the room. Buildings with high ceilings/roofs, such as factories, will result in greater heat losses through the roof as a result. A radiant heat emitter will therefore have a further running cost advantage as stratification will be reduced.

6.6 Safety issues

Where the emitter surface can be touched by the occupants, it must be ensured that such surface temperatures will not lead to burns. This is particularly important where the occupants are very young, sick or frail. For primary schools, hospitals and old people's homes there are regulations limiting the exposed surface temperature to about 45°C. Such a low surface temperature will have important impact on the heat output rate and therefore area of emitter needed.

Care also needs to be taken to avoid sharp obtrusive edges and finger traps, especially where children are present. Radiators are often enclosed in a decorative case with grilles to allow room air movement over the heated surface. This may allow a higher mean water temperature to be used but it still reduces total heat output rate and virtually reduces radiant heat output rate to zero.

6.7 Controlling heat output rate

The heat emitter must be sized to meet the peak heating load (at design conditions) of the room. This involves determining the design steady-state heat loss rate (equation 6.7 or 6.9) and applying a plant ratio factor to allow for the thermal capacitance of the structure where intermittent heating is used. This is discussed in detail in Chapter 10. The emitter will be required to provide full heat output at start-up, but solar gains and internal gains due to occupants, lighting and equipment will quickly reduce the heat load. Even in winter, shortly after occupation commences, heat gains in modern buildings can exceed fabric and infiltration heat losses. At other times of the year, the outdoor temperature will be substantially above the winter design temperature and fabric and infiltration heat loss will be significantly reduced. It is essential therefore that the heat output rate of the emitter can be efficiently controlled both to ensure thermal comfort and to reduce energy consumption. The latter is recognised in the current Building Regulations[8], which require adequate room temperature control (see section 6.7.3).

The heat output rate of any heat emitter can be achieved in one of two ways: modulating the water flow rate whilst maintaining the flow water temperature; modulating the flow water temperature whilst maintaining the water flow rate. (Fan convectors may be controlled by modulating the air flow rate, although water side control is more common.)

6.7.1 Modulation of water flow rate

This is usually achieved by means of a two-port valve which throttles the water flow rate in response to a room temperature sensor. For radiators and radiant panels, thermostatic radiator valves (TRV) are invariably used. These are low cost, self-actuating valves requiring no wiring: one TRV per emitter is used. For natural convectors a two-port valve connected to a room temperature sensor is common. Occasionally, three-port valves are used to divert water away from a group of emitters, controlled from a room sensor. Regardless of how the flow modulation is achieved, the response of the emitter is the same.

Figure 6.12 shows how heat output varies with water flow rate. Appendix 6.2 describes how this graph was produced. The axes used represent fractional heat output and flow rate, i.e. the ratio of actual heat output (or flow rate) to the design value at full duty. The curve is based on a flow

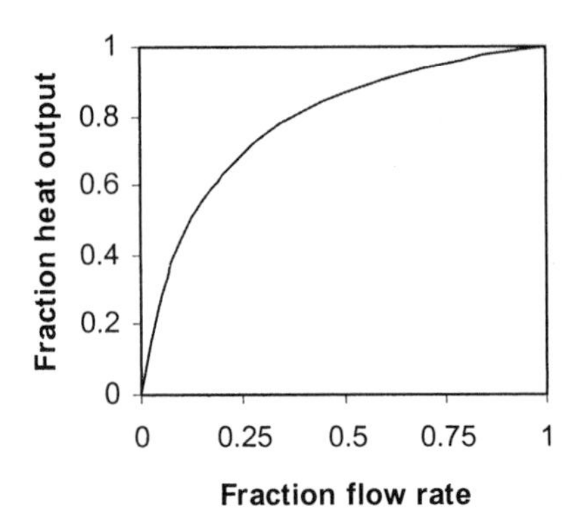

Figure 6.12 Variation of heat output rate with water flow rate.

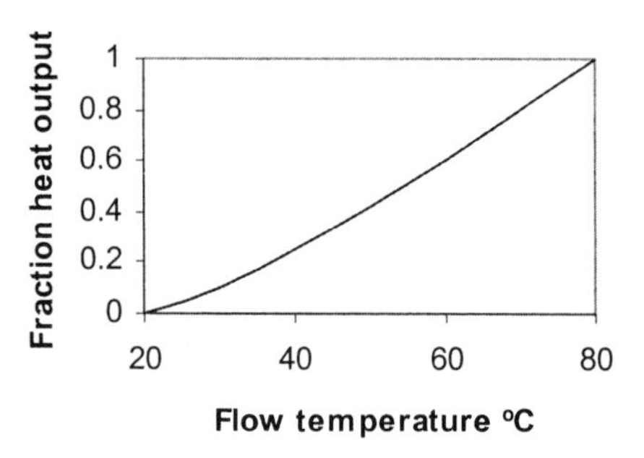

Figure 6.13 Variation of heat output rate with water flow temperature.

temperature of 80°C and a design return water temperature of 70°C. As can be seen, the response curve is highly non-linear: a large reduction in water flow rate results in only a small reduction in heat output rate. The curve becomes more linear as the design flow and return temperature difference is increased.

Such non-linearity requires a valve characteristic which will produce a large reduction in flow rate for a small valve stem movement. Such valves are available but are relatively expensive. They might be used for controlling a group of emitters but are not economically attractive propositions for individual emitters. Less expensive valves need to be virtually closed to give a significant reduction in heat output rate. Furthermore, as the two-port valve closes, the pressure drop across it increases and so the valve has to close further. This is discussed in Chapter 7. This results in poor controllability and possibly a valve that is unable to shut off completely.

6.7.2 Flow temperature modulation

Flow temperature modulation is achieved by the blending of return water with the flow using a three-port 'mixing' valve (see Chapter 7). The response of the heat emitter to a change in flow water temperature is shown in Figure 6.13. As can be seen, unlike the two-port throttling valve, the response is very nearly linear so good control is achieved. Unfortunately, such valves and control systems are relatively expensive. As such, this system is used for controlling large groups of emitters.

6.7.3 Combined flow rate and temperature modulation

Good control is achieved by combining the two methods. As the outdoor temperature increases above the design value used for sizing, so all the emitters in the building will experience a reduction in heat demand as the fabric and infiltration heat loss rate reduces. Flow temperature modulation can therefore be carried out centrally by scheduling it to outdoor air temperature. This is known as 'outdoor temperature compensation' and is discussed in detail in Chapters 7 and 10. Local (zone) two-port valves, controlled by the local room temperature (e.g. TRVs), then modulate the water flow rate to each emitter in response to changing solar and internal heat gains. The amount of throttling now required by each two-port valve is greatly reduced and overall control improved. This combination of central and local control is specified in the Building Regulations.

6.8 Underfloor heating

The construction of heated floors results in considerable differences to the performance and operation of other types of heat emitters:

- The surface temperature (exposed to the room) will be at a significantly lower temperature than the mean water temperature. The water flow temperature required will need to be carefully determined in order to provide the desired floor surface temperature, crucial for thermal comfort.
- There may be a significant downwards flow of heat which will not contribute to the heating of the room.
- The spacing of the hot water pipework and the temperature difference between flow and return must not result in excessive variations in temperature across the floor surface area.
- The thermal capacitance of the floor together with the small temperature difference between floor and room will result in a very slow response to control changes in water flow rate or temperature.

6.8.1 Upward heat flow

Under steady-state conditions (i.e. when the floor has been in operation for some time with the room and water flow temperatures constant), then the heat output rate from the finished floor surface to the room must equal the heat output rate upwards from the heating pipe through the screed and floor covering. If we assume that the temperature of the screed in the plane of the heating pipework is at a uniform temperature, θ_{pp}, then from equation 6.5:

$$\frac{Q_{up}}{A} = h(\theta_f - \theta_{room}) = \frac{(\theta_{pp} - \theta_f)}{R_{up}} \qquad (6.11)$$

where $R_{up} = \left[\frac{x_c}{\lambda_c} + \frac{x_s}{\lambda_s}\right]$

where Q_{up} (W) is the same as Q in equation 6.5 A is the floor area (m^2), x and λ are thickness (m) and conductivity (W/mK) and the subscripts c and s represent the floor covering and screed respectively. Then:

$$\theta_{pp} = R_{up}h(\theta_f - \theta_{room}) + \theta_f \qquad (6.12)$$

So the required temperature in the plane of the pipework is dependent on the resistance of the screed and floor covering. For example, for a 50 mm screed thickness above the pipework ($\lambda = 0.41$ W/mK) and a 25 mm marble floor tile ($\lambda = 2.0$ W/mK), with a desired floor surface temperature of 25°C and a room at 20°C, the temperature in the plane of the heating pipework is:

$$\theta_{pp} = \left[\frac{0.025}{2.0} + \frac{0.05}{0.41}\right] \times 10(25 - 20) + 25 = 32°C$$

If, instead of the marble, a 25 mm wood block floor was used (λ = 0.14 W/mK), then θ_{pp} would need to be 45°C.

There will be little resistance between the water in the pipework and the surrounding screed so the required mean water flow temperature will be only slightly higher than θ_{pp}. This analysis is based on one-dimensional heat flow and will only hold to acceptable accuracy if there is only a small temperature difference between the flow and return water temperatures and pipe spacing is not excessive. Circuits are normally set out using many short pipe lengths in parallel to prevent excessive temperature differences. The manufacturer's expertise is usually required to design these circuits.

6.8.2 Downward heat flow

With the temperature in the plane of the pipework known, the downward heat flow through intermediate floors can be found using an equation of the same form as 6.11:

$$\frac{Q_{dn}}{A} = \frac{(\theta_{pp} - \theta_{room})}{R_{dn}} \qquad (6.13)$$

where θ_{room} is now the temperature of the room or space below, R_{dn} represents the structure below the pipework plane. The latter must include a surface heat transfer coefficient for the underside of the floor (i.e. ceiling). This will be significantly less than the value used for upward heat flow from the floor surface as natural convection is greatly reduced. A typical value is 6 $Wm^{-2}K$. This is based on still air conditions.

For solid or suspended ground floors, this approach cannot be used. In the former, there is no definitive depth as the floor is in contact with the earth. In the case of the latter, the temperature in the floor void is not known. For solid floors, the 1986 edition of the CIBSE *Guide*[9] gives some guidance and suggests that a fictitious value of downward conductance be used based on outdoor air temperature. Such calculations indicate a downward heat loss of typically about 12 Wm^{-2}, assuming an insulation thickness of 50 mm (polystyrene) over the whole floor area.

6.8.3 Thermal capacitance

Solid floors will have a time constant of several hours and it may be worth considering continuous operation of the heated floor, particularly where the floor provides only background heating.

Where intermittent operation is used, there will be a swing in room temperature which can be estimated from the following equation[9]:

$$\tilde{\theta}_{room} = \frac{Q_{up}F}{\{\sum(AY) + 0.33NV\}} \tag{6.14}$$

where $\tilde{\theta}_{room}$ is the daily swing in room temperature (°C). For a solid ground floor, F is equal to 0.7. $\Sigma(AY)$ is the sum of the product of area (A) and admittance (Y) for all the surfaces in the room excluding the floor. This is similar to the fabric heat loss coefficient ($\Sigma(AU)$) but represents the flow of heat into the room surfaces rather than through them to outside. Note that all the room surfaces need to be considered (excluding floor), not merely those of the external fabric (walls, windows, etc.).

Appendix 6.1 Thermal Comfort

We humans (in common with other warm-blooded animals) need to maintain our deep-body core temperature (about 37°C) within tight limits. Just a few degrees below and we have hypothermia; a few degrees above and we have fever. This is a very small range compared to variations in outdoor temperature so, not surprisingly, our bodies have evolved a very sophisticated method of controlling core temperature. In comparison, skin and limb temperature can vary over very wide temperatures without physical harm, although this does lead to a feeling of thermal discomfort.

Heat and work energy is generated within our body by chemical reaction, our metabolic rate. About 120 W of heat is produced at rest, rising to perhaps ten times this amount during extreme exertion (e.g. sporting activities). In order to maintain our core temperature near constant we have to lose heat to our surroundings at very nearly the same rate as it is generated. If we lose heat too slowly we feel 'warm' and our body's mechanisms strive to increase heat loss rate by various means. Likewise, if we lose heat too quickly our body tries to compensate and we feel 'cold'.

We lose heat to our surroundings through our skin and breath by a mixture of sensible and latent heat exchange. Sensible heat is lost by radiation and convection from the skin and through breathing out warm air. Latent heat is lost through the evaporation from the skin of water vapour and through the moisture-laden breath. Most importantly, our bodies are able to change the ratio of latent to sensible heat, within limits, allowing us to live in a wide range of thermal environments.

The rate of latent heat loss depends on the partial pressure of water vapour in the air and to a lesser degree on air velocity. At typical room air temperatures, if the relative humidity remains between about 35% and 65%, the partial pressure of water vapour will be such that the latent heat loss rate is not significantly affected. Below about 30%, a sensation of dryness occurs as latent heat loss rates increase and the body will respond with a drop in skin temperature to compensate. In addition, static electricity often causes problems. At about 70% relative humidity, the rate of evaporation of water from the skin is reduced, leading to pools of water remaining and a sensation of clamminess. Skin temperature increases to compensate.

Fortunately, during the heating season, the %RH in most buildings will remain between 35% and 65%. Although the outdoor moisture content is low, provided there is not excessive ventilation for the number of occupants, the moisture given off by the occupants leads to acceptable relative humidity. Excessive ventilation can lead to too low a humidity, in which case humidification is required. Where moisture loss by occupants is high (e.g. squash courts), high rates of ventilation are required.

Heat loss rate by convection and radiation is much more sensitive to room conditions than latent heat, and as only the former is normally affected by the type of heat emitter used, this is of much greater interest to the heating system designer. Also, as heat transfer by convection and radiation occurs at the skin, this combined rate of heat loss affects skin temperature, which is one of our key thermal comfort indicators. However, as we will see, not only must the sum of convective and radiant heat loss be within limits to achieve comfort, but there are also limits on each of the two separate rates.

The convective heat loss rate depends on skin temperature, air temperature and local air velocity. Increasing the air velocity over the occupant's body increases the convection coefficient. For this reason a desk fan will provide a cooling effect in summer even though the air temperature is no different to that in the room. In winter, such an airflow would lead to discomfort. Certain parts of the body, in particular the neck and ankles, are especially sensitive to excessive air velocity. Local air velocities should normally be limited to about 0.1 m/s in winter to avoid draughts. At the same time, velocities below about 0.05 m/s can lead to sensations of stuffiness.

Radiant heat loss rate depends on skin temperature and room mean radiant temperature. This latter is a complex function of room surface temperatures and emissivities and the geometric relationship between occupant and each surface, including the heat emitter surface. This is discussed more fully in the section below.

All forms of heat loss rate depend, of course, on the amount and type of clothing worn.

Provided the room humidity is between 35% and 65%, when sitting at rest we feel comfortable when:

$$\frac{\theta_a\sqrt{10v}+\theta_r}{1+\sqrt{10v}} \approx 21^\circ C \qquad (6.15)$$

where θ_a is air dry bulb temperature (°C), θ_r the mean radiant temperature (°C) and v the local room air velocity (m/s). This is an empirical relationship and is referred to as the 'dry resultant temperature index' or more commonly 'comfort temperature' and given the symbol θ_c (°C). It is the comfort temperature index recommended by CIBSE. There are a number of other temperature indices used to measure thermal comfort. In North America 'operative temperature' is used.

As already mentioned, air velocity in rooms needs to be about 0.1 m/s during heating to avoid sensations of draught or stuffiness, in which case:

$$\frac{\theta_a+\theta_r}{2} \approx 21^\circ C \qquad (6.16)$$

So, if we are to feel comfortable (seated at rest and normally clothed) the heat emitters must maintain the average of the room air and room surface temperatures at about 21°C. Under other circumstances the desired comfort temperature will be different. See CIBSE (1999)[11] for a list of recommended comfort temperatures for different building types and activities.

Even so, there is a limit on the individual values of air and room surface temperatures that are acceptable. As a general rule, for sedentary occupants, the difference between air temperature and mean surface temperature should not be more than 5°C. In most modern buildings with even moderate amounts of insulation, the difference will be much smaller than this. Older buildings with either uninsulated solid walls or large expanses of single glazing can result in low mean radiant temperatures and thermal discomfort, which can be partly offset by raising the air temperature.

In addition, there are limits on the vertical variation of air temperature (stratification) and spatial variation in surface temperature (asymmetric radiation). Again, for sedentary occupants, vertical air temperature gradient should not exceed about 2 K/m, with the higher temperature at floor level (to avoid the cold feet warm head syndrome) while vertical and horizontal differences in surface temperature should not exceed about 5 K and 10 K respectively. Large areas of single glazing close to occupants can lead to excessive radiant asymmetry and the sensation of a draught even though there is little air flow. This problem is often referred to as 'radiation draught'.

The choice of heat emitter must therefore address the thermal comfort requirements of the occupants, which depend on their activity level and the thermal properties of the building structure. Heat emitters also have a large impact on both stratification and asymmetric radiation. For example, a high temperature radiant panel could lead to excessive radiant asymmetry in the wrong application, or a warm air system could lead to unacceptable stratification. Selecting appropriate heat emitters is essential if thermal comfort is to be achieved.

Appendix 6.2 Emitter Heat Output Rate Analysis

This appendix looks in detail at the heat output rate of emitters, using fundamental relationships for both natural convection and radiation, and so is applicable to natural convectors, radiators and radiant panels. The effect of changing flow water temperature and water flow rate is then analysed.

Fundamental equations

The radiation output can be expressed by the standard equation:

$$Q_{radiation} = A_R \sigma \Im (T_m^4 - T_r^4) \tag{6.17}$$

where A_R (m^2) is the area of the radiator relevant to radiation (usually the front only). T_m and T_r refer to the mean surface temperature of the radiator and the mean temperature of the room surfaces (also known as mean radiant temperature) respectively (K). The factor $\Im$ is a combined function of shape factor between the emitter and the room surfaces and the emitter and room emissivity (dimensionless). As the emitter is completely surrounded by the room and has a small surface area in comparison with the room, $\Im$ is equal to the emissivity of the radiator, typically 0.9.

The convection output is given by:

$$Q_{convection} = A h_c (\theta_m - \theta_a) \tag{6.18}$$

where A is the total surface area of the radiator (m^2), θ_m is the mean surface temperature of the radiator and θ_a the room air temperature (°C). The convective heat transfer coefficient h_c (W/m^2K) is determined by the equations for natural convection:

$$Nu = \frac{h_c x}{\lambda} = 0.36 Gr^{1.25} \quad \text{for laminar flow}$$

$$Nu = \frac{h_c x}{\lambda} = 0.1 (Pr . Gr)^{0.33} \quad \text{for turbulent flow} \tag{6.19}$$

$$\text{where } Pr = \frac{C_p \mu}{\lambda} \text{ and } Gr = \frac{\beta g \rho^2 x^3 \Delta\theta}{\mu^2}$$

where Nu is the Nusselt Number, Pr the Prandtl Number and Gr is the Grashof Number.

In perfectly still air, both laminar and turbulent heat transfer will occur. In practice, the room air movement in a heated room and the intentional design of the radiator will lead to most of the heat transfer being turbulent.

Using the turbulent natural convection equation and the radiation equation, the heat output rate for a typical pressed-steel double panel radiator (without fins) was calculated for a range of emitter mean surface temperatures. The radiator dimensions used were 1 m wide by 0.5 m high. The mean room surface temperature was assumed to be the same as the room air

Table 6.1 Calculation of heat output rate for a typical radiator.

Mean radiator temp °C	$Q_{convection}$ W	$Q_{radiation}$ W	Q_{total} W	% radiation	% convection	$Q = K\Delta\theta^{1.33}$
70	418	165	583	28	72	583
60	311	126	436	29	71	433
50	212	90	302	30	70	296
40	124	57	180	31	69	172
30	49	27	76	35	65	69
20	0	0	0			0

temperature, which was kept constant at 20°C. The results are shown in Table 6.1.

As can be seen, the % radiation decreases slightly with a rise in mean radiator temperature.

The extreme right-hand column is an attempt to fit the total heat output rate to a single equation where k is some constant, and $\Delta\theta = (\theta_m - \theta_i)$ where θ_i is the room temperature. As can be seen, there is a very good fit, at least at the higher temperatures. So, we may conclude that:

$$Q = k\{\theta_m - \theta_i\}^{1.33} \tag{6.20}$$

This is the empirical equation commonly used to describe radiators. The value of k depends on the design and installation of the radiator. It is the basis of the British Standard method of testing. As it is a much simpler approach than using the fundamental equations, and as it gives good agreement, this simplified equation has been used in the analysis which follows.

Mean temperature of radiator versus heat output rate

Expressing the part load as a fraction of design heat output (keeping room temperature, θ_i, constant), equation 6.20 can be written as the ratio of temperature differences in which the emitter constant, k, is eliminated:

$$\frac{Q}{Q_{design}} = \left\{\frac{\theta_m - \theta_i}{\theta_{m,design} - \theta_i}\right\}^{1.33} \tag{6.21}$$

where subscript 'design' represents the design (full load) value. From this, the mean surface temperature of the radiator required to produce any fraction of design heat output can be easily found from:

$$\theta_m = \theta_i + (\theta_{m,design} - \theta_i)\left\{\frac{Q}{Q_{design}}\right\}^{\frac{1}{1.33}} \tag{6.22}$$

Note that this mean temperature is required regardless of how the radiator is controlled.

The mean radiator surface temperature will be very close to the mean water temperature. It appears to be common practice that the mean water temperature is calculated as the arithmetic mean of the flow and return water temperatures. However, a simple examination shows that this cannot be the case. Table 6.1 shows that a mean surface temperature of 40°C gives about 30% of design heat output rate, based on a flow temperature of 80°C and a room temperature of 20°C. To achieve an arithmetic mean water temperature of 40°C, the return water temperature would need to be 0°C. Of course, the return water temperature can never fall below the room temperature. In reality, as the water flows through the radiator its temperature will not drop linearly but logarithmically. Accordingly, some form of log mean temperature difference (LMTD)[11] should be used instead, which is common in heat exchanger design. In addition, there is almost certainly some degree of mixing of the water within the radiator, particularly at low flow rates. For perfect mixing, the surface temperature of the radiator would be uniform over its entire area and would be equal to the return water temperature (a 'stirred tank' idealisation). It would seem reasonable therefore to assume that the mean surface temperature of the radiator was somewhere between the arithmetic mean of the water flow and return temperatures and the return water temperature. In the analysis which follows, we will look at the two extreme cases (designated 'no mixing' and 'perfect mixing'), and assume the truth lies somewhere between these two extremes. In the case of radiant panels and natural convectors there will be no mixing of the water streams as the water is contained within a small diameter serpentine pipe. However, a LMTD would still be more appropriate.

From an energy balance on the water-side (assuming steady-state):

$$Q = \dot{m}C_p(\theta_F - \theta_R) \tag{6.23}$$

where $\dot{m}$ is the mass flow rate of water (kg/s), C_p is the water specific heat (J/kgK), θ_F is the water flow temperature and θ_R is the return water temperature. This can be used to produce an alternative equation to 6.21 for the part load fraction:

$$\frac{Q}{Q_{design}} = \frac{\dot{m}(\theta_F - \theta_R)}{\dot{m}_{design}(\theta_{F,design} - \theta_{R,design})} \tag{6.24}$$

We can now examine the effect of changing either water mass flow rate, $\dot{m}$, or the flow water temperature, θ_F, the two principal methods of controlling heat output rate. In both methods the return water temperature (in equation 6.24) will change, so we need to eliminate θ_R from the equation. There are two possibilities, discussed above, regarding return water temperature. In the case of there being no mixing within the radiator, the mean surface temperature of the radiator can be taken as the arithmetic mean of the flow and return water temperature, in other words:

$$\theta_m = \frac{\theta_F + \theta_R}{2} \Rightarrow \theta_R = 2\theta_m - \theta_F \tag{6.25a}$$

In the case of perfect mixing then:

$$\theta_R = \theta_m \tag{6.25b}$$

For both cases, we can then substitute for θ_m using equation 6.22 to give:

$$\theta_R = 2\theta_i + 2(\theta_{m,design} - \theta_i)\left\{\frac{Q}{Q_{design}}\right\}^{\frac{1}{1.33}} - \theta_F \tag{6.26a}$$

$$\theta_R = \theta_i + (\theta_{m,design} - \theta_i)\left\{\frac{Q}{Q_{design}}\right\}^{\frac{1}{1.33}} \tag{6.26b}$$

These two equations can be substituted into equation 6.24 so as to eliminate θ_R. For the case of no mixing:

$$\frac{Q}{Q_{design}} = 2\left\{\frac{\dot{m}}{\dot{m}_{design}}\right\}\frac{\left[\theta_F - \theta_i - (\theta_{m,design} - \theta_i)\left\{\frac{Q}{Q_{design}}\right\}^{\frac{1}{1.33}}\right]}{(\theta_{F,design} - \theta_{R,design})}$$

and as $\theta_{m,design} = \dfrac{\theta_{F,design} + \theta_{R,design}}{2}$

$$\frac{Q}{Q_{design}} = 2\left\{\frac{\dot{m}}{\dot{m}_{design}}\right\}\frac{\left[\theta_F - \theta_i - \left(\frac{\theta_{F,design}}{2} + \frac{\theta_{R,design}}{2} - \theta_i\right)\left\{\frac{Q}{Q_{design}}\right\}^{\frac{1}{1.33}}\right]}{(\theta_{F,design} - \theta_{R,design})} \tag{6.27a}$$

A similar approach gives the equivalent equation for the case of perfect mixing:

$$\frac{Q}{Q_{design}} = \left\{\frac{\dot{m}}{\dot{m}_{design}}\right\}\frac{\left[\theta_F - \theta_i - (\theta_{R,design} - \theta_i)\left\{\frac{Q}{Q_{design}}\right\}^{\frac{1}{1.33}}\right]}{(\theta_{F,design} - \theta_{R,design})} \tag{6.27b}$$

Variable water flow rate, constant flow temperature

For variable flow rate control of the heat output from the radiator, the flow temperature is maintained constant, i.e. $\theta_F = \theta_{F,design}$. Equations 6.27a and 6.27b can then be solved for various fractional flow rates (m/m_{design}) to determine the fractional heat output rate (Q/Q_{design}). As these are implicit equations, they must be solved by a process of iteration.

Figure 6.14 (a) shows such a set of solutions, based on a design flow and return temperature of 80/70°C with a room temperature of 20°C. There is a significant deviation between the two extremes possible at around 50% of the design flow rate with the two extremes converging at 0% and 100% flow as would be expected. As suggested earlier, the actual profile lies somewhere between.

To get 40% of the design heat output we need only between about 5% and 10% of the design flow rate. The control response is highly non-linear. If the temperature difference between the design flow and return is increased, the response becomes more linear. The solutions have been reworked for 80/60°C and presented in Figure 6.14 (b), where it can be seen that to achieve the same 40% load, the fractional flow rate has increased to between about 10% and 20%. This implies that a larger design temperature difference between flow and return is to be preferred as control of the emitter is improved. The choice of flow and return temperatures will be discussed further in the next chapter.

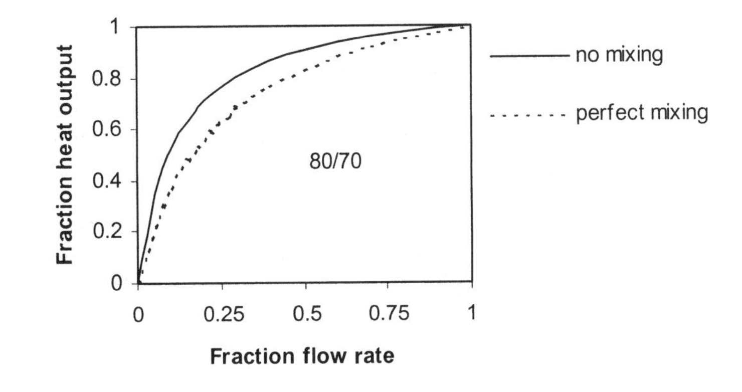

(a)

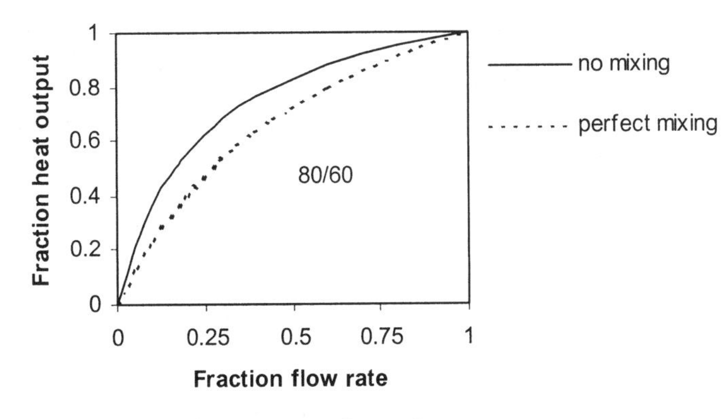

(b)

Figure 6.14 (a) and (b) Variation of heat output rate with mass flow rate.

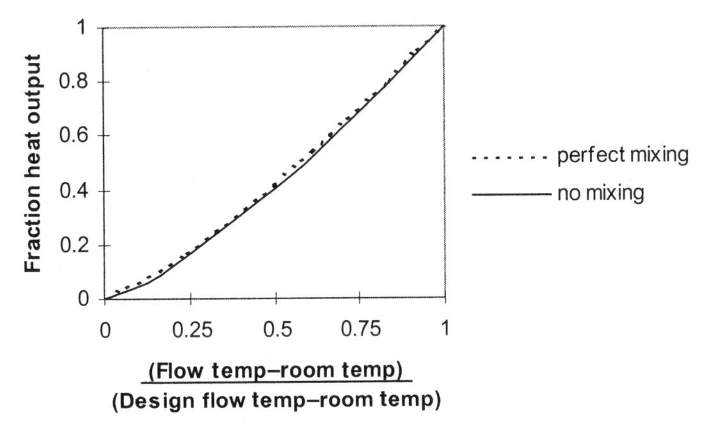

Figure 6.15 Variation of heat output rate with flow temperature.

Constant flow rate, variable flow temperature

In this case $(m/m_{design}) = 1$ and so disappears from equations 6.27a and 6.27b. These two equations can be solved (by iteration) for various values of flow water temperature as shown in Figure 6.15. This is based on a design flow and return temperature of 80/70°C. Note that to make the 'x' axis dimensionless and directly comparable with Figure 6.14, the flow temperature is represented as a fractional temperature difference using room temperature. The response is much more linear than for the flow rate control and there is negligible difference between the two extreme cases of perfect mixing and no mixing.

References

1. Recommended textbooks dealing with heat transfer theory: Kreith, F. (1976) *Principles of Heat Transfer*, 3rd edn. Harper International, New York. Incropera, F. & De Witt, D. (1996) *Fundamentals of Heat and Mass Transfer*. Wiley, New York.
2. Building Regulations (2002) Approved Document L, Conservation of Fuel and Power. The Stationery Office, London.
3. Chartered Institution of Building Services Engineers (CIBSE) (2002) Heating Systems, Section B1 of the CIBSE *Guide*. CIBSE, London.
4. Kreith, F. (1976) Chapter 11 in *Principles of Heat Transfer*, 3rd edn. Harper International, New York.
5. Chartered Institution of Building Services Engineers (CIBSE) (1999) Environmental Design, Section A1 of the CIBSE *Guide*. CIBSE, London.
6. British Standards Institution (BSI) (1995) BS EN ISO 7730: Moderate thermal environments – Determination of the PMV and PPD indices and specification of the conditions for thermal comfort. BSI, London.
7. Chartered Institution of Building Services Engineers (CIBSE) (1999) Environmental Design, Section A5 of the CIBSE *Guide*. CIBSE, London.
8. Building Regulations (2002) Approved Document J, Heat Producing Equipment.
9. Chartered Institution of Building Services Engineers (CIBSE) (1986) Heating Systems, Section B1 of the CIBSE *Guide*. CIBSE, London.
10. *ibid.*
11. Chartered Institution of Building Services Engineers (CIBSE) (1999) Environmental Design, Section A1 of the CIBSE *Guide*. CIBSE, London.
12. Kreith, F. (1976) Chapter 11 in *Principles of Heat Transfer*, 3rd edn. Harper International, New York.

7 Heating Circuits

This chapter deals with the design of the pipework circuits and the various items of equipment used to supply and regulate the heating water from the boiler(s) to the heat emitters, etc. We begin with a discussion of the selection of appropriate flow and return water temperatures and the impact this has on the design and performance of the system. This leads on to an analysis of insulation thickness. Various circuit arrangements are then considered, together with methods of control including variable volume pumping. The fundamentals of control and balancing are explained. The chapter then continues with a discussion of pipework expansion and the forces experienced by pipework.

7.1 Choice of flow and return water temperatures

The pipework has to carry sufficient hot water to meet the heating load at any instant in time. In practice, the circuits are designed for peak loads with this peak heat load related to the design mass flow rate of water and design flow and return water temperatures by the equation:

$$Q = \dot{m}C_{pw}(\theta_F - \theta_R) \qquad (7.1)$$

where Q is the peak load on the circuit, $\dot{m}$ is the mass flow rate of water in the circuit, C_{pw} is the specific heat of water and θ_F and θ_R are the design flow and return water temperatures. The usual procedure is to choose θ_F and θ_R and then determine $\dot{m}$. The choice of the flow and return temperature will be determined by the following:

(1) Safety – the surface temperature of the exposed pipework and heat emitters must not constitute a hazard to occupants – see Chapter 6. Also, HWS needs to be generated at sufficient temperatures to prevent growth of Legionella bacteria – see Chapter 8.
(2) Comfort – surface temperatures of floors must be limited to 25° or 28°C depending on application. Emitter surface temperature will also affect radiant and air temperature – see Chapter 6.
(3) Structural – embedded pipework or panels must not damage the structure.
(4) Boiler type – condensing boilers require special consideration of return water temperatures to promote condensation. Conversely, non-condensing boilers need to be protected from excessively low return water temperature – see Chapter 3.

Except where items 1 to 4 demand lower flow and return temperatures, it is all too common in the UK for 'standard' values to be used – invariably 82/71°C for LPHW systems. Yet these values are no more than a direct conversion from the 180/160°F preferred in the mid twentieth century. The following are all strongly influenced by choice of flow and return temperatures and have an economic impact which should be considered at the design stage:

(5) Emitter sizes – the heat output rate of emitters is a function of the mean surface temperature so that as θ_F and θ_R are increased so emitter size reduces – see Chapter 6.
(6) Pipe sizes – the greater the difference *between* θ_F and θ_R the smaller the mass flow rate of water required and so the smaller the pipework – this is discussed later in this chapter.
(7) Boiler control – the difference *between* θ_F and θ_R (differential) has an impact on the control of boilers, particularly with sequence control of multiple boilers – see Chapter 10.
(8) Efficiency – the higher the values of both θ_F and θ_R the greater the heat loss rate from the pipework.

This last aspect is heavily dependent on the performance of the insulation and the location of the pipework as discussed below.

7.2 Insulation of pipework

Insulation of hot water pipework to a minimum standard is required by the Building Regulations[1] in order to reduce heat loss and consequential CO_2 emissions from fossil fuels. Strictly, the current regulations exempt pipework which will provide useful heating to the surrounding space at all times. However, such situations would be rare. In any event, there are sound financial reasons to insulate hot water pipework.

The heat loss rate per unit length of insulated pipe is given by the equation:

$$\frac{Q}{L} = \pi D_o U(\bar{\theta}_w - \theta_a) \tag{7.2}$$

where D_o is the outside diameter of the insulation, $\bar{\theta}_w$ is the mean temperature of the water in the pipe and θ_a is ambient temperature. The U value is similar to that used for a plane wall, but due to the geometry involves logarithmic function of the insulation thickness and pipe diameter. It also depends to some extent on the external surface temperature of the insulation. This is explained in more detail in Appendix 7.1.

Figure 7.1 shows how the heat loss rate varies with insulation thickness for various nominal pipe sizes. The heat loss rate (Q/L) is given per metre

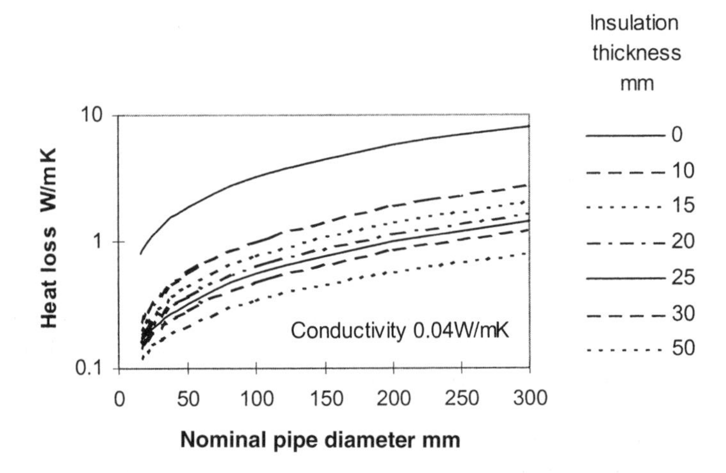

Figure 7.1 Heat loss rate and insulation thickness.

length of pipe per °C difference in water–ambient temperature difference. A logarithmic scale has been used for Q/L due to the large range involved. It can be seen that an uninsulated 300 mm diameter pipe loses about 8 Watts for every °C difference in temperature per metre length. So a 20 m section carrying water at 75°C with an ambient temperature of 20°C would lose (8 × 20 × (75 – 20) W) 8.8 kW of heat. Were it insulated with only 25 mm of insulation the heat loss rate would be reduced to about 1.4 W/m°C (an 84% reduction). Using twice the thickness of insulation would reduce the heat loss rate to about 0.8 W/m°C, not quite halving the loss. The law of diminishing returns applies.

The values in Figure 7.1 were calculated using the rather complex heat transfer equations described in Appendix 7.1. Using this data with curve fitting methods has produced a reasonable approximation which is much faster and easier to use:

$$\frac{Q}{L} = \left[0.0613\frac{D_i}{X^{0.82}} + 0.1\right](\bar{\theta}_w - \theta_a) \qquad (7.3)$$

where X is the insulation thickness and D_i the inside diameter of the pipe, both in mm. As with Figure 7.1, this only applies to insulation with a conductivity of 0.04 W/mK. This cannot be used to determine heat loss rate for uninsulated pipes and applies only to insulation more than 25 mm thick.

Alternatively, tables for both insulated and uninsulated pipes are given in CIBSE[2].

7.2.1 Economic insulation thickness

The question arises as to what thickness of insulation should be used. Obviously, thicker insulation reduces heat loss rate – although at a diminishing rate – and so reduces annual fuel costs (assuming that the heat lost from the pipe is not usefully used to heat the building). At the same time installation costs increase. Is there an economic optimum thickness which results in a lowest overall cost over some agreed time period?

The installation cost will depend on thickness and type of insulation and diameter of pipe. Annual fuel costs due to heat loss will depend on fuel tariff, boiler efficiency, hours of operation, hot water temperature, ambient temperature, insulation thickness and conductivity, and pipe diameter. But the annual fuel costs need to be converted into an equivalent cost based on some definite time period (e.g. 5 years). To do this, the Present Value concept is used which, in addition to a specified time period, needs to take into account interest rate and fuel inflation rate. Equations used are set out in Appendix 7.1.

BRECSU has produced tables of optimum economic thickness for hot water pipework insulation for a wide range of conditions[3]. However, given the large number of variables involved, it is reasonable to ask how relevant these figures are to a particular project.

Figure 7.2 has been produced for a 50 mm pipe for the conditions shown. It is clear that any thickness of insulation between about 15 mm and 40 mm would give close to the minimum total costs. Further, when some of the factors are varied, although the absolute value of total costs changes, there is little difference in the shape of the curve, suggesting that the optimum is not particularly sensitive to most factors. The greatest effect is due to operating time per year and the number of years. The installation cost had little impact provided a period of at least three years was used.

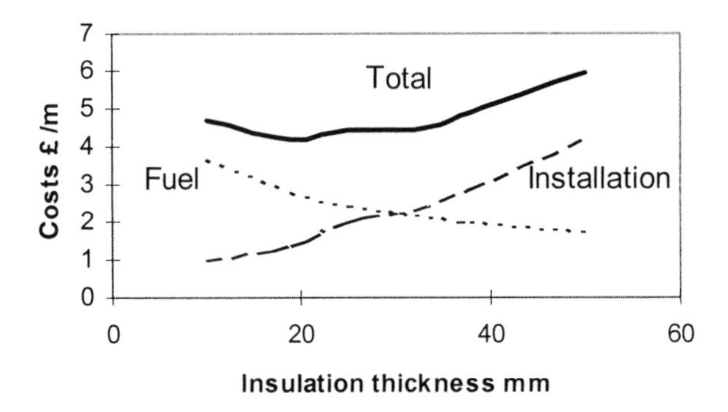

Figure 7.2 Heat loss rate from insulated and uninsulated pipes.

Table 7.1 Criteria used in Figure 7.2.

Time period	10 years
Fuel costs	1.5 p/kWh
Boiler seasonal efficiency	70%
Hours operation per year	1000
$\theta_w - \theta_a$	55 K
Interest rate	10%
Fuel inflation rate	3%
Installation costs for insulation	£1/m for 10 mm thickness £4.2/m for 50 mm thickness

7.2.2 Environmental considerations

It is relatively easy to estimate the annual CO_2 emissions attributable to heat loss from pipework. This is given by the equation:

$$M_{co_2} = \left[\frac{Q}{L}\right](\bar{\theta}_w - \theta_a)\frac{tm}{\zeta_{bs}} \times 10^{-3} \tag{7.4}$$

where t is the operating time per year (hours) and m is the mass of CO_2 produced per kWh of fuel consumed giving M_{co_2} in kg of CO_2 per metre length of pipe per year. Values of m are given in Table 7.2. ξ_{bs} is the seasonal boiler efficiency.

Given the lack of sensitivity in the total cost to the thickness of insulation used, it would appear to be an opportunity for heating design engineers to make substantial reductions in CO_2 emissions by choosing to install more rather than less insulation, i.e. working to the right-hand side of the total cost curve.

7.2.3 Other considerations

The above analysis is based on the assumption that the heat lost from the pipework does not contribute towards the useful heating of the building. Clearly in many cases it will to some extent (though perhaps not at all times as required by the Building Regulations). Even so, unlike heat emitters, there is no control over the heat output from pipework, which could lead to overheating in some instances.

Table 7.2 Carbon Dioxide Emissions for a range of Fuels used for Heating.

fuel	CO_2 emissions kg/kWh (useful heat output)
Natural gas	0.18
LPG (butane)	0.26
Gas oil	0.29
Light fuel oil	0.30
Electricity	0.46 (average)

Boiler duty includes an allowance for pipe heat loss rate: a figure of 10% appears to be common. Economically optimised thicknesses of insulation (or minimum thicknesses as specified by the building regulations) are unlikely to lead to such high percentages. Although these need to be established for each case, they are unlikely to exceed 2% or 3%.

As discussed in section 7.1, the choice of flow and return temperatures will have an effect on size of pipes required. The mass flow rate is directly proportional to the difference between flow and return temperature. Ignoring any changes in water density (which would be negligible) we can say that the volume flow rate required is also directly proportional to this temperature difference. So, if the temperature difference was doubled, for the same water velocity, the pipe cross-sectional area needed would be halved. As cross-sectional area is proportional to the square of the internal diameter, then the new internal diameter required would be reduced by a factor of $2^{0.5}$, or $D_{new} = 0.707D_{old}$. Pipes come in standard sizes: for pipes smaller than about 40 mm, the above would normally only allow a reduction of one size, whereas above 40 mm it would be more likely that the pipe could be reduced by two sizes.

Reduced flow rates also result in reduced running costs for the pumps. The pump power is proportional to the product of the volume flow rate and pressure drop through the circuit. Whilst the flow rate is proportional to the flow/return temperature difference, the pressure drop may increase or decrease slightly as pipe sizes are reduced.

7.3 Pipework arrangements

The following discussion is concerned only with two-pipe flow and return systems for LPHW. In the past one-pipe systems have been employed but the lack of balancing and control has made this method obsolete.

The complete heating circuit can be thought of as comprising two principal circuits: the primary and the secondary. The primary circuit contains the boiler(s) and the secondary circuit distributes the hot water throughout the building. This arrangement allows the primary to function, to some extent, independently of the secondary. In particular, the flow rate through the secondary may vary, but maintaining a constant flow rate through the primary

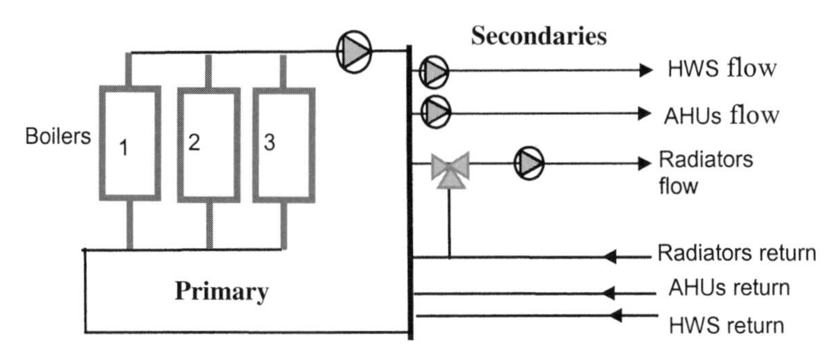

Figure 7.3 Primary/secondary pipework arrangement.

is often preferred. This aspect is discussed in detail in Chapter 10. The present chapter will be concerned with the secondary circuit which supplies the various heat emitters and heating coils.

7.3.1 Secondary circuit

The secondary circuit is itself often broken down into a number of independently operated parallel circuits. In a building containing a wide range of heating systems there may be a separate circuit for each of the following:

- Radiators
- Underfloor heating
- Air handling units
- Terminal re-heat coils
- Hot water services.

Further, the radiator circuit might be broken down into two or more circuits.

There are several reasons for this. Some circuits will be required to operate at different times of the day or year. For example, the underfloor heating will probably run 24 hours/day in winter whilst the radiator system will run intermittently. Hot water services and terminal re-heat will be required all year round. Some circuits will have different flow temperature requirements, for example the underfloor heating will operate at about 45°C compared to 80°C for the others. Some circuits will have specific control requirements. The load on the radiators will vary considerably with outdoor temperature and will be better controlled if the flow temperature is reduced in warmer weather (outdoor temperature compensation – see Chapters 6 and 10). Other systems will require constant flow temperature. Finally, there may be advantages in zoning the building to allow independent operation, for example north and south wings or landlord's and tenant's services.

Both the primary circuit and each secondary circuit now require a dedicated circulation pump. A typical arrangement is shown in Figure 7.3. In practice, rather than use a single pump, a duty standby pump set would be used.

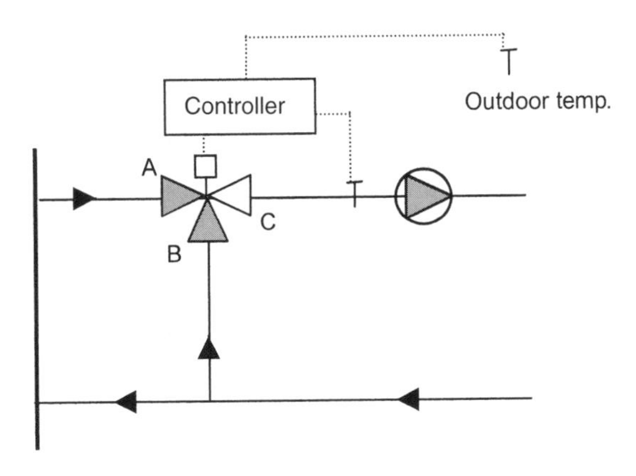

Figure 7.4 Flow temperature control.

7.3.2 Compensated temperature circuit

As discussed in Chapter 6, radiators, radiant panels and natural convectors are best controlled using zone controls combined with flow temperature regulation. The flow temperature is modulated downwards as the outdoor temperature rises above the design condition, leaving the zone control valve to respond to changing incidental gains in the particular zone. This allows a relatively inexpensive zone control valve, such as a thermostatic radiator valve, to be used. Figure 7.4 shows how flow temperature control is achieved. A mixing valve is installed upstream of the circulation pump, allowing cooler return water to mix with the constant temperature flow from the primary circuit. The valve contains three ports. Ports A and B are designed so that as A opens B closes in such a manner as to maintain a constant flow rate. Port C remains fully open at all times. The instruction to modulate A and B comes from the controller. This controller contains the function relating outdoor temperature to required flow temperature (known as the *schedule*) and is provided with two input signals: the current values of outdoor and flow temperature. The controller calculates the required flow temperature from the outdoor temperature input signal and compares this with the actual flow temperature. It then sends a signal to the valve actuator to correct any error. The flow temperature sensor provides a feedback signal. If the primary flow and secondary return temperatures were always constant, then this sensor would not be required. However, in practice the secondary return water temperature will vary with the heating demand on the circuit; the primary flow temperature will vary with total system load depending on how the boiler sequencing is controlled (see Chapter 10). In order for the mixing valve to be able to achieve acceptable control, it is important that the valve is correctly sized and has the appropriate characteristics. This is discussed in section 7.5.

7.4 Variable volume flow

Any form of two-port (*throttling*) zone control valves, including TRVs, will result in a variation in flow rate around the radiator circuit despite the use

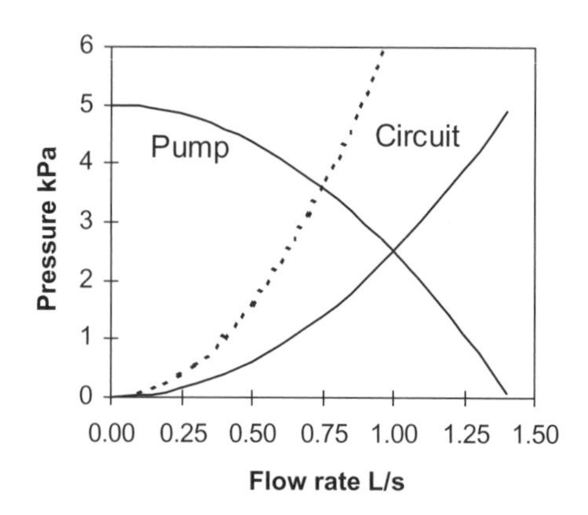

Figure 7.5 Idealised pump and circuit curves.

of a dedicated circulation pump and temperature compensation. It is the zone control valve which must respond to incidental gains. As was seen in Chapter 6, to reduce the heat output of a radiator by even a small percentage requires a considerable reduction in flow rate. As virtually every zone will have some incidental gains (lighting, people, equipment, solar), which are not taken into account when sizing the heat emitters, there will always be appreciable flow rate variation. The impact of this is most easily understood by reference to the pump characteristic curve as shown in Figure 7.5. A mathematical treatment is given in Appendix 7.2.

Any type of pump will have a pressure/flow rate curve similar to the idealised curve shown. The pump will produce maximum pressure at no flow and minimum pressure at maximum flow rate. The idealised curve can be represented by the equation:

$$P = A - B\dot{V}^2 \tag{7.5}$$

where P is the pressure produced by the pump, $\dot{V}$ is the flow through the pump and A and B are constants for a particular pump. In this example, A = 5 (kPa) and B = 2.5 (kPas^2/L^2). Note the units of B.

The pressure drop through the pipework circuit can be expressed by a similar quadratic equation. (This is, strictly speaking, only true when the flow is sufficiently turbulent (fully-rough) that the friction factor is no longer a function of the Reynolds Number but only of the relative roughness of the pipe wall. Water flow in heating systems closely approaches this ideal.)

$$\Delta P = R\dot{V}^2 \tag{7.6}$$

where ΔP is the pressure drop through the circuit due to the flow rate $\dot{V}$, and R is a measure of the resistance to flow offered by the circuit. The value of R will vary depending on the degree of opening of the zone control valves. As the valves close, so R will increase. The diagram shows two circuit curves. The lower curve (full line) has a value R = 2.5 representing the circuit with all valves fully open.

The pump must generate a pressure which exactly balances the pressure drop through the circuit under all conditions. As the flow rate through both the pump and the circuit must also be the same, the operating point of the pump must be where the pump and circuit curves cross. For the case where all valves are open this occurs at a flow rate of 1.0 L/s and a pressure (or pressure drop) of 2.5 kPa. If the circuit now required a flow rate of only 0.75 L/s, then the valves would have to partially close until the 'R' value increased as shown by the upper, dotted, circuit curve. There are two possible disadvantages to this. Firstly, as can be seen from the graph, the pressure developed by the pump increases as the flow rate reduces, in this case to about 3.6 kPa. Some valves, particularly those close to the pump, may be exposed to an excessive pressure and may not be able to close, with subsequent loss of control, and water velocities through these valves can be very high, leading to noise and erosion. Secondly, the pump is wasting energy by producing more pressure than is required.

One (partial) solution is to install a bypass in the circuit. This consists of a pipe joining the flow and return pipework, which includes a pressure-actuated two-port valve. As the pressure across the valve increases, so the valve opens to try to maintain a near constant flow rate, in effect cancelling out the increase in resistance value R arising from the closing of the zone control valves. The pump pressure developed is then maintained at close to design value. For large circuits it is preferable to include a bypass in each main branch in order to limit the local pressure rise. The bypass system will improve control and prevent excessive water velocities through valves. However, this is still wasteful of energy: the pump is delivering a greater flow rate than is required. Heating circuits must be sized to meet the design maximum demand. This design demand will only occur for a fraction of the total operating time, mostly at start-up in the mornings. Even on a cold day there will still be incidental gains. Variable volume pumping is required to capitalise on this potential saving in pump running cost and offers improved control by preventing overpressure.

7.4.1 Variable volume pumping

The input power required to a pump motor is given by:

$$\dot{W} = \frac{P\dot{V}}{\zeta} \tag{7.7}$$

where ζ is the overall efficiency of the pump and motor. When P is in kPa and V in m^3/s then W is in kW. For the moment we will assume that the efficiency of the pump is constant. As the required flow rate reduces, so the pressure drop through the circuit reduces if R can be held constant and as a result the pump power reduces. Using the suffix '1' to represent the full flow condition and '2' the reduced flow for condition, for equations 7.6 and 7.7 we have:

$$\Delta P_1 = R\dot{V}_1^2 \quad \text{and} \quad \Delta P_2 = R\dot{V}_2^2 \quad \text{then} \quad \Delta P_2 = \Delta P_1\left\{\frac{\dot{V}_2}{\dot{V}_1}\right\}^2$$

$$\dot{W}_2 = \frac{\Delta P_2\dot{V}_2}{\zeta} = \frac{\Delta P_1\left\{\frac{\dot{V}_2}{\dot{V}_1}\right\}^2\dot{V}_2}{\zeta} = \frac{\Delta P_1\dot{V}_1}{\zeta}\left\{\frac{\dot{V}_2}{\dot{V}_1}\right\}^3 = \dot{W}_1\left\{\frac{\dot{V}_2}{\dot{V}_1}\right\}^3 \tag{7.8}$$

So, if the flow rate can be reduced without increasing R, pressure drop will fall with the square of the flow rate, and the pump power with the cube of the flow rate. For example, if the required flow rate is only half that at full load, then the pressure drop through the circuit is only one quarter of the pressure drop at full flow rate, and the input power is only one eighth of the full flow input power. This is a potentially very large saving in running cost if the pump can be controlled to match the circuit requirements without the need to increase the system resistance.

7.4.2 Variable speed drives

The speed at which a pump turns is related to the volume flow rate and pressure developed by the following equations, known as the fan/pump laws (being identical for both fans and pumps):

$$\frac{\dot{V}_2}{\dot{V}_1} = \frac{N_2}{N_1} \quad \frac{P_2}{P_1} = \left\{\frac{N_2}{N_1}\right\}^2 \quad \frac{\dot{W}_2}{\dot{W}_1} = \left\{\frac{N_2}{N_1}\right\}^3 \tag{7.9}$$

where N is the rotational speed. As with equation 7.8, suffixes '1' and '2' represent different operating conditions. Again, the power equation assumes that the efficiency remains constant. Thus, pump speed control allows the pump to exactly (in theory) match the change in pressure required in the circuit for a change in volume flow rate: half the design volume flow rate is achieved by running the pump at half the design speed, resulting in one quarter of the pressure being developed (and only one eighth of the power consumed). The pump curve of Figure 7.6 show the pressure-flow rate characteristic at a number of speeds for the same pump used in Figure 7.5. If the full speed characteristic of the pump is given by equation 7.5, then it can be shown (see Appendix 7.2) that the characteristic at any fractional speed N (i.e. $0 < N < 1$) is given by:

$$P_N = AN^2 + B\dot{V}^2 \tag{7.10}$$

In the past, a number of methods of speed control have been used. These included gears and hydraulic and magnetically coupled clutches. These were relatively expensive, bulky and inefficient due to mechanical losses. The modern solution is to use a frequency inverter on the electrical supply to the motor. Pump motors, as with most motors in (non-domestic) building services, are often 3-phase induction. The speed at which these motors rotate is directly proportional to the frequency of the electrical supply. This is

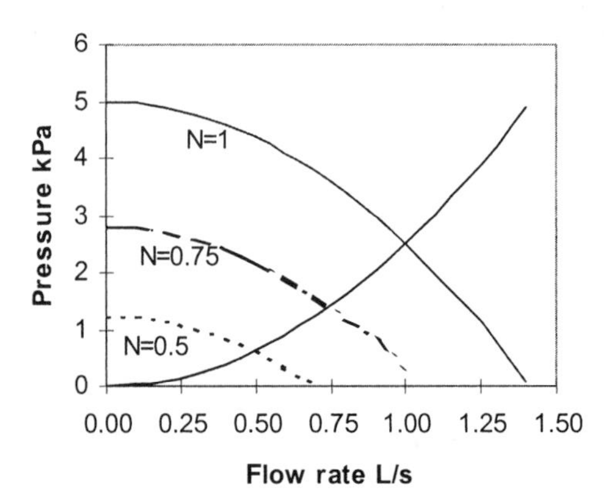

Figure 7.6 Variable speed pump curves.

easily and inexpensively varied using a solid state device with very little energy losses. Some pump manufacturers now supply their pumps equipped with built-in frequency inverters.

Equation 7.8 gives the theoretical power saving assuming that the pump efficiency remains constant. Pump efficiency will vary with rotational speed as will the efficiency of the electric motor. The change in efficiency of the pump with speed will depend on the design of the pump and its suitability to the design flow rate and pressure drop. The manufacturer should be consulted. The motor efficiency, and power factor, will also reduce as the load is reduced. Modern developments in inverter technology have resulted in this being minimised. A best practice publication[4] suggests that for a typical LPHW heating system in an office building, pump energy consumption could be reduced by two-thirds using variable volume speed control.

7.4.3 Control of variable speed pumps

The aim of pump speed control is to minimise pump energy consumption whilst ensuring that all zones are provided with sufficient heating water at all times. Unfortunately, as the heating demand of each zone varies independently and in an unpredictable manner, the only way of ensuring adequate water flow to each zone is to design the system such that the design flow rate is available at all times to each zone. This means overprovision and so wasted pump energy with zone valves having to increase system resistance unnecessarily. Alternatively, a compromise between energy savings and heating provision can be considered, as discussed below.

The conventional method of modulating pump speed in response to changing heating demand involves sensing static pressure in the circuit. This is a simple and as a consequence, inexpensive, method of control. As the heating demand falls within a controlled zone, so the two-port valve serving the emitter(s) in that zone will start to throttle back, causing a drop in flow rate. This fall in flow rate will result in an increase in static pressure. The control system responds to an increase in static pressure by reducing pump speed. The static pressure varies throughout the circuit and with the pump speed and zone valve positions. The question arises therefore where to locate the pressure sensor. (A similar problem arises in the control of variable air volume air conditioning systems[5].) This requires an analysis of the system hydraulic behaviour under various load conditions.

Figure 7.7 shows a simple heating circuit with three identical heat emitters, each controlled by a two-port valve. The design water flow rate of each

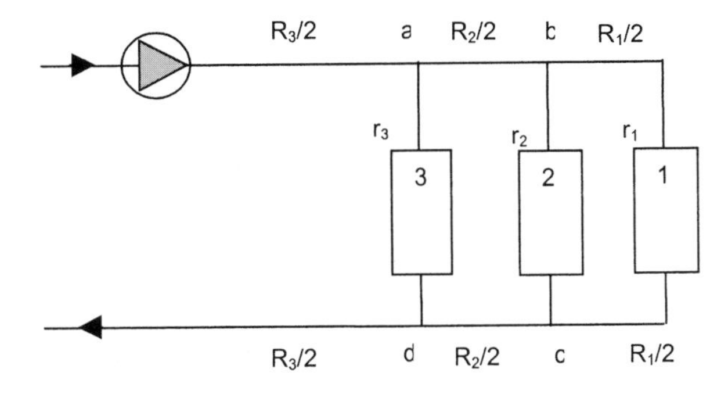

Figure 7.7 Simple circuit with three heat emitters.

emitter is the same as is the design pressure drop. Assume that emitter 1 forms the index circuit. When the system is balanced, additional resistance will need to be added to branches 2 and 3 in order to achieve equal flow rates through each emitter when the zone valves are all fully open. In Figure 7.7 the resistance of each section is represented by letters $r_{1,2,3}$ and $R_{1,2,3}$. $r_{1,2,3}$ represent the combined resistance of emitter, zone valve, regulating valve and pipework between the flow and return pipe when the system has been balanced and all zone valves are fully open. When the zone valves start to close, these resistances will increase. $R_{1,2,3}$ represent the fixed resistance of the sections of pipework connecting the branches. These values remain constant at all times. The relative values of r_2 and r_3 needed to balance the system are easily found by equating the pressure drop at design flow rate. The pressure drop between points B and C can be expressed using equation 7.6 for branches 2 and 3:

$$\begin{aligned}&\Delta P_{bc} = r_2\dot{V}_2^2 = (R_1 + r_1)\dot{V}_1^2\\&\text{And, as } \dot{V}_2 = \dot{V}_1\\&r_2 = R_1 + r_1\end{aligned} \tag{7.11}$$

In other words, if the flow rates through each identical branch are to be equal, then the regulating valve on branch 2 must be partially closed to provide an additional resistance equal to R_1. By considering the pressure drop between A and D, a similar method leads to:

$$\begin{aligned}&r_3\dot{V}_3^2 = R_2(\dot{V}_1 + \dot{V}_2)^2 + r_2\dot{V}_2^2\\&\text{And, as } \dot{V}_1 = \dot{V}_2 = \dot{V}_3\\&r_3 = 4R_2 + r_2 = 4R_2 + R_1 + r_1\end{aligned} \tag{7.12}$$

This shows that the added resistance needed to balance the branch increases quickly moving towards the pump.

The total pressure drop around the circuit, and the pressure the pump must develop, at design flow rate will be given by:

$$\begin{aligned}P = \Delta P &= R_3(\dot{V}_1 + \dot{V}_2 + \dot{V}_3)^2 + R_2(\dot{V}_1 + \dot{V}_2)^2 + (R_1 + r_1)\dot{V}_1^2\\&= (9R_3 + 4R_2 + R_1 + r_1)\dot{V}_1^2\end{aligned} \tag{7.13}$$

and the pump power (equation 7.7) by:

$$\dot{W} = \frac{P(\dot{V}_1 + \dot{V}_2 + \dot{V}_3)}{\zeta} = \frac{3(9R_1 + 4R_2 + R_1 + r_1)\dot{V}_1^3}{\zeta} \tag{7.14}$$

We now look at two possible positions for measuring and controlling pump speed by static pressure. To simplify the comparisons, it would be more convenient to simplify the above equations by putting $R_1 = R_2 = R_3 = r_1$ giving at design flow rate:

$$P = 15r_1\dot{V}_1^2, \quad \dot{W} = \frac{45\dot{V}_1^3}{\zeta} \tag{7.15}$$

Across the pump

The differential static pressure needs to be set at a value given by equation 7.13. As zonal demand varies, so the pump speed is adjusted to maintain this

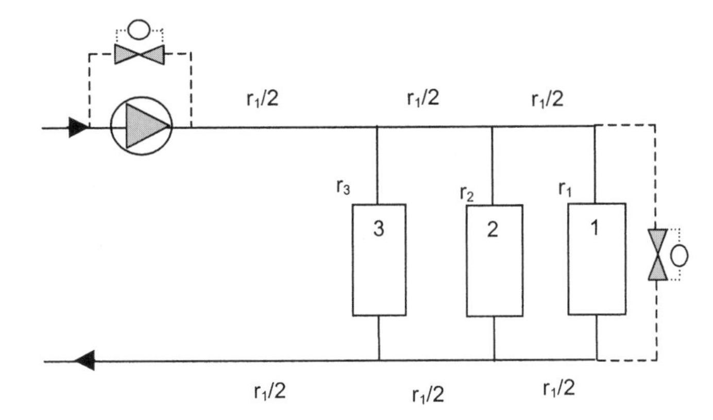

Figure 7.8 Differential pressure sensor location.

pressure across the pump. If heat demand in zone one was to fall to zero while zones 2 and 3 remain at full load, the pump speed would be reduced to give this lower flow rate but at the same pressure. Assuming no change in pump efficiency, the pump output power would be given by:

$$\dot{W} = \frac{P\dot{V}}{\zeta} = \frac{15r_1\dot{V}_1^2 \cdot 2\dot{V}_1^2}{\zeta} = \frac{30r_1\dot{V}_1^3}{\zeta} \tag{7.16}$$

i.e. the pump power is reducing in proportion to the speed, not the speed cubed.

The pressure difference across branch 3 will be given by:

$$P - r_1 \cdot 2\dot{V}_1^2 = 9r_1\dot{V}_1^2 \tag{7.17}$$

an increase of 50% above the value at full circuit design flow rate. Thus zone 3 will have more than sufficient pressure available to provide design flow rate, and the zone control valve will be partly closed.

At end of circuit

Going to the other extreme, the differential static pressure could be sensed across the index branch, 1. The set point for the sensor would now be the design pressure drop for this part of the circuit, i.e. $\Delta P = r_1\dot{V}_1^2$. The design flow rate through branch 1 is therefore always guaranteed. Under the same conditions as before, with zone 1 having zero demand and zones 2 and 3 requiring full load, we can examine how the pump behaves. In theory, it is delivering two-thirds of the design flow rate. The pressure developed will equal the pressure drop around the circuit given by:

$$P = \Delta P = R_3(\dot{V}_2 + \dot{V}_3)^2 + (R_2 + r_2)\dot{V}_2^2 = 4r_1\dot{V}_1^2 + (r_1 + r_2)\dot{V}_1^2$$

But, the pressure drop across r_2 is that of the senor equal to $r_1\dot{V}_1^2$ giving:

$$P = 6r_1\dot{V}_1^2 \quad \text{so that pump power, } \dot{W} = \frac{6r_1\dot{V}_1^2 \cdot 2\dot{V}_1}{\zeta} = \frac{12r_1\dot{V}_1^3}{\zeta} \tag{7.18}$$

This is a considerable saving compared to controlling with the sensor located across the pump. At full design load the pump power is given by equation

7.15, a saving of 73% at two-thirds flow rate. Unfortunately, the pump fails to deliver the required flow rate to zones 2 and 3.

With no flow in branch 1, the pressure drop across r_2 will be the same as the set point for the sensor. But during balancing, r_2 was increased to allow for the resistance R_1, (equation 7.11), in fact it was doubled. As a result, zone two, with its control valve fully open, would receive only about 70% of the required flow rate. Zone 3 would also suffer from lack of heating water. In fact, equation 7.18 overestimates pump power as the required flow rate is not achieved.

This analysis appears to suggest that the designer must install the differential static pressure sensor across the pump to ensure that all zones receive sufficient heating water at all times, in the process sacrificing pump energy savings. The situation is not as drastic as it at first seems, however. As was seen in Chapter 6, the heat output of a heat emitter, such as a radiator, falls only slightly with a large reduction in water flow rate. In addition, it is unlikely that one zone will be at zero load whilst others are at full load. Of course, the sensor could be placed at some intermediate point between pump and end of index circuit. This will be a compromise between pump energy savings and performance. The designer should consider the effect of locating the sensor at various distances from the pump and assess the resulting risk of underheating, balancing risk against potential energy savings. This is a considerable design task and one that many designers would not have the time or resources to carry out. Some assistance is available in a BSRIA software package.[6]

If the sensor is located at the end of the index circuit, the set point could be set higher than the design pressure drop across it. For large circuits, for example circuits serving several floors, it may be advisable to use several differential pressure sensors, one per major branch. The pump would then be controlled from the worst case, i.e. any sensor recording a differential pressure less than the set point for that particular sensor. Some manufacturers have developed control systems to try to overcome the above problems related to achieving good energy savings whilst providing acceptable system performance.

The above does not make a particularly strong case for using variable speed pumps. The reality appears to be far removed from the apparent simplicity and potential savings indicated by equation 7.9. There are, however, further good reasons why variable speed pumps should be considered.

When selecting a pump for a particular circuit, the pump will invariably be oversized. The design pressure loss through the circuit has probably been overestimated (to be on the 'safe side'), and pumps are only available in certain sizes. At commissioning, the circuit will be balanced and then finally regulated to give the design flow rate. For a fixed speed pump this is usually done by partially closing the pump discharge valve. With a variable speed pump, this is easily – and efficiently – achieved by reducing the maximum speed. Reducing the flow rate by only 10% by this method results in a 33% saving. In comparison, throttling at the discharge valve will save only a few %.

Two-port control valves will not operate correctly when subjected to excessive differential static pressures with resulting overheating of the space. The manufacturer will specify this maximum differential pressure. This can be quite low for the small, and inexpensive, valves used commonly on heating systems, particularly TRVs. The maximum differential pressures can easily be exceeded as valves start to close throughout the circuit. Pump speed control helps to reduce the risk of this happening, particularly if the sensor(s) is located well away from the pump.

As will be seen later, control valves need to provide a reasonably high pressure drop even when fully open in order to have sufficient control authority. This has substantial pump running cost implication. In constant speed pumping, the pressure drop across an open two-port valve needs to be equal to about 40% of that of the index circuit. With variable speed pumping, this can be relaxed somewhat.

Finally, inverter driven variable speed pumps remove the need for starting current control circuitry such as star-delta or autotransformer. The pump can be started at low speed and gradually speeded up to give a 'soft' start. This facility is usually provided as standard by the inverter manufacturer.

7.5 Control valves

Control valves are used to vary either the flow rate or the temperature of water to a circuit or individual heat emitter. They do this in response to a control signal from a controller. The controller and valve must be carefully selected if good control is to be achieved. By good control, we usually mean that the heating system will maintain the space temperature (or, for example, water temperature in a storage cylinder) within the specified limits but without rapid fluctuations. This section deals with the requirements of control valves to achieve this objective. Controllers have additional requirements which are outside the scope of this book. There are innumerable text books on control theory, a few of which are directed specifically at heating systems or building services[7,8].

Although control valves may be used to control flow temperature, they do so indirectly by regulating water flow rate, for example the mixing valve used in temperature compensation (see section 6.7.2). To achieve good control, the implication for all types of control valves is that they are able to adjust water flow rate over the entire range from zero to design flow rate. Further, the signal from the controller should result in a reasonable linear change in the controlled property – room temperature – rather than the water flow rate. The control signal is used to raise or lower the valve spindle; there is no actual measurement of flow rate (feedback) so the effect of the spindle movement on the controlled item is important.

For example, in a temperature compensated circuit, it is the water temperature which is the controlled item. The water temperature leaving the three-port control (mixing) valve is given by the simple equation:

$$\theta_{w,out} = \frac{\dot{m}_F\theta_F + \dot{m}_R\theta_R}{(\dot{m}_F + \dot{m}_R)} \tag{7.19}$$

where subscripts F and R represent the flow and return water entering the valve. Assuming that the flow and return temperatures remain constant, a linear relationship between valve spindle movement and outlet water temperature requires a linear relationship between valve spindle position and the flow rate of water entering the flow (and return) port, as shown in Figure 7.9. The relationship between water flow rate and valve spindle position is an important property of the valve and is referred to as its 'characteristic'.

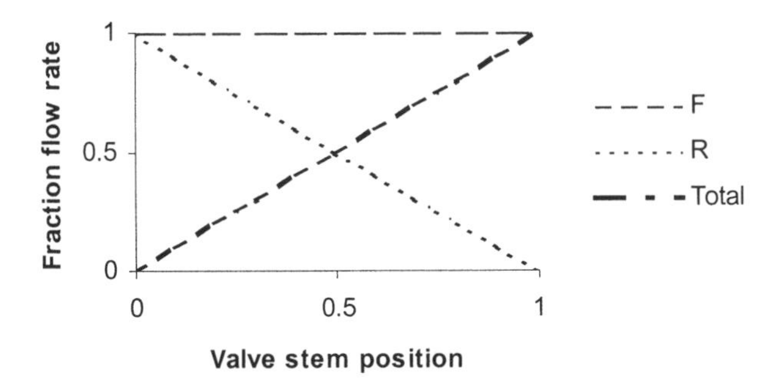

Figure 7.9 Three-port mixing valve ideal response.

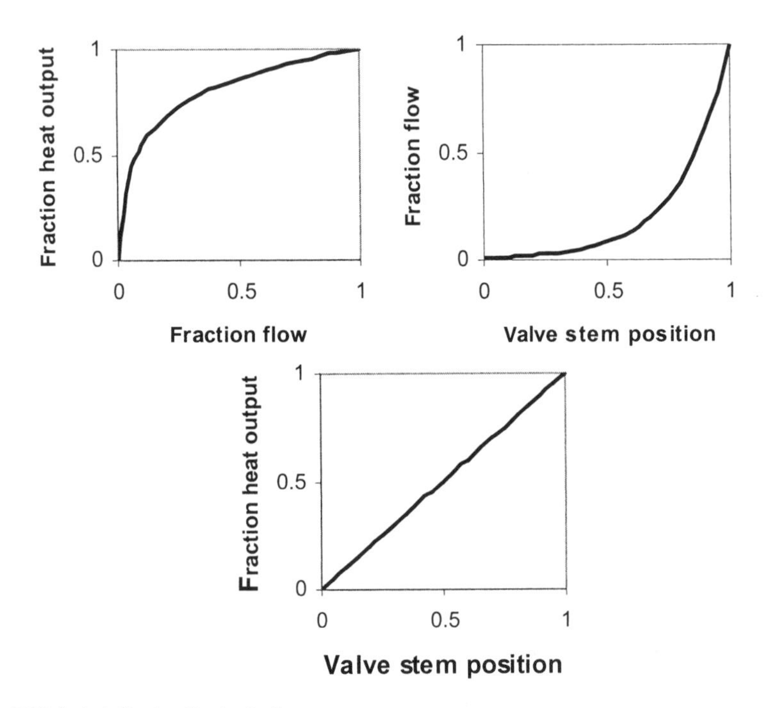

Figure 7.10 Control of heat emitter to give linear response.

In contrast, a two-port valve used to control heat output from a heat emitter would not benefit from a linear characteristic, as can be seen in Figure 7.10. As was seen in Chapter 6, the heat output rate of a heat emitter varies logarithmically with water flow rate. So to achieve a linear relationship between valve spindle position and heat output rate, an inverse characteristic to the emitter is required.

However, the flow rate through a valve depends not only on valve spindle position but also on pressure drop across it. As was seen in the previous section, as a two-port control valve begins to close, so water flow rate through it and the circuit reduces. This results in a rise in pump static pressure. The resistance of the circuit, excluding the control valve, is constant, so as the flow rate through it reduces, the pressure drop through it reduces rapidly, with the square of the flow rate. The pressure drop across the valve there-

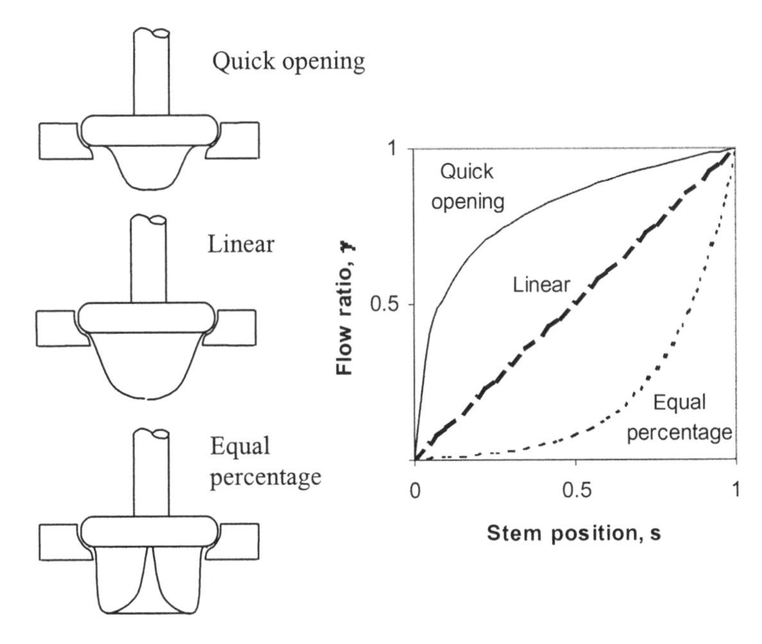

Figure 7.11 Plug shapes and inherent characteristics.

fore increases significantly and the valve has to close further in response. A pump with a steep pressure/flow curve will result in a large rise in pressure for a small reduction in flow rate. The effect of the drop in pressure across the rest of the circuit will depend on the relative resistance of it to the control valve. Manufacturers need a method of describing the characteristic of a control valve independent of the pump and circuit characteristics in which it is installed. The heating system designer can then use this information to determine the actual performance within the circuit. This is similar to the way pumps are characterised. The manufacturer tests the pump under standard conditions to produce a pressure/flow curve. The designer then uses this pump characteristic to determine the pump performance when installed, knowing the circuit characteristics. The following describes how the 'inherent characteristic' of the valve is produced and how it can be used to determine the 'installed characteristic'.

7.5.1 Valve inherent characteristic

This is the relationship between valve spindle position and water flow rate when the pressure drop across the valve is maintained constant, i.e. unaffected by the pump and circuit. This can easily be determined by installing the valve in a test rig containing a manual regulating valve, a flow meter and static pressure tappings at the inlet and outlet to the control valve. The flow rate is measured for various valve spindle positions, using the manual regulating valve to maintain a constant pressure across the control valve. Agreed standard values of pressure drop are used[9]. The shape of the characteristic depends on the shape of the plug and plug seating, as shown in Figure 7.11.

The inherent characteristic can be represented algebraically by the function:

$$\gamma = \frac{\dot{V}_s}{\dot{V}_o} = fn(s) \qquad (7.20)$$

where s represents valve spindle position between 0 and 1, (0 = closed, 1 = fully open). The subscript s refers to the flow rate at spindle position s and subscript o the flow rate through the fully open valve.

The 'quick opening' valve provides most of the flow during the initial opening of the valve. This is a very simple valve design and of little use in the control of heating systems.

The linear valve characteristic is used in three-port mixing valves where a linear relationship is required. Each inlet port would have this characteristic, but acting in opposition to each other to maintain a constant leaving flow rate, i.e.:

$$\gamma = \frac{\dot{V}_s}{\dot{V}_o} = s \quad (\text{or } -s) \qquad (7.21)$$

The logarithmic, or equal percentage, valve characteristic is used in the design of two-port valves used to regulate heat output rate from emitters. It has a characteristic of the form

$$\gamma = \frac{\dot{V}_s}{\dot{V}_o} = e^{a\left(1-\frac{1}{s}\right)} \qquad (7.22)$$

7.5.2 Installed characteristic

The installed characteristic depends, as described above, on the pump and circuit characteristics. The pump characteristic can be represented by the standard pump pressure/flow equation:

$$P = A - B\dot{V}^2 \qquad (7.5)$$

which is applicable to constant speed, unregulated pumps. The circuit characteristic, excluding the control valve, can be represented by an equation based on equation 7.6:

$$R = \frac{\Delta P_{c,o}}{\dot{V}_o^2} \qquad (7.23)$$

where R is the resistance of the circuit, excluding the control valve, which can be calculated from the pressure drop at the design flow rate when the control valve is fully open.

However, it is convenient to define a value known as the valve authority (N) given by the ratio of the pressure drop across the control valve to the pressure drop across the whole circuit, including the control valve, when the control valve is fully open, i.e. at full flow rate:

$$N = \frac{\Delta P_{v,o}}{\Delta P_{v,o} + \Delta P_{c,o}} \qquad (7.24)$$

As previously discussed, the higher this ratio, the less the increase in pressure drop across the valve when it starts to close. As will be seen, it is a useful measure of the control valve's ability to regulate flow rate, hence the use of the word 'authority'. (In determining the valve's inherent characteristic, the control valve has an effective authority of unity.)

It can be shown[10] that the installed characteristic (γ') can then be represented by the equation:

$$\gamma' = \sqrt{\frac{BN + R_{v,o}}{\left[BN + R_{v,o}\left(1 - N + \frac{N}{\gamma^2}\right)\right]}} \tag{7.25}$$

i.e. it is a function of the control valve's inherent characteristic, authority and resistance when fully open, and the pump constant B.

Should the pump be speed controlled, as described in section 7.4.3, by placing the differential pressure sensor across the pump, then the pump pressure will be maintained constant. This is equivalent to setting the value of B to zero, and equation 7.25 reduces to:

$$\gamma' = \sqrt{\frac{1}{\left(1 - N + \frac{N}{\gamma^2}\right)}} \tag{7.26}$$

And if the idealised condition of unity authority could be achieved, this would give an installed characteristic equal to the inherent characteristic. This, of course, occurs in the index branch of a circuit utilising variable speed pumping when the differential pressure sensor is located here rather than at the pump.

If the installed characteristic as given by equation 7.25 is plotted, the result of various values of valve authority and pump characteristic is immediately clear, as shown in Figure 7.12a for an equal percentage valve. Figure 7.12b shows the effect this has on the heat output rate of a typical heat emitter.

7.5.3 Recommended valve authorities

A high authority will lead to a more linear response between spindle lift and the controlled variable. However, this requires a greater pressure drop across the control valve and hence higher pump running costs. As usual, a compromise is required between performance and cost. CIBSE[11] recommend a minimum valve authority of 0.5 for two-port throttling valves (or three-port mixing valves used to divert water) and a minimum of 0.3 for mixing valves (used in a mixing application). However, the pump characteristic also needs to be considered. A pump with a low value of B ('flat curve'), or controlled to maintain or, better still, reduce pressure with fall in flow rate (for example speed control), will lead to improved control performance, and a lower valve authority may be worth considering.

7.6 Forces acting on pipework

Various forces act on pipework and these need to be accommodated in the design of both the pipework and supports. These forces arise from gravity, fluid static pressure, fluid flow and thermal expansion.

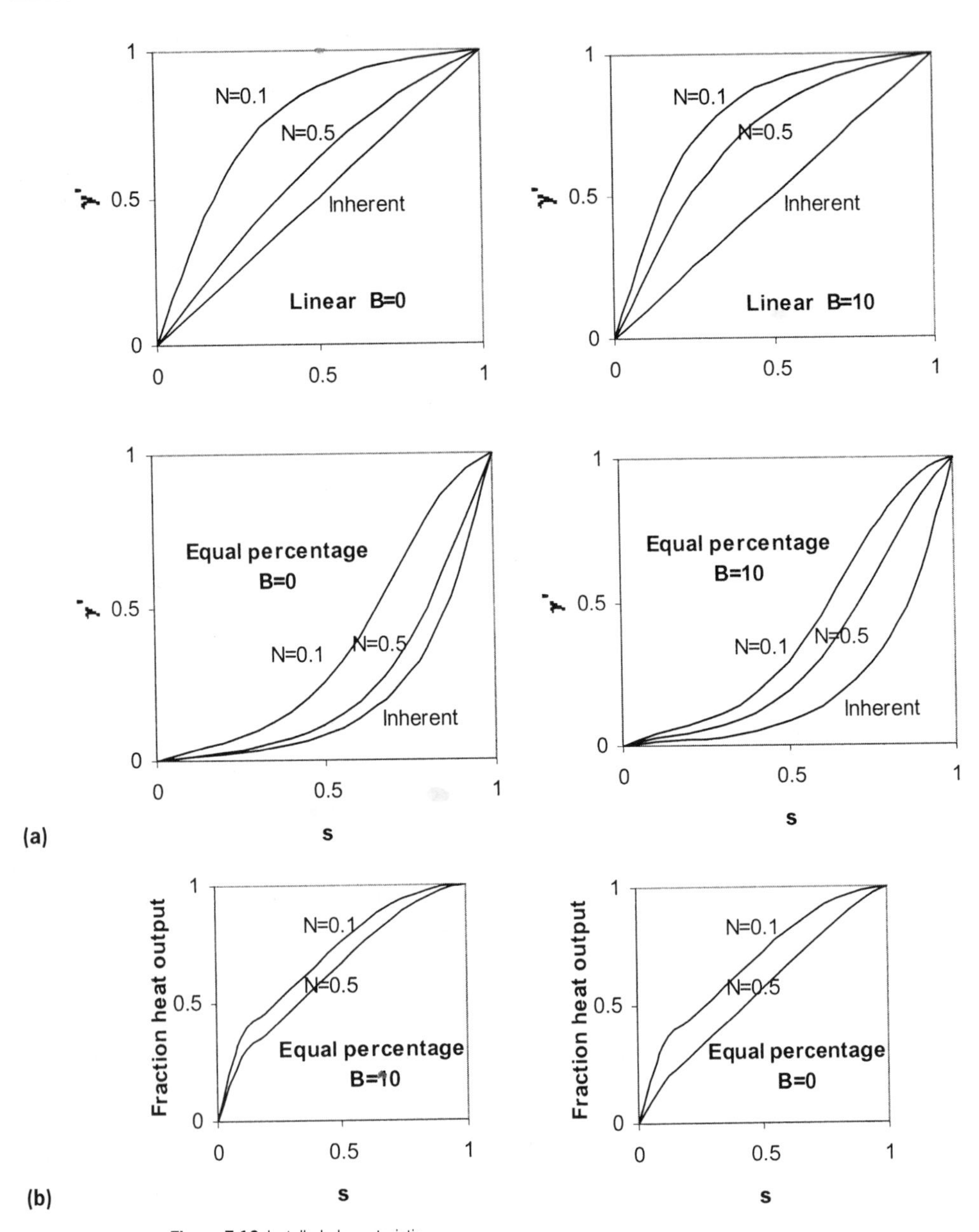

Figure 7.12 Installed characteristic.

Gravity will cause (horizontal) pipes to sag due both to the weight of the pipe and the weight of the water in the pipe. This will result in bending stresses within the pipework, which could lead to failure of the joints. The spacing of pipe supports must be such as to limit the resultant stresses.

The static pressure of the water in the pipe will tend to push pipe sections apart so that pipe joints are vulnerable to failure. At bends, particularly at the base of risers, the static pressure can lead to very large forces, requiring the use of substantial 'anchors' to prevent pipe movement and failure.

Closely related are forces exerted on the pipe due to the flow of the water, characterised by a change in static pressure and, at bends, a reaction force due to the inertia of the water. In most cases, these tend not to be as significant as those in the previous paragraph, but nevertheless they still need to be checked.

Finally, each time the heating system is started up, the water will be raised in temperature by about 60 K, as will the pipework. Just as the expansion of the water must be accommodated by the design of the system (expansion vessel or f & e tank), so too must the expansion of the pipework. Over this temperature lift, each metre length of pipe will increase its length by about 1 mm if free to do so. If prevented from doing so by the pipework supports, large compressive forces will arise which could easily cause failure.

There is therefore a conflict between the need to rigidly fix the pipework to resist the forces arising from gravity, static pressure and fluid flow, and to prevent damaging forces arising from restricting expansion. As will be seen, appropriate pipework layout and a combination of rigid and semi-rigid or sliding supports are required.

7.6.1 Types of support

There are basically two types of pipework support – those that are designed to prevent movement in any direction and those that allow movement in a selected direction(s).

Anchors fix the pipework rigidly, preventing all movement. They incorporate a clamp which is firmly tightened around the pipe. To prevent excessive heat loss by conduction and to spread the load imposed by the clamp over a larger area, a sleeve of an insulating material is used. The anchor is then secured to the building structure. On large pipes, the anchor may be welded directly to the pipe.

Guides allow axial movement of the pipe whilst supporting its weight and preventing lateral movement. For larger pipes, Teflon pads, rollers and tubular guides are available. These are designed to reduce friction between the pipe and support so that negligible axial force due to pipe expansion results. For small pipes, clips are used.

Hangers are used to support pipework from ceilings, etc. The drop rod, depending on its length, will allow some axial and lateral movement.

7.6.2 Spacing of supports

For horizontal pipes, the spacing between supports will depend on the pipe diameter and the pipe material. Larger diameter pipes will have a greater cross-sectional area of pipe wall to carry the weight-induced forces so that stresses will be less. Also, as pipe diameter increases, there is generally an increase in wall thickness. So, even though the weight of pipe and water will be greater, larger diameter pipes can cope with greater spacing between supports. CIBSE[12] recommend pipe spacings for a range of pipe sizes and materials. These are based on the resulting stress in the pipework not exceeding 10% of the yield stress (or in some cases ultimate tensile stress) of the particular materials. The reason for using a figure of 10% is not clear. However, it is not usually the pipe material which would fail, but rather the pipe joint.

7.6.3 Expansion and contraction

The amount of axial (i.e. along the pipe length) expansion is given by the equation:

$$\Delta L = \alpha\, L\, \Delta\theta \qquad (7.27)$$

where α is the coefficient of linear expansion of the material from which the pipe is made (K^{-1}); L is the original length of the pipe (m); $\Delta\theta$ is the temperature change of the pipe, (K). This gives ΔL in metres. Note that as α is a property of the material, the axial expansion is independent of pipe diameter.

Pipes are normally installed at room temperature. When the heating system becomes operational, the pipe will be heated to the water temperature and so expand. It will contract when the heating system switches off and will go through many cycles of expansion and contraction throughout its life. This repeated cycling is one reason for eventual failure. The coefficients of linear expansion for mild steel and copper are $11.3 \times 10^{-6} K^{-1}$ and $16.9 \times 10^{-6} K^{-1}$ respectively. Thus, a steel pipe 10 m long and heated from 20° to 85°C would expand, if unrestrained, by ($11.3 \times 10^{-6} \times 10 \times 65$) 0.0075 m or 7.5 mm, whilst a copper pipe under same conditions would expand by about 11 mm.

If, conversely, the pipes were fully restrained so that no expansion could occur, then they would be subjected to a compression stress given by:

$$\sigma = \alpha\, \Delta\theta\, E \qquad (7.28)$$

where E is the Young's modulus of the material (N/m^2) and the stress, σ, is in N/m^2. Again, this stress is independent of the pipe diameter but it is also independent of the pipe length. E for copper is about $100 \times 10^9 N/m^2$ so the stress in the fully restrained copper pipe would be ($16.9 \times 10^{-6} \times 65 \times 100 \times 10^9$) $110 \times 10^6 N/m^2$ or $110 MN/m^2$.

The force (being given by the product of stress times area) will vary with diameter:

$$F = \sigma\, A_c = \sigma\, \pi\{D_{op}^2 - D_{ip}^2\}/4 \qquad (7.29)$$

For a 50 mm (nominal) copper pipe, the outside and internal diameters are 54.0 and 51.6 mm respectively, giving a cross-sectional area of 0.199×10^{-3} m^2 and hence the force would amount to ($110 \times 10^6 \times 0.199 \times 10^{-3}$) 21 900 N. This is equivalent to the gravitational force on a mass of 2200 kg. The anchors at each end of the pipe would need to be able to share this axial force without deforming. In comparison, the mass of the pipe and water would only be about 39 kg for the 10 m section. But note that the axial force calculated is independent of length. Of course, if there were no guide supports between the two anchors then in practice the pipe would bend under this axial force. In doing so the pipe length would increase so the stress in the pipe and the force acting on the anchors would be reduced. This is the principle behind the practical management of expansion in pipework.

Dog legs and loops

Figure 7.13 shows a simple method for allowing for expansion between two anchors, known as a dog leg or set or pulled bend. Both a single and double are shown. The solid line represents the pipe when cold and the dotted when

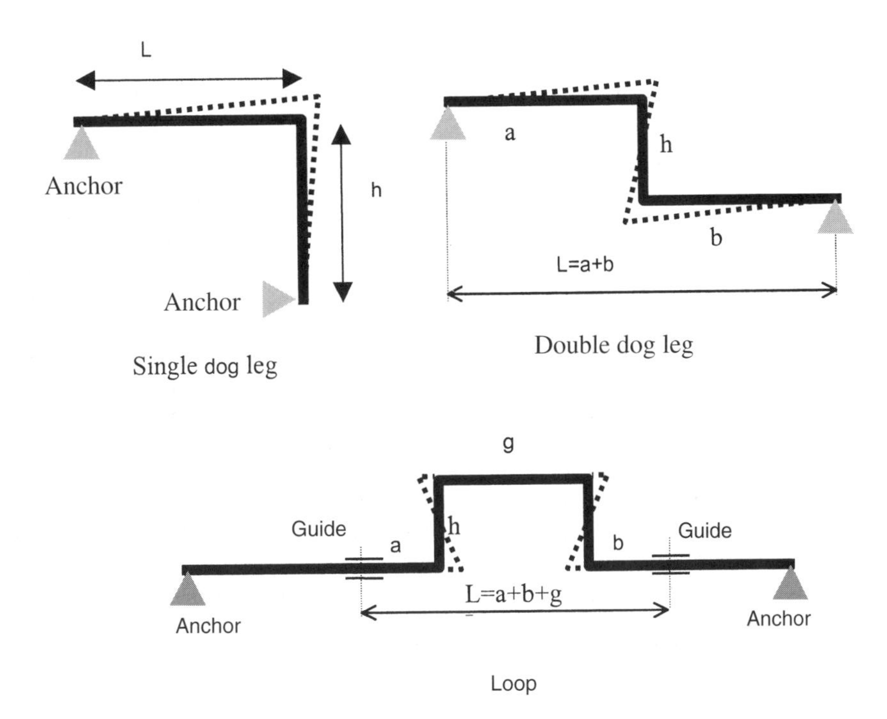

Figure 7.13 Dog legs and loops.

hot following partially restrained expansion. For the stresses in the pipe to be contained within an acceptable limit, the lengths of the set, L and h, must permit sufficient expansion. CIBSE[12] give minimum recommended values of h relative to L for steel pipes undergoing various temperature changes (no figures are given for copper). Although the stress induced is independent of pipe diameter, the flexibility of the pipe is not. Therefore L and h are given for various pipe diameters. For a 40 mm steel pipe undergoing an 80 K rise in temperature and for L = 10 m, the single set requires a minimum value of h of 1.9 m. For the double set this reduces to 0.48 m – a significant reduction. Values for the double set assume that the set is symmetrical (a = b). When this is not the case, CIBSE give correction factors.

Alternatively, a pipe loop may be used, as shown also in Figure 7.13. Repeating the above example with a loop, CIBSE recommend a minimum value of h of 0.17 m on the basis that the loop is symmetrical and that g = L/2. Correction factors are available for asymmetric loops. Notice that guides may be used in conjunction with a loop but not within the loop itself. Guides must not be used anywhere between the anchors of dog legs.

In addition to giving recommended minimum values of h, CIBSE also provide tables and equations for determining the forces resulting from thermal expansion acting on the anchors.

Dog legs and loops are useful solutions for overcoming expansion problems, although they do take up a considerable amount of space. Fortunately,

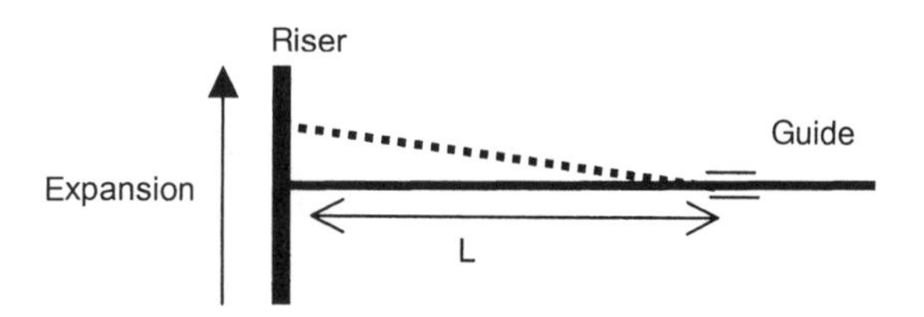

Figure 7.14 Branch connection.

it is often possible to integrate dog legs into the natural layout of the pipework, provided adequate clearance for the expansion at the corners is available.

A common cause of pipe failure in heating systems occurs at branch connections, as shown in Figure 7.14, for example from a riser. Expansion in the main pipe can be considerable and unless the branch pipe has sufficient length between the connection to the main and the first anchor, it will not have the flexibility to take up the lateral movement. This can result in leakage at the branch connection or at the heat emitter. Again, CIBSE give recommendations on minimum branch length. For suspended pipework, hangers will normally have sufficient in-built flexibility to allow for this movement of the branch pipe. A more serious problem is connection to heat emitters, for example when several emitters are connected to a long straight F + R mains running at skirting level or emitters on each landing of a stairwell are connected to a F + R riser. In each case the branches will be very short and could result in failure of the pipe connections at the branch or emitter.

Expansion bellows

Expansion bellows allow lateral expansion without inducing high stresses. The wall is made from corrugated steel so as to form a bellows which can elongate and shorten. Placed between anchors, these bellows can take up all of the expansion without a need for dog legs or loops. Guides should be installed each side of the bellows to prevent lateral movement. Alternatively, these may be incorporated into the design of the bellows by the manufacturer. In selecting bellows, the static pressure of the water must be taken into account as this will both stiffen the bellows and cause it to elongate, so reducing its ability to deal with thermal expansion.

7.6.4 Forces arising from the fluid

The static pressure of the water in the pipe will impose a force on the pipe wall, as will the inertia of the moving water. This is most noticeable at a pipe bend. Appendix 7.3 shows how the net force acting on a 90° bend, as shown in Figure 7.15, has a magnitude equal to:

$$F = \sqrt{2}A_c(P + \rho v^2) \qquad (7.30)$$

and acts at 45° away from the bend. A_c is the cross-sectional area of the pipe (m^2), P the static pressure in the pipe (Pa, gauge), ρ the density of the water, (kg/m^3) and v the water velocity (m/s) giving F in Newtons.

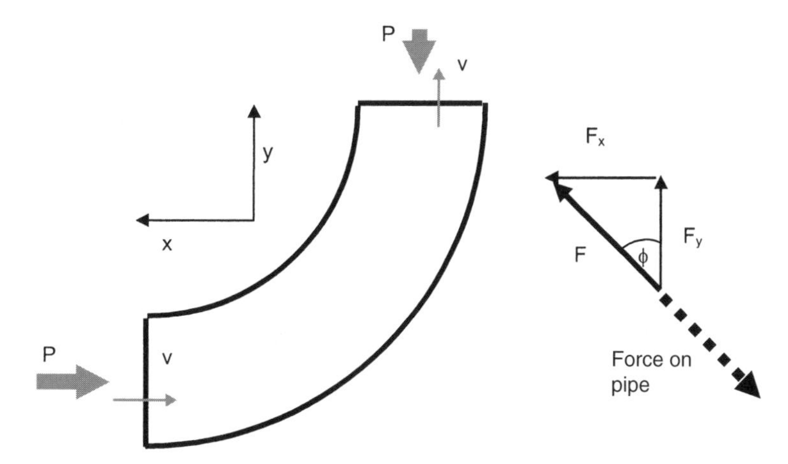

Figure 7.15 Forces acting on a bend with no flow.

This equation demonstrates the relative importance of static pressure (P) compared to inertia (ρv^2). Static pressures are usually in excess of 1 bar (gauge), that is 100 000 Pa. Even at a velocity of 4 m/s, which would be appropriate only in very large pipes[13], the inertia term is only equivalent to about 16 000 Pa, or 16%. On pipes below 50 mm, when velocities are normally about 1 m/s, the inertia accounts for only about 1% of the total force on the pipe bend.

Appendix 7.1 Heat Loss from Pipework

The equation to determine the heat loss rate from an insulated pipe is given by the equation:

$$Q = AU(\theta_i - \theta_a) \qquad (7.31)$$

where θ_i is the mean water temperature in the pipe and θ_a the ambient temperature. The term AU represents a combined surface area times 'U' value. Unlike a plane wall which has a constant cross-sectional area at all points throughout its thickness, the surface area of an insulated pipe increases with distance outwards from the inside surface of the wall. Any surface area can be used provided that the same surface is used in the calculation of the U value. When the *outside* surface of the insulation is selected as the reference area then:

$$A = 2\pi\, r_o L$$

$$U = \frac{1}{\left[\dfrac{r_o}{r_i h_i} + \dfrac{r_o}{\lambda}\mathrm{Ln}\left(\dfrac{r_o}{r_i}\right) + \dfrac{1}{h_o}\right]} \qquad (7.32a)$$

and when the inside surface area of the pipe is used then:

$$A = 2\pi\, r_i L$$

$$U = \frac{1}{\left[\dfrac{1}{h_i} + \dfrac{r_i}{\lambda}\mathrm{Ln}\left(\dfrac{r_o}{r_i}\right) + \dfrac{r_i}{r_o h_o}\right]} \qquad (7.32b)$$

where r_i is the inside radius of the pipe and r_o the external radius of the insulation, λ is the conductivity of the insulation and h_i, h_o are the internal and external surface heat transfer coefficients. The numerical value of the product AU is the same in both cases.

Both equations for U ignore the resistance of the pipe wall itself as this is generally insignificant, being relatively thin and made of highly conductive material.

For turbulent flow, the inside surface heat transfer coefficient may be determined from the general empirical heat transfer relationship:

$$Nu = \frac{h_i D_i}{\lambda_f} = 0.023 Re^{0.8} Pr^{0.33}$$

$$\text{where} \quad Re = \frac{\rho v D_i}{\mu} \quad \text{and} \quad Pr = \frac{C_p \mu}{\lambda} \qquad (7.33)$$

In practice, h_i is also normally very high relative to the other terms and can be safely neglected. In such cases, the outside surface temperature of the pipe is very nearly the same as the water in the pipe.

The external surface heat transfer coefficient is problematic in that it depends not only on the surface finish but also on the surface temperature and ambient conditions. Both radiant and convective heat loss take place at the outside surface so that h_o is in reality a combined radiant and convective transfer coefficient. The radiant heat transfer coefficient depends on the (absolute) surface temperature and the emittance (see Chapter 6). Low emittance is achieved with the use of metallic finishes. The convective heat transfer coefficient is affected by wind (air) speed and, being natural convection – i.e. driven by the increased bouyancy of the ambient air heated by the warm surface – depends on the surface temperature. The value of h_o can therefore vary considerably. For substantial amounts of insulation the resistance afforded by the insulation will dominate the overall heat loss rate, and errors in selecting an appropriate value of h_o will not be great. As the thickness of insulation increases so the external surface area will increase so that the significance of the surface heat transfer coefficient will increase. Even so, for conventional thicknesses of insulation the error introduced will not be significant. However, for little, or no, insulation, the choice of h_o is critical.

The following equations[14] may be used to estimate h_o:

$$h_o = h_{oC} + h_{oR}$$

$$h_{oC} = 13.5\left[\frac{T_o - T_a}{d_o}\right]^{0.25} \qquad h_{oR} = 5.67 \times 10^{-8}\,\varepsilon\left[\frac{T_o^4 - T_a^4}{\theta_o - \theta_a}\right] \tag{7.34}$$

The expression for h_{oC} assumes natural convection for horizontal pipes in still air conditions. For other cases, appropriate heat transfer relationships should be used, usually of the standard form Nu = f (Re, Pr).

The economic thickness of insulation for a particular pipe size is determined by estimating the financial cost of producing the heat lost through the insulated pipe over an agreed time period and adding this to the cost of installing the insulation to give the total cost. This calculation is repeated for several thicknesses of insulation to determine the lowest total cost and therefore the economic thickness. A financial method, such as Present Value, is used where the time period is greater than three or four years to discount future savings:

$$\text{Total cost} = C + cf\,F \int^{year} \frac{Q}{\zeta_b}\,dt \tag{7.35}$$

where C is the cost of installing the insulation (£), t is time (hours), year is the time per year that the heating system is operating (hours), Q is the heat loss rate from the pipe (kW), F is the cost of fuel (£/kWh), ζ_b is the boiler efficiency. cf is the cummulative discount factor based on the number of years over which the assessment is to be made (N) and the real interest rate (i) which in turn is given by:

$$cf = \frac{1-(1+i)^{-N}}{i} \tag{7.36}$$

It is common practice to use a constant value for boiler efficiency: seasonal efficiency, ζ_{bs}. Then, where the water flow and ambient temperatures are constant (or nearly so), the integral in equation 7.35 can be replaced by $Q \times year/\zeta_{bs}$. Where the flow water temperature varies (e.g. as a function of outdoor temperature) then a degree-day method is required to solve the integral (see Chapter 11). As discussed in section 7.2, the heat loss rate from insulated pipes can be approximately expressed by the equation:

$$\frac{Q}{L} = \left[0.0613\frac{D_i}{X^{0.82}} + 0.1\right](\theta_i - \theta_a) \tag{7.3}$$

It is reasonable to assume that the installed cost, C, can be expressed in the form:

$$\frac{C}{L} = (A + BX)\pi D_i \tag{7.37}$$

where A and B are constants. That is, the installed cost is a linear function of insulation thickness and pipe circumference. Equation 7.35 can then be written as a general function of X which can be differentiated to find the minimum total cost:

$$\text{Total cost} = L(A + BX)\pi D_i + L\left[0.0613\frac{D_i}{X^{0.82}} + 0.1\right](\theta_i - \theta_a)tcf\frac{F}{1000\zeta_{bs}}$$

(note the factor of 1000 is needed to convert W to kW)

$$\frac{d(\text{Total cost})}{dX} \approx \left\{B\pi - 5\times10^{-5}X^{-1.82}(\theta_i - \theta_a)tcf\frac{F}{\zeta_{bs}}\right\}LD_i \tag{7.38}$$

giving a minimum total cost when

$$X = \{1.6\times10^{-5}(\theta_i - \theta_a)K\}^{0.55}$$

$$\text{where } K = tcf\frac{F}{B\zeta_{bs}}$$

implying that the same thickness of insulation should be used for all pipe sizes operating at the same water-ambient temperatures. This is in disagreement with published tables of economic insulation thickness[15], which suggest that insulation thickness should increase as pipe diameter increases. This is most probably due to the assumption made that installation cost was a linear function of insulation thickness. It is more likely that the value of B used in equation 7.37 will fall slightly as X increases, in which case it may be better to use different values of B for ranges of pipe diameters.

Appendix 7.2 Idealised Pump Characteristics

The idealised characteristics allow a simple mathematical analysis to be carried out when considering control of flow rate. The pump and circuit characteristics can both be represented as quadratic functions of flow rate:

$$P_{pump} = A - B\dot{V}^2_{pump} \qquad \Delta P_{circuit} = R\dot{V}^2_{circuit} \tag{7.39}$$

The flow rate through both pump and circuit must be identical. Also, the pressure produced by the pump must exactly match the pressure loss through the circuit. Thus:

$$\dot{V}_{pump} = \dot{V}_{circuit} = \dot{V}$$

$$A - B\dot{V}^2 = R\dot{V}^2 \tag{7.40}$$

$$\text{So } \dot{V} = \sqrt{\frac{A}{B+R}}$$

For a real pump, the values of A and B can be quickly found from the manufacturer's data tables or graphs:

$$P = A - B\dot{V}^2 \quad \text{when} \quad \dot{V} = 0, A = P \tag{7.41}$$

so, A is equal to the pressure at no flow. Likewise, when $P = 0$:

$$A = B\dot{V}^2 \quad \Rightarrow \quad \dot{V} = \sqrt{\frac{A}{B}} \tag{7.42}$$

so, knowing A, B can be found at the zero pressure point where the maximum flow rate occurs. This is sometimes called 'free delivery' volume flow rate.

For variable speed pumps, a characteristic for any speed can be derived from the characteristic at full speed:

Let $P = A - BV^2$ at full speed and $P_N = A' - B'\dot{V}_N^2$ at speed N

where $N = \dfrac{\text{pump speed}}{\text{design pump speed}}$ i.e. $0 \le N \le 1$

At zero flow rate $P_N = A'$ and $P = A$

But from the pump laws, $P_N = N^2P$

So $P_N = A' = N^2P = N^2A$ i.e. $A' = N^2A$

And, at zero pressure, $\dot{V}_N = \sqrt{\dfrac{A'}{B'}}$ and $\dot{V} = \sqrt{\dfrac{A}{B}}$

But from the pump laws, $\dot{V}_N = N\dot{V}$

$$\text{So, } \sqrt{\frac{A'}{B'}} = N\sqrt{\frac{A}{B}} \Rightarrow \frac{A'}{B'} = N^2\frac{A}{B} \tag{7.43}$$

But, $A' = N^2A$, so that $\dfrac{N^2A}{B'} = N^2\dfrac{A}{B} \Rightarrow B' = B$

Finally giving $P_N = N^2A - B\dot{V}_N^2$

Appendix 7.3 Forces Due to Fluid Flow

From Newton's Law (force = mass × acceleration):

$$F = m\frac{\Delta v}{\Delta t} = \frac{m}{\Delta t}\Delta v = \dot{m}\Delta v \qquad (7.44)$$

F is the net force acting on the body of mass m, v is the velocity and t is time. So, for a moving fluid, this is the same as Force = mass flow rate × change in velocity. This is also the same as Force = rate of change of momentum and is a much more useful form of Newton's Law when dealing with fluid flow. Notice that velocity, in this context, is a vector quantity (i.e. comprises both magnitude and direction). So, water flowing around a pipe bend will experience a change in velocity as the direction changes, even though the magnitude (speed) remains the same.

The net force, F, acting on the water is equal and opposite to the force experienced by the wall of the pipe, so we proceed by determining F. However, F is also a vector quantity so we need to find both magnitude and direction. We can do this by considering F to be equivalent to two component forces acting at right angles to each other, Fx and Fy, and resolving the forces and rate of change in momentum in these two directions.

Consider a right-angled pipe bend of constant cross-sectional area as shown in Figure 7.15 and take the directions X and Y to be in the directions of the arms of the bend:

In the X direction, net force acting on the fluid is given by $PA_c + F_x$. The rate of change of momentum in the X direction is given by $\dot{m}(0 - v)$, that is, the fluid ends up with no velocity in the X direction. So, equating the net force with the rate of change of momentum:

$$\begin{aligned} &PA_c + F_x = \dot{m}(0 - v) \\ &\Rightarrow F_x = -(PA_c + \dot{m}v) = -(PA_c + \rho\dot{V}v) = -(PA_c + \rho vA_c v) = -A_c(P + \rho v^2) \end{aligned} \qquad (7.45)$$

This is negative, which implies it is acting in the negative X direction.

We can do the same in the Y direction:

$$\begin{aligned} &-PA_c + F_y = \dot{m}(v - 0) \\ &\Rightarrow F_y = (PA_c + \dot{m}v) = A_c(P + \rho v^2) \end{aligned} \qquad (7.46)$$

This is the same magnitude as the X component and is acting in the positive Y direction.

The resultant force F is found by combining F_x and F_y graphically, as shown in Figure 7.15 to give:

$$\begin{aligned} &F = \sqrt{2}A_c(P + \rho v^2) \\ &\phi = \tan^{-1}\left(\frac{F_x}{F_y}\right) = 45^\circ \end{aligned} \qquad (7.47)$$

References

1. Building Regulations (2002) Approved Document L, Conservation of Fuel and Power. The Stationery Office, London.

2. Chartered Institution of Building Services Engineers (CIBSE) (1986) Heat Transfer, Section C3 of the CIBSE *Guide*, Tables C3.14, 15 and 20. CIBSE, London.
3. Building Research Energy Conservation Support Unit (BRECSU) (1997) *Economic Thickness of Pipe Insulation.* Fuel Efficiency Booklet FEB008. BRECSU, Garston.
4. Building Research Energy Conservation Support Unit (BRECSU) (1996) *Variable Flow Control.* General Information Report 41. BRECSU, Garston.
5. Shepherd, K. (1999) *VAV Conditioning Systems.* Blackwell Publishing, Oxford.
6. Parsloe, C. (1999) Pipework analysis software in Application Guide 14/99, *Variable volume in heating and cooling circuits.* Building Services Research and Information Association (BSRIA), Bracknell.
7. Leatherman, K.M. (1981) *Automatic Controls for Heating and Air Conditioning.* Pergamon Press, Oxford.
8. Chartered Institution of Building Services Engineers (CIBSE) (1985) *Automatic Controls and their Implications for Systems Design.* Applications Manual AM1. CIBSE, London.
9. British Standards Institute (BSI) (1999) BS EN 1267: Valves: Tests of flow resistance using water as a test fluid. BSI, London.
10. Legg, R.C. (1991) *Air Conditioning Systems – design, commissioning and maintenance.* Batsford, London.
11. Chartered Institution of Building Services Engineers (CIBSE) (1986) Automatic Controls, Section B11 of the CIBSE *Guide.* CIBSE, London.
12. Chartered Institution of Building Services Engineers (CIBSE) (1986) Miscellaneous Equipment, Section B16 of the CIBSE *Guide.* CIBSE, London.
13. Chartered Institution of Building Services Engineers (CIBSE) (1986) Heating Systems, Section B1 of the CIBSE *Guide.* CIBSE, London.
14. Chartered Institution of Building Services Engineers (CIBSE) (1986) Heat Transfer, Section C3 of the CIBSE *Guide.* CIBSE, London.
15. Building Research Energy Conservation Support Unit (BRECSU) (1996) *Economic Thickness of Pipe Insulation.* Fuel Efficiency Booklet FEB008. BRECSU, Garston.

8 Hot Water Services

Hot water service, HWS, is the generation and delivery of hot water for consumption at outlets such as sinks, baths, showers, dishwashers, etc. Although the water supplied is not required to be potable, it will still be used for washing and so must be delivered, at the point of use, in a hygienic and pleasant condition. In particular, great care needs to be taken to prevent bacterial growth such as legionella. As biocides cannot be added high temperatures need to be maintained, though not so high as to risk scalding. The design of the system must also allow for the expansion of the water so as to prevent explosions. These safety requirements are covered by regulations and codes of practice. HWS systems consume significant quantities of both water and energy and these are also dealt with in various regulations.

Regulations place restrictions on the design of HWS, yet there are still many decisions that the designer must make. Should an instantaneous or a storage system be used? Should the hot water be produced centrally or locally? Should it be part of or separate from the space heating system? Should it be vented or unvented?

The sizing of HWS also raises particular problems. Unlike a space heating system, demand for hot water is erratic, being entirely at the vagaries of the occupants.

Finally, water hardness is also a major issue. Unlike closed systems, the water is continuously replaced so scale can quickly build up. Water softening is one way of preventing this but it is expensive in terms of salt consumption and maintenance. In recent years magnetic 'de-scalers' have become common.

This chapter looks at some of these issues, particularly those not covered by regulations, such as the choice of system to use, the sizing of the system and the energy consumption, where the designer needs more guidance. Solar hot water is also discussed.

8.1 Instantaneous versus storage

The heat (energy) required to raise the temperature of water is given by the well-known equation:

$$E = MC_{pw}\Delta\theta \qquad (8.1)$$

where M is the mass of the water, (kg); C_{pw} is the specific heat of water, (4.18 kJ/kgK); $\Delta\theta$ is the rise in water temperature,(K), when the energy, E, is given in kJ. The power (rate of energy supplied) to achieve this in a given time period, Δt is then:

$$Q = \frac{MC_{pw}\Delta\theta}{\Delta t} \qquad (8.2)$$

Equation 8.2 is directly applicable to the heating of a fixed mass of water in a storage vessel. However, it can easily be modified for a flow of water by noting that the fraction M/dt is the same as mass flow rate. So, for an instantaneous HWS system:

$$Q = \dot{m} C_{pw} \Delta\theta \qquad (8.3)$$

where $\dot{m}$ is the mass flow rate of the water (kg/s).

It is revealing to compare the power requirements of a storage and an instantaneous system. For example, a 22 mm tap serving a bath has a maximum flow rate of up to about 0.3 kg/s (note that 1 kg of water is approximately 1 litre). A typical delivery temperature is 55°C. In winter the water may enter the system at 10°C. This gives an instantaneous demand (equation 8.3) of some 56 kW. Yet domestic storage systems are quite capable of supplying a hot bath using a mere 3 kW electric immersion heater. Obviously, an electric instantaneous water heater to serve a bath is out of the question. Such high heat output rates would also be uneconomic for domestic gas-fired water heaters. Domestic combination boilers are limited to a HWS output of about 20 kW. Spray taps are frequently used for wash hand basins in non-domestic buildings because of their very low rate, only about 0.015 kg/s. As well as helping to conserve water, a 3 kW instantaneous electric heater is just sufficient to heat the water to an acceptable temperature for a single tap. Even so, peak electrical demand would be high in a building containing many such taps.

One disadvantage of the storage system is that it takes a finite time to reach design temperature once all the hot water has been run off. This is known as the recovery time and depends on the mass of water stored and the power of the heating element. Equation 8.2 can be arranged to give the typical recovery time of a domestic hot water cylinder (125 kg water capacity) based on a 3 kW immersion heater for a 50 K lift in temperature (storage temperatures will be discussed later): Δt = 8700 s or about two and a half hours. This is rather a long time. Domestic immersion heaters are limited to 3 kW for practical reasons regarding the design and cost of domestic electrical distribution installations. When the HWS cylinder is served by the boiler, the heat exchanger is rated at about 6 kW to give a recovery time of nearer one hour. Instantaneous systems do not suffer from this problem. Although the water flow rate is limited, it is always available.

A second drawback of storage systems is that they occupy more space. The designer could decide to install a smaller storage vessel. For the same heat exchanger duty this would result in shorter recovery times but would increase the number of times each day when hot water was not available. Alternatively, a larger vessel could be installed of sufficient capacity to allow it to meet the maximum demand during the day with continuous recovery. The size of domestic hot water cylinders has become very much standardised, being a compromise between physical space requirements and hot water availability. For non-domestic HWS, the range of sizes of storage vessels and LPHW heat exchangers available is considerable. In the past some designers have opted for large storage volumes, but an increasingly common system nowadays is the 'hot water generator'. This has a small water storage volume to meet maximum demand for a few minutes combined with a direct, gas-fired heater to give very short recovery times. Guidance on how to select appropriate storage volumes and heat exchanger duties is given in section 8.2.

8.2 Central versus local systems

Instantaneous systems, other than domestic, tend to be localised since the power demand is very great for even a few hot water outlets. Storage systems may be centralised or localised.

Central systems offer potential capital cost savings compared to local systems with duplicate plant distributed throughout the building. Conversely, central systems require extensive pipework, longer and of larger diameter than local systems. This added cost may outweigh savings in plant. As to space requirements, central systems tend to be located in areas of low rental value, for example basements, but do require distribution space. Only a detailed cost analysis will determine which is the cheapest and this is likely to depend on the value attached to space.

A more complete cost analysis should include running costs. Central systems are likely to be less efficient due to heat loss from the more extensive pipework. On the other hand, maintenance costs should be less.

In sizing plant and pipework, diversity is usually applied. Central systems will be able to make greater use of diversity. This is discussed next.

8.3 Sizing

In order to design and size a HWS system, it is essential to be able to make a realistic estimate of the hot water demand, both in the short and long term (e.g. peak demand, average hourly, average daily). This is impossible to do with any certainty for most buildings as we are dealing with the behaviour of its occupants. The CIBSE provide empirical data for a range of buildings types, and various sizing methods have evolved.

8.3.1 Peak instantaneous demand

Peak demand is needed to size instantaneous systems, and also the maximum demand imposed on storage systems and the distribution pipework. This is determined by the number of hot water outlets in use simultaneously. Binomial probability theory has been used to develop simple calculations which do this. In essence, it calculates the maximum number of outlets that are *likely* to be in use at any one time. For example, in a typical office building toilet containing say four wash hand basins, it is unlikely that the hot taps to all four would be open at the same time. There is a diversity in use. There is a possibility that all four could be used but only a very small probability. In building services, we do not usually design for the worst case as this is not economic. Instead we design to an acceptable level of risk.

The following equation[1] is an approximation to the binomial theorem which determines the maximum number (m) of taps in simultaneous use out of a total of n installed:

$$m = np + x[2np(1-p)]^{0.5} \qquad (8.4)$$

where p is the 'usage ratio' of each tap. The usage ratio represents the probability of any one tap being in use at any instant in time. It is given by the simple ratio:

$$p = \frac{\text{average time in use(s)}}{\text{average time between successive use(s)}}$$

The value of x in equation 8.4 depends on the level of risk that the system is to be designed for. CIBSE lists values of this factor for a number of risk levels. The commonly accepted level of risk for the design of hot water services is 1% for which x = 1.8.

This equation is not restricted to hot water taps. It is applicable to any situation where there are a number of identical items operated independently of each other and which may be in only one of two possible states. It is just as applicable to the number of personal computers switched on at any one time in a building or the number of lamps in a lighting scheme which will fail on any one day.

Taking the example of the office toilet, let us assume that each tap is, on average, in use for 30 seconds every 300 seconds ($p = 0.1$). Then the number of hot taps in use simultaneously for sizing purposes is:

$$m = 4 \times 0.1 + 1.8[2 \times 4 \times 0.1(1 - 0.1)]^{0.5} = 1.9 \approx 2$$

i.e. would size for two taps in use. There is only a 1% probability that more than two out of the four taps would be in simultaneous use. The design hot water flow rate can then be calculated knowing the design flow rate for an individual tap. Further, if the office contained, say, 10 such toilets served by the same HWS system, this same approach could be taken to determine the total demand from all 40 hot water taps:

$$m = 40 \times 0.1 + 1.8[2 \times 40 \times 0.1(1 - 0.1)]^{0.5} = 8.8 \approx 9$$

This represents a significant diversity. For large, central systems this process is repeated for each branch with the diversity increasing towards the central plant.

However, great care does need to be used in determining the usage ratio. The value of 'average time of use' and 'average time between successive use' should be based on busy periods of the day, not the entire day necessarily. A washroom in a factory will be very busy during relatively short breaks; all the showers in a leisure centre might be in use at the end of a football match, etc. A well known case is toilets at a theatre during intermissions. Clearly, the designer must use some discretion in determining the appropriate usage ratio.

This raises the question of the true risk of underestimating the peak demand. In the first place, the theory assumes the tap is either on or off. In reality, the tap is likely to be on at some intermediate position rather than fully on (depending on the tap used). So demand will be less than design flow rate. Secondly, should the number of taps in simultaneous use be underestimated, water will still be received at each tap but at a reduced flow rate which users may not notice. Although it is common practice to use a risk factor of 1%, there may well be a good economic case for designing to a higher risk.

Demand units

A major drawback with the above approach is that it can only be used with identical hot water outlets. Where there is a mixture of fittings (different usage ratios and different flow rates – for example wash hand basins and baths) on a distribution network, an alternative approach is required. This involves producing a 'standardised hot water fitting' and treating all other fittings as a simple multiple of the standard. This procedure is known as demand units and is used also for cold water outlets and discharge from sanitary fittings ('discharge units'). To simplify the procedure, sets of tables giving total demand units against design flow rate are provided.[2]

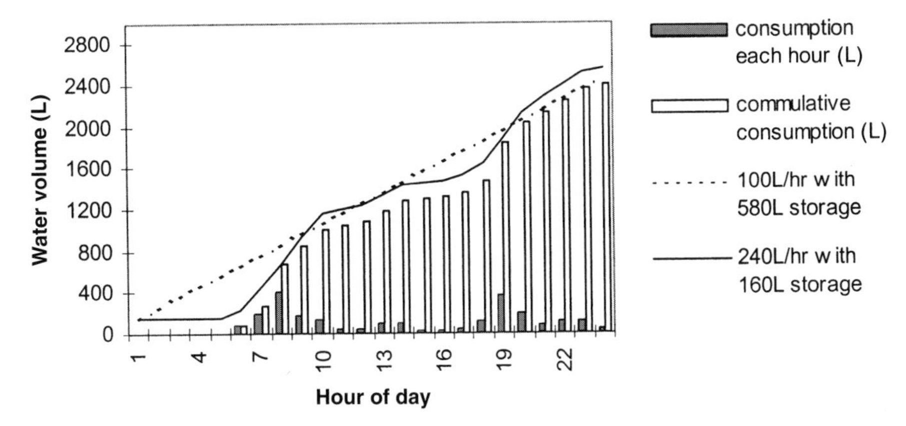

Figure 8.1 Hot water consumption and options for storage system sizing.

8.3.2 Time-averaged demand

Time-averaged demand is used to assess the storage capacity (volume) and heating power requirements of storage systems.

Consider the example shown in Figure 8.1 which shows the average hourly hot water consumption rate of an imaginary 12-bedroom hotel (for simplicity ignoring catering). As expected, there is a peak in the morning and evening arising from guests running baths. These figures have been arrived at simply by estimating the number of baths that might be run each hour and allowing for other uses, e.g. wash hand basins, showers. The time interval of one hour was used purely for convenience. From this, cumulative consumption is produced. This totals 2400 litres per day. Using this cumulative demand curve, we will examine several hot water generation rates and determine the minimum storage volume needed to just satisfy demand.

Begin by considering a generation rate which exactly matches the average rate for the day, that is 2400 litres in 24 hr or 100 litres/hr. This is represented in Figure 8.1 by a straight line of gradient 100 litres/hr which has been positioned on the graph so that it just touches the cumulative demand curve but never falls below it. With this system the hot water production rate must be constant (otherwise it will fail to produce 2400 litres in 24 hr). When demand is less than 100 litres/hr water is stored. When demand exceeds 100 litres/hr then water is taken from the store. The storage volume required is equal to the greatest difference between the cumulative demand curve and the amount generated. This is about 580 litres and occurs at about 6 AM. Such a system would just meet demand. At 11 PM the store is completely emptied. To produce 100 litres/hr would require a heat exchanger of approximately 6 kW. Recovery time would be nearly 6 hours.

Now consider a system with a generation rate of 240 litres/hr. It is assumed that the heat exchanger is controlled so as to maintain the desired storage temperature. By trial and error it can be determined that a storage volume of 160 litres is needed to just satisfy demand. Its output curve then closely follows the cumulative demand curve. The heat exchanger would require a duty of about 14 kW. Recovery time would be less than one hour.

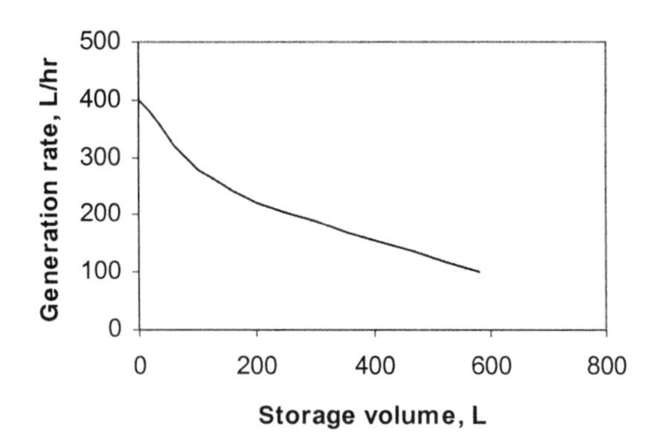

Figure 8.2 Hot water generation rate versus storage volume.

Higher hot water generation rates would fit the demand curve more closely and require still smaller storage volumes. A perfect match between the production curve and demand curve would require an instantaneous system with no storage. This is 400 litres/hr (which occurs at 8 AM), about 24 kW. These options have been plotted on a graph to produce Figure 8.2 and produce a classic sizing curve. This allows intermediate values of generating capacity to be paired with minimum required storage volume. Note, however, that the graph cannot be extrapolated below 100 litres/hr. No matter how great the storage volume, the system would never be able to generate sufficient hot water during the course of 24 hours.

These options represent the system sizes which would just meet the predicted cumulative demand. But how good is the prediction? There may be times when peak hourly demand is greater than estimated, even though daily demand is not exceeded. The designer will need to allow for some margin of error. This could be done by increasing hot water generation rate, storage volume or both. At the same time, excessive oversizing will increase capital costs and is likely to increase running costs. In the past, the evidence suggests that hot water systems were grossly oversized. There have been a number of studies in the past[3] which have attempted to provide design data using actual installations and computer models. CIBSE has produced a series of sizing curves[4] for various building types which attempt to avoid excessive oversizing.

8.3.3 CIBSE sizing charts for storage systems

These are based on studies of installations in a variety of building types where the system was known to satisfy demand. Paired values of installed generating capacity and storage volume were plotted for each type of building. A curve was then drawn which would just accommodate these data points. This implies that the CIBSE curves should satisfy the requirements in most cases. To normalise the results, generating capacity and storage volume are given per person, with generation capacity in kW rather than litres/hr and referred to as 'boiler power'. Hot water for catering and non-catering uses have been separated.

Figure 8.3 is based on the CIBSE sizing chart for hotels. This gives a minimum generating capacity of about 6 kW (0.25 kW/person) with a corresponding storage volume of 480 litres (20 litres/person). This is remarkably

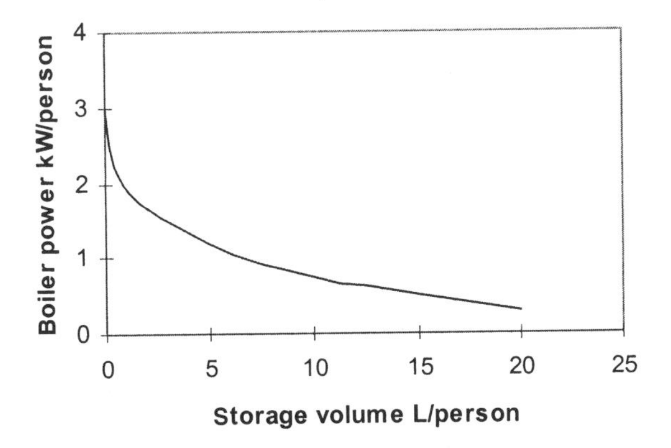

Figure 8.3 CIBSE sizing chart for hotels.

close to the prediction in the previous section. However, at lower storage volumes it recommends much greater generating capacity. This should not be taken as a criticism of the CIBSE sizing curves but as a warning of the difficulty in predicting hot water consumption rates.

8.4 Systems

8.4.1 Domestic storage systems

A typical size of domestic size hot water cylinder is 125 litres with a 6 kW heat exchanger and a 3 kW electric immersion heater for use in summer. The heat exchanger is fed from the boiler. Recovery times are about 1 hour for the heat exchanger and 2 hours for the immersion heater. A typical system is shown in Figure 8.4. This is a vented system, i.e. the system pressure is produced by a tank of water open to the atmosphere.

Whenever a hot water tap is opened, water flows by gravity from the feed and expansion tank into the bottom of the hot water cylinder. An equal volume of water flows from the top of the cylinder. This inlet and outlet arrangement combined with the tall, narrow shape of the cylinder allows stratification to occur. Below the heating element the water will remain close to entry temperature; only water above the element will be heated. By preventing any mixing of the cold incoming water, the temperature of the water in the upper part of the cylinder remains more or less constant and ready for use. If mixing were to occur, this would not represent an energy loss; no energy is lost in the mixing process but the temperature of the hot water is 'diluted' and may not be sufficiently hot for the intended use.

The primary heat exchanger is fed from the boiler. Control is through a simple solenoid valve with a temperature sensor mounted on the cylinder. Where an electric immersion heater is fitted, this comes as a complete package of heating element and thermostat with high temperature cut-out. The latter is important. The feed from the boiler is limited by the boiler cut-out thermostat to about 90°C so that it cannot heat up the water in the cylinder above this temperature. There is no limit to the temperature which could be reached in the cylinder with an electric heater. As a further safety feature, the outlet from the tank is connected to an open vent. This allows for expansion of the water when it is heated. Under normal operating conditions, no water

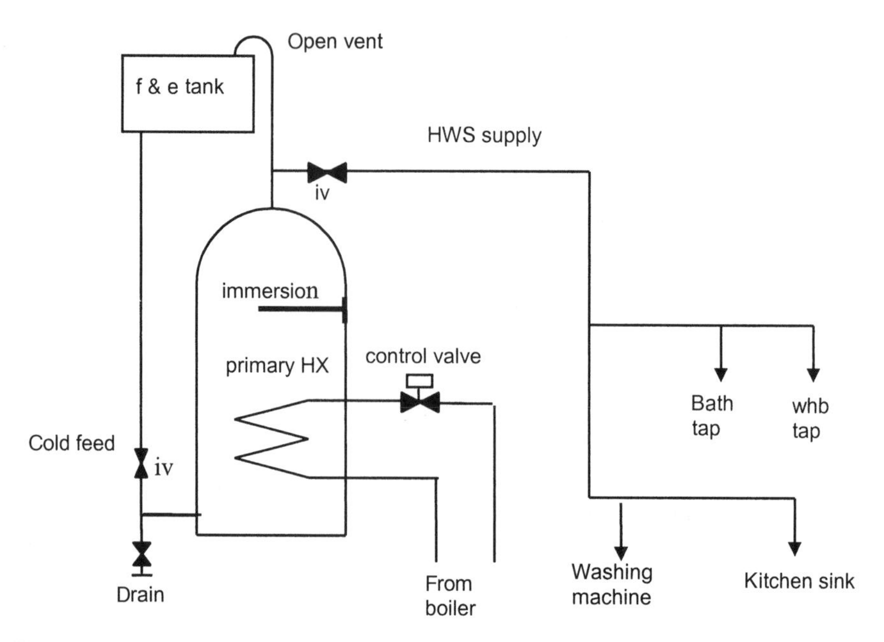

Figure 8.4 Typical domestic HWS Storage installation.

would be forced into the f & e tank; there would merely be a rise in water level within the vent pipe. The vent pipe must be of a minimum size (typically 28 mm) and must not be fitted with any isolating valves.

Modern cylinders are pre-insulated and comply with British Standards to reduce standing heat losses. These losses are relatively small and can be estimated very easily by assuming a constant water temperature and ambient temperature:

$$\text{annual energy standing losses} = AU(\theta_f - \theta_a)t_{op} \tag{8.5}$$

where A is the external surface area of the cylinder (m^2), U is the effective unit thermal transmittance – U value (W/m^2K), of the cylinder, θ_f is the temperature of the water in the cylinder (°C), θ_a is the ambient temperature (°C) and t_{op} is the operating time per year (hr) giving the annual energy in Wh. As an approximate calculation, U can be assumed to be the same over the entire surface of the cylinder and is typically about 0.8 W/m^2K. The surface area of a 125 litres cylinder is about 1.4 m^2. Typical design water temperature is 65°C and an ambient temperature of 20°C. Assuming continuous operation all year round (8760 hr per year) this equates to an annual heat loss of approximately 440 kWh. Assuming a gas-fired boiler is used with a seasonal efficiency of 65% and a tariff of 2 p/kWh for domestic gas, this leads to an annual cost of only about £13. Nevertheless, this represents about 10% of the annual cost of HWS for an average dwelling, thought to be about £100/year. Of course, not all this heat loss from the cylinder is necessarily wasted as it may contribute to the space heating.

Heat is also lost from distribution pipework. Current building regulations require that distribution pipework is insulated. This is both to save energy

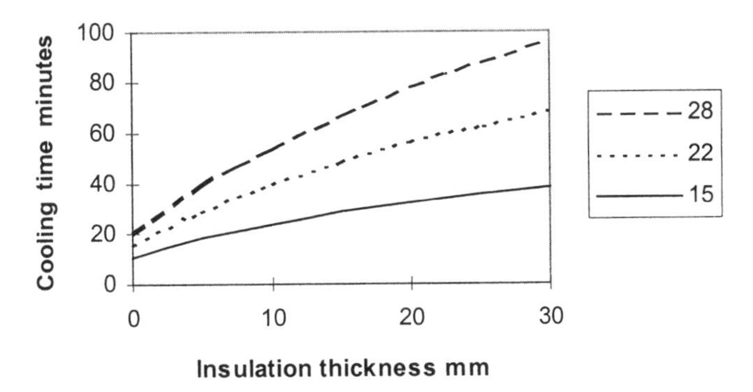

Figure 8.5 Effect of insulation thickness on cooling rate.

and water. Insulation slows down the heat loss rate but the water will still eventually fall below an acceptable temperature. Users will run the tap until sufficiently hot water is obtained, wasting significant amounts of lukewarm water. Figure 8.5 shows the heat loss rate for various nominal pipe sizes and various thicknesses of insulation for the time taken for the water temperature, initially at 65°C, to fall to 50°C (when the ambient temperature is 20°C). Uninsulated 15 mm pipes require only about 10 minutes for the water to cool down whereas adding commonly used 10 mm thick DIY insulation extends this to about 24 minutes. Yet increasing the thickness to 15 mm (which current building regulations stipulate) would only slow down the process by a few minutes.

Estimating the quantity of energy (and water) wasted in this way is very difficult as it depends on the frequency of use, amount of water which needs to be run off (dead-leg), and the behaviour of the occupants. A simple model suggests a cost of about £30/year for a domestic system. So total losses are about £43 (for an insulated system) in total annual costs of around £100. In other words, even a well-insulated system is only about 55–60% efficient.

8.4.2 Non-domestic storage system

The waiting time at taps for acceptably hot water to reach the user has a further economic implication in commercial buildings: loss of workforce productivity. In order to reduce waiting time, water must be available within the pipework at acceptable temperatures at all times (during occupancy). Further, legionella is a major concern and maintaining a high water temperature in all parts of the pipewok will reduce the risk. As insulation will only slow down the cooling rate of the water and not stop it, other measures are needed. There are two approaches: trace heating of the pipework or a continuous circulation of water. In both cases, good standards of insulation are still required. Other than this, the design is essentially the same as for domestic systems.

Figure 8.6 shows a typical system utilising a return circuit. The system is designed to provide a continuous flow of hot water around the circuit just sufficient to prevent the water temperature at the furthest tap from the cylinder falling below the required value. Flow rate is generated by a small pump. The return pipe is connected to the end of each supply branch so that there is a continuous circulation of water through all branches. It would be

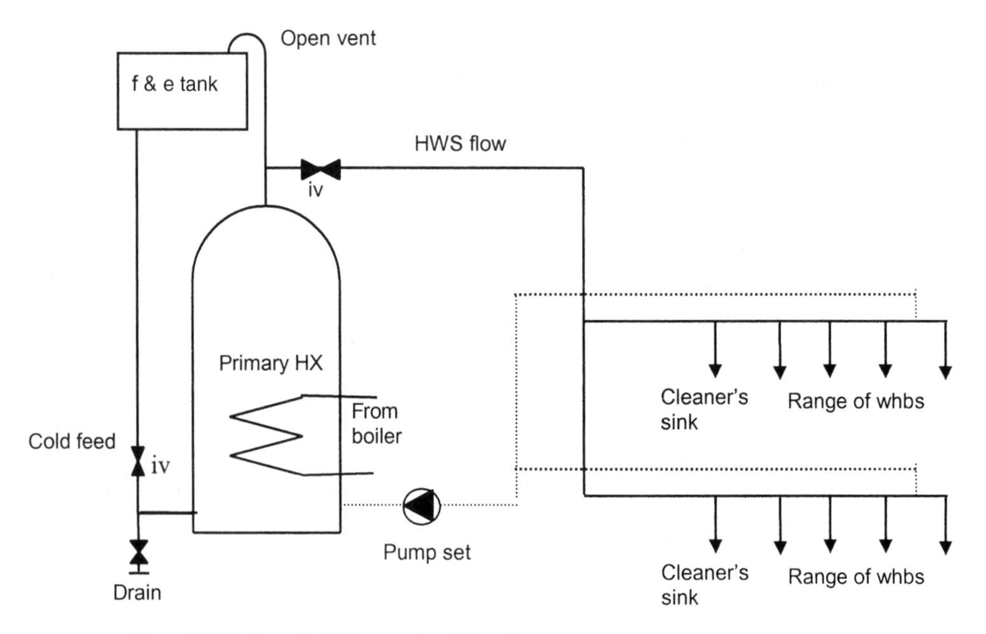

Figure 8.6 Non-domestic HWS storage installation with return circuit.

impracticable to provide a continuous circulation through pipes serving an individual tap. Such pipes are called 'dead-legs'. Current regulations[5] allow a maximum length of dead-leg (depending on pipe diameter). The purpose of this was originally to reduce water wastage but it has obvious benefits in terms of reducing energy consumption and risk from legionella.

In order to determine the flow rate required for the continuous circulation, the heat loss rate through the whole of the supply distribution pipework system, Q, is determined, based on a mean water temperature of about 55°C, as described in section 7.2. Q can then be equated to the heat lost by the circulating water using a form of equation 8.1:

$$Q = \dot{m}_{circ} C_{pw} (\theta_s - \theta_{min}) \tag{8.6}$$

where $\dot{m}_{circ}$ is the mass flow rate of water required (kg/s), θ_s is the temperature of water in the storage cylinder and θ_{min} is the minimum acceptable water temperature within the supply pipework (°C).

The circulating pump is sized to produce this flow rate only. It is not intended to supply any of the flow to the taps. For this reason it is important to locate the pump such that it will not restrict flow of water when taps are opened. The return pipe is one obvious position. Alternatively, some designers prefer to locate it in the main flow pipe from the cylinder with a bypass around it. In this position the pump assists the flow of water to the taps but results in a more complex arrangement of pipework and fittings. In most cases, the return pipework will be only 15 mm diameter. This method of maintaining a high flow temperature also helps to reduce stratification in the cylinder, so reducing the risk of legionella bacteria growing.

Trace heating is considerably simpler and probably has a lower capital cost than circulating water. However, its use of electricity will increase running costs and have a much greater environmental impact. It may be cost effective to increase the insulation thickness, using the optimisation method discussed in section 7.2.

The extensive distribution pipework combined with maintaining a high water temperature, by either method, will result in the non-domestic storage system having an even lower efficiency than the domestic system. Indeed, if no hot water is drawn off, the efficiency becomes zero. For certain types of buildings, such as offices, the hot water demand will be very low and can be further reduced by the use of spray taps. This will in effect reduce the system efficiency even though water and energy consumption are reduced. Efficiency, then, is not a particularly useful indicator of the effectiveness of the non-domestic hot water services system. A better indicator would be annual primary energy consumption per person or per unit of floor area.

The CIBSE Energy Code[6] bases HWS targets on a floor area basis. At the same time, the HWS demand is based on daily hot water consumption per person and surprisingly high HWS system efficiencies – 80% for a storage system.

8.4.3 Unvented systems

Unvented systems are becoming very popular for both domestic and non-domestic systems. The cold feed can still be from a feed tank if required so as to maintain a constant head pressure, although the trend for domestic seems to be for it to be served directly from the incoming cold water mains. This has the advantage of a higher head than can normally be achieved in domestic premises but does mean that hot water service is lost should the cold mains fail. In some areas of the country mains water pressure can be very low during peak hours.

The unvented system must prevent excessive build-up of pressure for whatever reason. To do this, a pressure activated safety valve is installed near the cylinder. Where an electric immersion is installed, a high-temperature cut-out is required in addition to the normal immersion heater controls. Water regulations require a double check valve, with test drain, to be installed to prevent back flow of hot water into the cold water supply pipe. A typical arrangement is shown in Figure 8.7.

8.4.4 Pumped storage

In designing the storage system above, gravity (or in unvented systems, mains pressure) provides the flow to the taps. Pipework must therefore be sized with this available pressure in mind. On very large systems the available head per metre of pipe can be quite small, resulting in large diameter pipework being required if the desired flow rates are to be achieved. This increases both capital costs and heat loss rate. In such instances, it may be cost effective to utilise a pumped system with more economically sized pipework. The pump will need to be able to provide the peak demand flow rate while for much of the time having to operate at greatly reduced duty. For trace heated systems, a boosted pump set complete with a small expansion vessel (as is used on boosted cold water systems) may be worth considering.

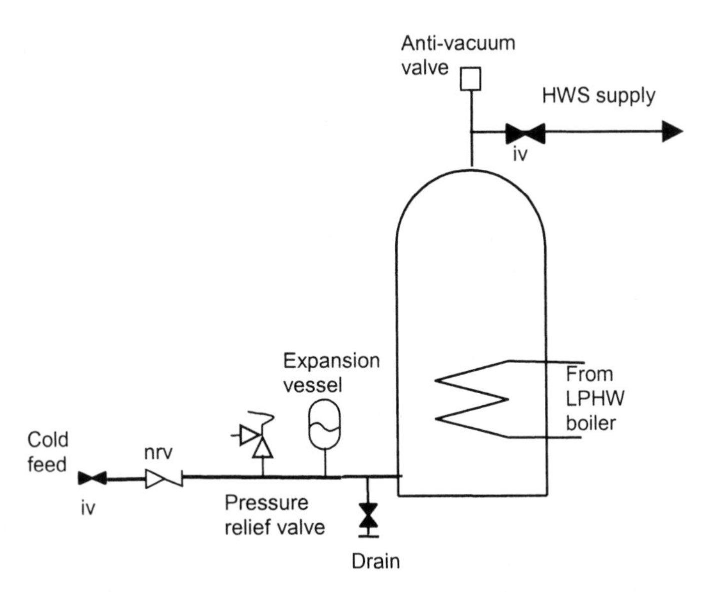

Figure 8.7 Unvented HWS storage system.

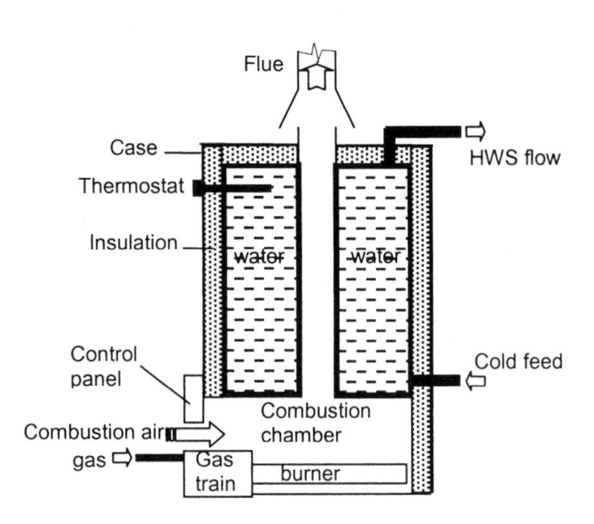

Figure 8.8 Hot water generator.

8.4.5 Hot water generators

The hot water generator is a direct-fired storage system, as shown in Figure 8.8. This has the advantage that it is completely independent of the space heating boilers and so does not compromise efficiency in summer (see Chapter 10). The burner may be oil or gas-fired and allows a high generation rate in comparison to the heat exchangers used in storage cylinders. At the same time, these systems have relatively small storage volumes. In this respect, they approach instantaneous systems with very fast recovery times (in the order of minutes). A major drawback can be rapid scaling of the heat exchanger in hard water areas unless water softening is used.

8.4.6 Instantaneous

Instantaneous systems for domestic buildings usually mean a combination boiler. As discussed in section 8.1, despite their relatively high heat output rate for a domestic boiler – typically about 20 kW, this still puts severe restriction on the hot water generation rate. The efficiency of a combination boiler is difficult to assess. Although there are no standing losses from a hot water cylinder, there will still be losses from the pipework and this will depend on how extensive the pipework is and the usage pattern, not on the efficiency of the boiler.

For non-domestic applications with low water consumption rates, such as offices, a popular solution is to use small, local electric heaters. Although this involves a high cost fuel with a high primary fuel ratio, the much higher efficiencies (probably about 90%) compared to the distribution losses of the centralised or localised storage system would suggest this has lower energy costs. It may also have lower capital costs. This system is also popular because it removes many of the problems associated with legionella (see section 8.6).

For higher consumption rates, such as sports centres, hotels and hospitals, a system finding increased application is the flat plate heat exchanger served by either the space heating or a dedicated boiler. This reduces the rate of scaling suffered by the hot water generator as the temperature on the heat exchanger is greatly reduced. These systems are now available as a package, complete with pumps and controls. The distribution pipework is as for a storage system.

8.5 Solar hot water

Solar hot water, at least for domestic buildings, is slowly becoming more popular in the UK. This, it must be said, is because of environmental concerns and the desire to save primary energy and reduce CO_2 emissions. The current cost of solar hot water, even most DIY systems, is still far too high to be economically viable. This is a pity as despite our often cloudy weather conditions in the UK, solar panels can easily contribute 50% or more of the annual hot water requirements of dwellings[7]. There are a number of useful web sites for detailed information on solar panels, including manufacturers and environmental organisations[8].

Despite the solar energy being free, the efficiency of the panel is still important as it reduces the area of panels needed and therefore capital cost. There may also be limitations on the roof area available and suitable for siting the panel. This needs to be predominantly south facing and at an angle of between about 30° and 60° to the horizontal. It should also not be shaded by trees and other buildings.

8.5.1 Integration with HWS system

Figure 8.9 shows a typical installation. A water-glycol solution is pumped through the panel and a heat exchanger in a hot water cylinder. This cylinder preheats the water flowing into the second cylinder which contains the conventional heat exchanger served by the boiler. The second cylinder then acts as a top-up as necessary. This means that useful heat can be obtained from the panel even on cloudy days and, just as importantly, the panel does not have to be designed to achieve a high outlet temperature.

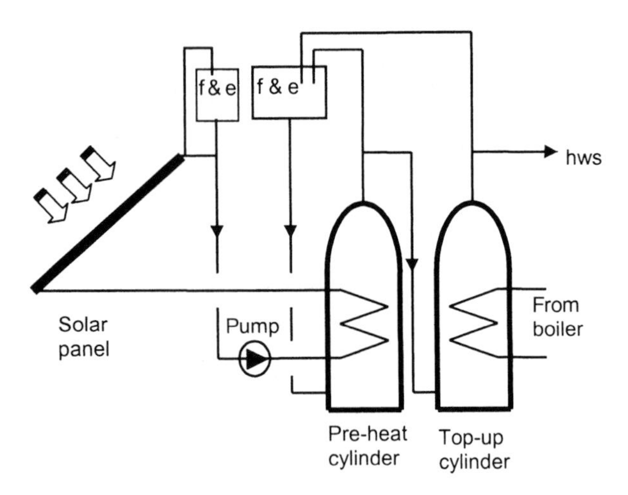

Figure 8.9 Solar hot water system.

The two cylinders improve energy recovery by increasing the temperature difference between the water in the store and the water leaving the panel. It is also a useful method of retrofitting solar water heating to an existing system. For new systems, it is now common practice to use a single, larger cylinder with the heat exchanger from the solar panel installed in the lower half and the heat exchanger served by the boiler in the upper half. Stratification maintains a good temperature difference. This uses less space than two smaller cylinders but requires a specialised cylinder.

The pump is controlled via a temperature sensor in the panel so that should the water in the panel be cooler than in the cylinder, the pump stops. The solar panel is not only a good collector of solar radiation but also a heat emitter at night or in cold and cloudy weather. As an alternative, an inherently simpler arrangement, if the panels can be arranged below the cylinders then water flow by thermo-syphoning will be achieved and the pump and pump control can be omitted. There will only be water circulation when the water in the panel is above that in the cylinder.

The panel temperature can drop well below 0°C at night, even in mild weather, due to radiation to the sky. A water-glycol mixture is recommended. Some systems have been designed to be drained down in cold weather, but this will increase maintenance requirements, mean no solar water heating in winter and still not guarantee that freezing will not occur during clear sky conditions.

8.5.2 Efficiency and output of panel

The efficiency of a solar collector is not merely a function of the construction of the panel, but also of operating conditions. As it loses heat to the surroundings as well as captures some of the incident solar energy, the efficiency depends on this balance between gain and loss. This can be represented by a simple graph, as shown in Figure 8.10. This graph is for a conventional flat plate collector. Much higher efficiencies can be achieved with more sophisticated (but more expensive) designs. Solar energy captured depends on solar intensity (W/m^2) whilst the rate of heat lost depends on the panel tempera-

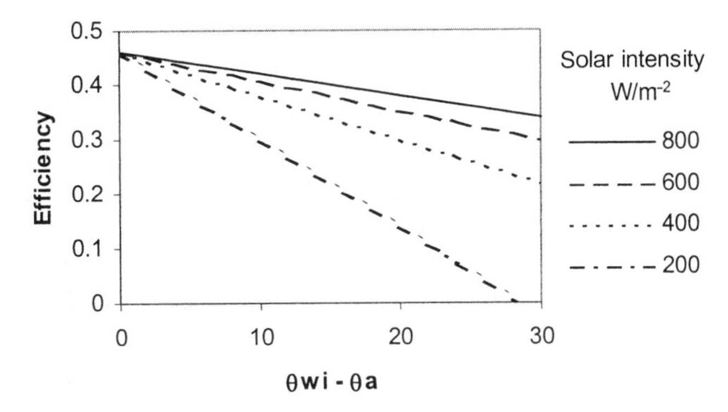

Figure 8.10 Flat plate solar panel efficiency.

ture which in turn depends on the temperature of the water entering the panel from the store ($\theta_{w,in}$) and the ambient temperature (θ_a). As can be seen, the highest efficiency occurs when the water enters at the same temperature as ambient. In practice, it will be higher of course. Higher solar intensities produce higher efficiencies.

The equation for the graph is given by:

$$\zeta = C_1 - C_2 \frac{(\theta - \theta_a)}{I_t}$$

$$\text{where:} \quad C_1 = \frac{\alpha}{1 + \dfrac{A_f(U_f + U_b)}{2\dot{m}C_{pw}}} \quad \text{and} \quad C_2 = \frac{(U_f + U_b)}{1 + \dfrac{A_f(U_f + U_b)}{2\dot{m}C_{pw}}} \qquad (8.7)$$

where α is the solar absorptance, U_f and U_b are the unit thermal transmittance for the front and back of the panel respectively, A_f is the face area of the panel, $\dot{m}$ the water mass flow rate and C_{pw} the specific heat of water. Appendix 8.1 shows how this equation can be derived. Understanding this equation allows us to be able to predict how the efficiency could be improved.

Reducing the U values will increase the efficiency. The back (and sides) of the panel are relatively easy to insulate with conventional insulation. The front presents more of a problem in that it must let in the solar radiation but reduce heat loss both by convection and radiation. A glass cover is usually employed as this has high transmittance in the short wave but low in the long range (infra-red). The energy in solar radiation is concentrated at short wave (around the visible wavelengths of about 0.5 μm); the warm panel will radiate at much longer wavelengths (peaking at about 9 μm for a panel temperature of about 50°C). Double glazing may be used, but this will increase cost and weight. For panels operating at only about 50°C, this may not be economically viable. Acrylic is lighter and more durable but has a higher transmittance in the long wave than glass.

Very high efficiency panels employ a vaccum to reduce convection heat losses through the front. This is achieved by using evacuated glass tubes through which the pipes run – tubes having greater strength than flat glazing units and so able to withstand atmospheric pressure. Such systems can give water leaving temperatures well in excess of 100°C.

It should be noted that even a perfectly insulated panel ($U_f = U_b = 0$) will have a maximum efficiency equal to α. This is a function of the transmittance of the glass, etc. and the absorptance of the panel surface. The transmittance of the glass, etc. has already been discussed above. The same argument applies to the absorptance of the panel surface. It will ideally be highly absorbing at short wavelengths but have a low emittance at long wavelengths. (Absorptance varies with wavelength. At any one wavelength, absorptance = emissivity.) Such surfaces are called selective emitters. Absorptance at short wavelength is more important than emittance at long wavelength when glass covers are used, so the panel is usually a matt black finish.

The flow rate per unit area of panel should be high. This will reduce the temperature of the panel and so reduce heat losses. However, there is usually a minimum water outlet temperature requirement. There is therefore a compromise between efficiency and water outlet temperature. Inlet water temperature should be as low as possible, which implies a large storage volume in the cylinder. The panel should obviously be installed to receive the maximum solar energy. It is possible to track the sun, but most use a fixed panel. The position is then a compromise as the sun's position will vary with time of day and time of year. A rule of thumb is to position it facing due south at an angle to the ground roughly equal to the angle of latitude of the site. For the UK this is about 50°.

8.5.3 Sizing of the system

Manufacturers have rule of thumb methods for determining appropriate sizes of panel and store required for dwellings. For other applications, a simulation model should be considered as the relationship between panel area, store volume and consumption pattern is critical. A doubling of panel area will rarely produce double the solar energy captured as the store temperature will increase, depending on consumption. This is complicated by the variations in solar availablity and outdoor temperature. Some computer models are available, one (cost free, subject to use) via the internet[9]. Alternatively, it is relatively simple to write a spreadsheet application using, for example, hourly values of hot water consumption, solar intensity and outdoor temperature with equation 8.7.

8.6 Legionella

The bacterium *legionella pneumophilia* is present throughout the natural environment, requiring warm, moist conditions and nutrients to multiply. An ideal environment can be created in a hot water services installation. If water droplets containing sufficient concentrations of the bacteria are inhaled, a pneumonia like infection (legionnaires diseases) can occur which is often fatal for the elderly and people with reduced immune systems. Showers in HWS systems create this ideal method of infection (as do cooling towers in air conditioning systems). Hospitals are naturally more prone to fatal cases and have very strict rules on the design and maintenance of hot water systems.

The ideal temperature for the bacterium is about 37°C (human body temperature) which is easily created in parts of a HWS system. Below about 20°C and above about 55°C the bacterium is dormant and above 70°C is killed. As the bacterium is almost certainly in the water supply, the objective is to prevent multiplication to dangerous concentrations. This is achieved by a combination of good design practice and regular maintenance. Guidance and regulations on both aspects are available from a number of sources[10].

Good design practice includes appropriate design of the cold water feed tank, where used, as this can easily reach temperatures above 20°C in the summer; prevention of stratification in hot water cylinders, so allowing a layer of water at about 37°C to form; maintenance of water temperatures above about 50°C in distribution pipework (see section 8.4.2); and the facility to regularly increase the store temperature to 70°C ('pasteurisation').

Appendix 8.1 Solar Panel Efficiency

Where moderately efficient solar panels are required, a common solution is the flat plate collector as shown schematically in Figure 8.11. The panel has a U value for the front and back of U_f and U_b respectively. The face area is A_f. It is subjected to an intensity of solar radiation normal to the face of I_t of which the fraction α is absorbed. There is a flow rate of water, m, entering at $\theta_{w,in}$ and leaving at $\theta_{w,out}$. The ambient temperature is θ_a. The system is operating at steady state and the mean temperature of the panel is θ.

Efficiency

We can define efficiency in the usual way, i.e. useful output/input:

$$\zeta = \frac{mC(\theta_{w,out} - \theta_{w,in})}{A_f I_t}$$

But we do not know the water leaving temperature $\theta_{w,out}$. If we can eliminate this, we can determine efficiency for stated inlet temperatures and solar intensities.

Consider the energy balance, where the heat loss to ambient is given by q_{loss}:

$$\alpha A_f I_t = q_{loss} + mC(\theta_{w,out} - \theta_{w,in}) \approx A_f(U_f + U_b)(\theta - \theta_\alpha) + mC(\theta_{w,out} - \theta_{w,in})$$

$$\theta_{w,out} = \theta_{w,in} + \frac{\alpha A_f I_t - A_f(U_f + U_b)(\theta - \theta_a)}{mC}$$

simplify by writing as:

$$\theta_{w,out} = \theta_{w,in} + XI_t - Y(\theta - \theta_a)$$

where $X = \dfrac{\alpha A_f}{mC}$ and $Y = \dfrac{A_f(U_f + U_b)}{mC}$

But the inside temperature of the panel, θ, is itself dependent on water outlet temperature. If we assume that there is no mixing of water in the panel then we can say that the mean temperature of the panel is equal to the average of the water entering and leaving temperatures:

$$\theta_{w,out} = \theta_{w,in} + XI_t - Y((\theta_{w,out} + \theta_{w,in})/2 - \theta_a)$$

$$\theta_{w,out} = \frac{\theta_{w,in}(1 - Y/2) + XI_t + Y\theta_a}{(1 + Y/2)}$$

We now have the water outlet in terms of inlet water temperature plus the solar intensity and ambient temperature. Now substitute this expression into the equation for efficiency and so eliminate the water leaving temperature:

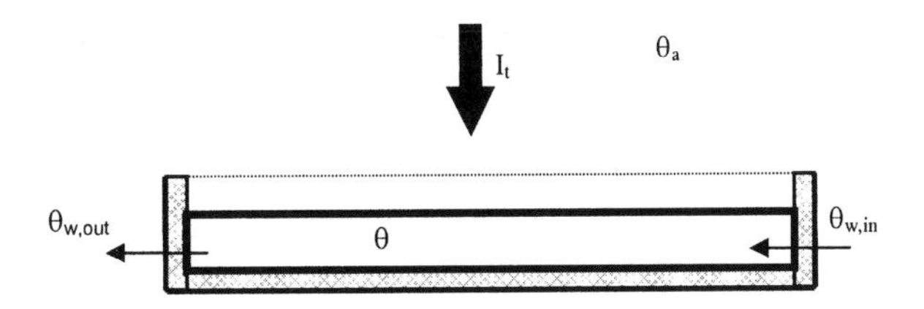

Figure 8.11 Schematic of flat plate collector.

$$\zeta = \frac{mC}{A_f I_t}\left\{\frac{\theta_{w,in}(1 - Y/2) + XI_t + Y\theta_a}{(1 + Y/2)} - \theta_{w,in}\right\}$$

which simplifies to:

$$\zeta = \frac{mC}{A_f(1 + Y/2)}\left\{Y\frac{(\theta_a - \theta_{w,in})}{I_t} + X\right\}$$

If we now replace the ratios X and Y:

$$\zeta = \frac{1}{1 + \frac{A_f(U_f - U_b)}{2mC}}\left\{(U_f + U_b)\frac{(\theta_a - \theta)}{I_t} + \alpha\right\}$$

This can be written in the form of a straight line equation:

$$\zeta = C_1 - C_2\frac{(\theta - \theta_a)}{I_t}$$

where: $C_1 = \frac{\alpha}{1 + \frac{A_f(U_f + U_b)}{2mC}}$ and $C_2 = \frac{(U_f + U_b)}{1 + \frac{A_f(U_f + U_b)}{2mC}}$

References

1. Chartered Institution of Building Services Engineers (CIBSE) (1986) Miscellaneous Piped Services, Section B6 of the CIBSE *Guide*. CIBSE London.
2. *ibid.*
3. Wise, A.F.E. & Swaffield, J.A. (1995) *Water, Sanitary and Waste Services for Buildings*, 4th edn. Longman, Harlow.
4. Chartered Institution of Building Services Engineers (CIBSE) (1986) Water Service Systems, Section B4 of the CIBSE *Guide*. CIBSE, London.
5. Water Supply Regulations 1999.
6. Chartered Institution of Building Services Engineers (CIBSE) (1998) *Energy Demands and Targets for Heated and Naturally Ventilated Buildings*. CIBSE, London.
7. Duffie & Beckman (1992) *Solar Engineering of Thermal Processes*, 2nd edn. John Wiley & Sons, New York.
8. www.ises.org (International Solar Energy Society which hosts WIRE – Worldwide Information systems for Renewable Energy); www.iea.org (International Energy Association); www.greenbuilder.com/sourcebook (information on the design of solar heating and other renewables); www.thermomax.com (solar panel manufacturer).

9. www.retscreen.gc.ca (software available free for a number of renewable energy sources)
10. CIBSE (2002) *Minimising the Risk of Legionnaires' Disease*. Technical Memorandum TM13. CIBSE, London. Health and Safety Commission (2001) *Approved Code of Practice – Legionnaires' Disease: the control of legionella bacteria in water systems*. HSC Books, London.

9 Sizing Central Boiler Plant

This chapter looks at the factors that affect the choice of boiler plant size (and the emitters to some extent), as well as the options available for switching the plant on and off in order to satisfy comfort conditions and protect the building fabric while ensuring minimum energy consumption. Sizing is not an exact science and there are contrasting opinions about the best approach. Some of these differing approaches are presented here, together with a more structured methodology that takes life-cycle cost into account. Discussions are included about the benefits and drawbacks of oversizing and how practical considerations may ultimately dictate the installed boiler duty. The term 'oversize' is used to describe plant that is sized larger than the calculated steady state design heat load; it is slightly misleading since some degree of oversize is desirable and even necessary, as will become clear in this chapter. Oversizing of boiler plant is therefore not necessarily bad design practice, but excessive oversize should be avoided.

There are two key components in boiler sizing:

(1) Obtaining the design steady state heat output
(2) Selecting the amount of oversize.

Both contain plenty of scope for assumptions, cautious choices and error. Steady state design conditions will be reviewed before looking in detail at the oversize issue.

9.1 Design heat output

The equations governing the design heat loss, Q_{design}, from a space were discussed in Chapter 6. These are:

$$Q_{design} = \sum(UA)(\theta_{ei} - \theta_{ao}) + 0.33NV(\theta_{ai} - \theta_{ao}) \qquad (6.7)$$

or in terms of dry resultant temperature

$$Q_{design} = [F_{1cu}\sum(UA) + 0.33F_{2cu}NV](\theta_c - \theta_{ao}) \qquad (6.9)$$

Note that for typical room geometries and low infiltration rates F_{1cu} and F_{2cu} are very nearly equal to 1, and $\theta_c \cong \theta_{ei} \cong \theta_{ai}$. For the purposes of simplicity this chapter will consider a single representative indoor temperature, θ_i, although the reader is advised to use the appropriate temperature for the systems and spaces under consideration. In certain types of application, e.g. a warehouse or factory, there may be large spaces with high infiltration. In these cases convective systems have a very high F_{2cu} because they are effectively dealing with a large air point load; on the other hand highly radiant systems are advantageous as they have low F_{2cu} values (<0.8), and while the air temperature is kept low, comfort is maintained by high radiant temperatures. For a more detailed discussion see Chapter 6.

The type of emitter can therefore affect the value of Q_{design}, but so can each of the other terms in equation 6.7. The U values of the walls, windows and doors, etc. are effectively dictated by the building regulations, but the designer does have scope to improve on these if he/she wishes. This is predominantly an economic decision, although the cost of insulation is a relatively minor one compared to the rest of the building. The U values do contain some small uncertainties, for example the surface heat transfer coefficients change with temperature – but these are insignificant in comparison to other uncertainties.

Perhaps the most uncertain factor is the infiltration air change rate, N, which will depend on how well a building has been built, and to some extent on the behaviour of the occupants (who may like windows open, or who do not shut loading bay doors). While the CIBSE *Guide* 1999[1] suggests typical values of N for different building types – typically 1 for offices – this is dependent on wind velocity and other variables which are difficult to predict. This highly circumspect value can have a significant impact on Q_{design} as modern buildings now have low fabric heat loss coefficients. The move to highly airtight buildings with controlled mechanical ventilation may remove some of this uncertainty, but there is still scope to overestimate the air infiltration rate. Note that mechanical ventilation loads should be assessed separately from space heating loads for clarity of the design process; these must be added to the heating load to assess boiler size, although strictly speaking this should be done after boiler oversize capacity has been determined. Note that infiltration load is only experienced by those rooms on the windward side of the building, and central plant size should be calculated with this in mind.

The other factor in equation 6.7 that greatly determines Q_{design} is the outdoor design air temperature θ_{ao}. The 1986 CIBSE *Guide* A2[2] suggested that the local 24-hour mean outdoor temperature that is exceeded for only one day in the year should be chosen for lightweight buildings, and the 48 hour mean exceeded for no more than two days for heavyweight buildings. The 1999 *Guide* A2 is less prescriptive and introduces the concept of design risk. The choice will also depend on application; for example, a hospital with critical care areas will need to cope with the worst weather situations – especially if all-air heating is used.

The lower the choice of outdoor design temperature the less the risk of underheating, but the greater the effective oversize of the plant will be for most of the heating season. The latter can be dealt with by good system design and effective control, but it may prove expensive. Where the building is to be heated intermittently this inherent oversize may be beneficial, but the CIBSE guidance does not completely identify the most economic size of boiler. The decision whether to choose a low outdoor temperature with small oversize, or higher temperature with larger oversize, is still an open question. The methods described in the sections that follow attempt to deal with this.

Other factors also come into play in plant sizing. It is common to add some margin for pipe losses – typically 10%; this may be unwarranted, particularly when pipes run within the building and contribute to the heating (see section 7.2.3 for a discussion of this). Pipe losses should certainly be considered in group or district heating systems and where pipes run in external ducts, but otherwise they add to the inherent oversize of the central plant. Another issue is coincidence of different loads, for example space heating and hot water service (HWS). HWS can be a significant load, which is unlikely to be called upon during preheat periods, and is effectively extra capacity for most of the time. Finally, there is the issue of plant size availability. Boiler sizes increase in steps and it is unlikely that a precise boiler load can be matched from a manufacturer's catalogue. Erring on the side of caution will invariably lead to larger plant than calculated. In many instances all of this leads to vastly oversized plant which is more expensive and underutilised. These considerations should be borne in mind when reading the following sections.

9.2 Traditional sizing approaches

For continuously heated buildings it is normally sufficient to size the central boiler plant according to the design steady-state heat loss. Overcapacity may be installed as stand-by to deal with any boiler failure, but this is normally only necessary in buildings where maintaining the level of service is essential (e.g. hospitals). In such cases the economic case for installing overcapacity is related to the risk of failure (and the likely coincidence of failure with very cold conditions). The traditional approach of installing two boilers each of $\frac{2}{3}$ design capacity aimed to cover most eventualities in this respect, without incurring excessive capital costs. In many cases this $2 \times \frac{2}{3}$ approach was also sufficient for intermittent operation, where extra boiler capacity is required to heat the building to the required temperature in a suitable time. However, the whole heating system design process contains opportunities for guesswork, assumptions and fudge factors (e.g. high infiltration rates, low design outdoor temperatures, just-in-case factors), normally coupled with a pathological fear of underdesign. The result can be a steady-state design value well in excess of anything the building will normally experience, so any additional capacity added for intermittency or failure risk will be in excess of what is intended. Large capacity plant will normally give the building occupier the service required, but with subsequent lack of attention given to the efficient running of the plant.

There are issues surrounding the relationship between efficiency and plant size which are discussed at the beginning of Chapter 10; these notwithstanding, it should be borne in mind that overcapacity of central plant requires extra capital cost, which is wasted if that extra capacity is never used (so called stranded capacity). On the other hand, if plant is too small there will be periods when there is a loss of service (causing doubts about the designer's credibility). Sizing of plant therefore needs to be put on a more rigorous footing that takes account of building thermal properties, boiler output, energy costs and relative capital costs.

9.2.1 CIBSE guidance

Past guidance has been given in the form of tables[2] which show typical preheat times in hours for different plant ratios, p, as well as expected energy savings over the heating season when compared to continuous heating. The definition of plant ratio is given by:

$$p = \frac{\text{Maximum plant output}}{\text{Design steady-state heat loss}}$$

This approach is somewhat limiting, and the designer has little control over the selection of p, or knowledge about the origins of the tables. The 1999 edition of the CIBSE *Guide* A[1] acknowledges this lack of transparency by introducing a formula for calculating p, or F_3 as it is referred to in the *Guide*:

$$F_3 = \frac{24f_r}{hf_r + (24 - h)} \tag{9.1}$$

where h is the sum of the preheat and occupancy times and f_r is the building response factor given by:

$$f_r = \frac{\sum(AY) + \frac{1}{3}NV}{\sum(AU) + \frac{1}{3}NV} \tag{9.2}$$

where A is area, Y is the fabric admittance values, U is the fabric unit thermal transmittance, N is the number of infiltration air changes per hour and V is the volume of the internal space. This can be used to assess different plant ratios for different practical preheat times, although it cannot be used directly to make assessments about the energy performance of an installation. We will see later that preheat energy is reduced using larger boilers with shorter preheat times, but equation 9.1 is unable to be used for such an analysis. Equation 9.1 can be rearranged, with the heating period split into its component parts (preheat and occupancy periods), to give the preheat time, t_{ph}, for a given plant ratio:

$$t_{ph} = \frac{24\left(\frac{f_r}{F_3} - 1\right)}{(f_r - 1)} - t_{occ} \tag{9.3}$$

where t_{occ} is the length of the daily occupied period in hours. However, this can lead to negative values of preheat time, even when using realistic values of plant ratio and response factor (for example when t_{occ} = 14 hours and $f_r = 5$, if $F_3 > 1.5$ then t_{ph} is always less than 0). This raises questions about its validity and makes it unworkable as the first step in carrying out an economic assessment of plant ratio.

9.2.2 Plant size and intermittent operation

To conduct an economic appraisal of plant size it is necessary to calculate the preheat energy consumption for a given plant output. This depends on building thermal characteristics, plant output capacity and external ambient conditions. Figure 9.1 shows the relationship between heat up time and plant size for different outdoor temperatures, and Figure 9.2 shows the same for two different building thermal capacities.

To calculate the preheat time it is necessary to have a model for the building thermal response. As we have seen above, the admittance based model cannot reliably do this; other approaches include dynamic simulation, or simplified lumped capacitance models. Dynamic simulation is too time consuming to justify, unless the building is already being simulated for other reasons,

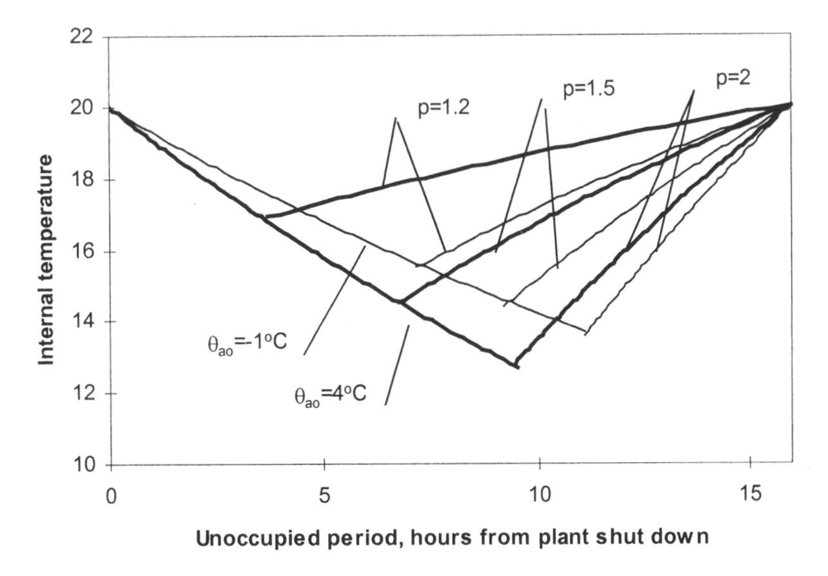

Figure 9.1 Building and plant response for different outdoor temperatures and plant ratios.

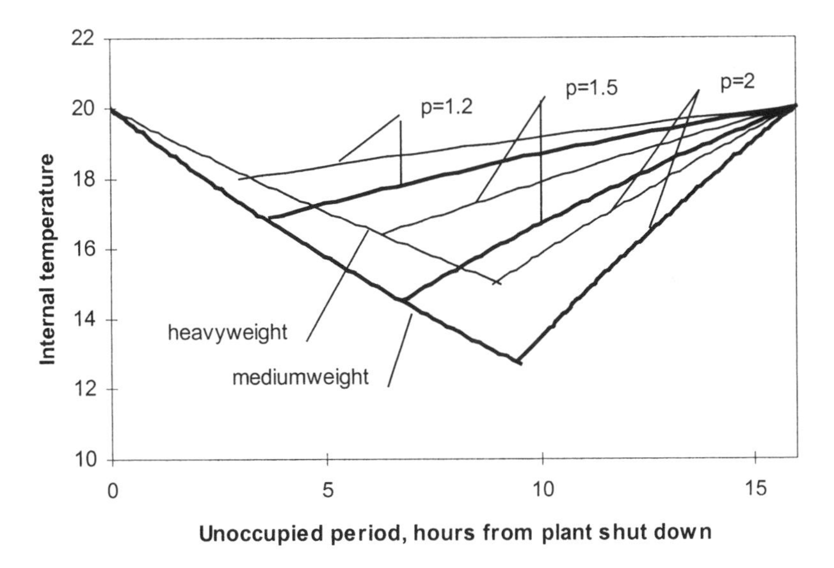

Figure 9.2 Building and plant response for different building thermal capacities and plant ratios.

in which case data may be available. What is really required is a simple, robust model that requires a small number of inputs, but that can be easily manipulated to assess the impact of various design assumptions on the size of the plant. We will adopt a first order capacitance approach, which combines all the thermal capacitances together.

The cooling of a structure to a constant temperature heat sink, θ_o, can be represented by:

$$-C\frac{d\theta_i}{dt} = U'(\theta_i - \theta_{ao}) \tag{9.4}$$

Where C is the thermal capacity of the structure in kJK^{-1} and U′ is the building heat loss coefficient in kWK^{-1}, defined as follows:

$$U' = \frac{\sum UA + 0.33NV}{1000}$$

Rearranging equation 9.4 and integrating between the limits t_1 and t_2 gives:

$$\int_{t1}^{t2} dt = -\frac{C}{U'}\int_{\theta i1}^{\theta i2} \frac{d\theta_i}{(\theta_i - \theta_{ao})}$$

with the solution being (in hours):

$$t_2 - t_1 = \frac{C}{U' \times 3600} \ln\left[\frac{(\theta_{i,2} - \theta_{ao})}{(\theta_{i,1} - \theta_{ao})}\right] \tag{9.5}$$

which is the equation of the cooling curve for the building. Similarly the heating of a structure can be represented by:

$$C\frac{d\theta_i}{dt} = Q_p - U'(\theta_i - \theta_{ao}) \tag{9.6}$$

where Q_p is the plant output, with the solution being:

$$t_3 - t_2 = \frac{C}{U' \times 3600} \ln\left[\frac{Q_p - U'(\theta_{i,1} - \theta_{ao})}{Q_p - U'(\theta_{i,2} - \theta_{ao})}\right] \tag{9.7}$$

Equations 9.5 and 9.7 can be combined as follows:

$$t_3 - t_1 = t_{unocc} = -\frac{C}{U' \times 3600} \ln\left[\frac{(\theta_{i,2} - \theta_o)(Q_p - U'(\theta_{i,1} - \theta_o))}{(\theta_{i,1} - \theta_o)(Q_p - U'(\theta_{i,2} - \theta_o))}\right] \tag{9.8}$$

where t_{unocc} denotes the unoccupied period.

Equation 9.8 can be solved for $\theta_{i,2}$ to find the optimum switch-on temperature, θ_{so}:

$$\theta_{i,2} = \theta_{so} = \frac{e^{-\left(\frac{t_{unocc}}{\tau}\right)} Q_p(\theta_{sp} - \theta_o)}{Q_p - U'(\theta_{sp} - \theta_o)\left(1 - e^{-\left(\frac{t_{unocc}}{\tau}\right)}\right)} + \theta_o \tag{9.9}$$

where θ_{sp} is the building set point temperature (which replaces $\theta_{i,1}$), and

$$\tau = \frac{C}{U' \times 3600} \tag{9.10}$$

which is also called the building time constant.

Time constant

When $\tau = t_2 - t_1$, then equation 9.5 can be re-written using:

$$\theta_2 - \theta_{ao} = (\theta_1 - \theta_{ao}) - (\theta_1 - \theta_2)$$

to give:

$$-1 = \ln\left[1 - \frac{(\theta_1 - \theta_2)}{(\theta_1 - \theta_{ao})}\right]$$

Thus:

$$\frac{(\theta_1 - \theta)}{(\theta_1 - \theta_{ao})} = 1 - e^{(-1)}$$

$$= 1 - \frac{1}{e}$$

$$= 0.63$$

This statement says that the temperature fall experienced by the building in the time equal to the time constant is 63% of the maximum possible temperature fall. Or the time constant is the time for the building to fall by 63% of the indoor to outdoor temperature difference (under steady conditions).

Having established the switch-on temperature, the switch-on time, t_2, can be now found from equation 9.5 or 9.7. When t_2 is the switch-on time (i.e. when $\theta_{i,2} = \theta_{so}$) then $t_3 - t_2$ is the preheat time, denoted by t_{ph}. This process can be used to make assessments about expected savings of intermittent operation against continuous operation using:

$$\%\ \text{saving} = \frac{U'(\theta_{sp} - \theta_{ao})t_{unocc} - Q_p t_{ph}}{U'(\theta_{sp} - \theta_{ao})t_{unocc}} \times 100 \qquad (9.11)$$

Thermal capacity

The thermal capacity of the building is found from:

$$C = \sum_{1}^{n} c_p \rho V$$

where c_p is the specific heat of an element, ρ is the material density and V is the volume of the element, for n elements.

The main weakness in this thermal response model is the choice of the total thermal capacity of the building. Not all of the mass of the building is actively involved in the thermal response, and different elements have different time constants. This is a problem that the admittance method (the basis of the CIBSE F_3 factor) does not suffer from, although the admittance method is more sensitive to the various input assumptions.

Some work has been done on this and a detailed discussion can be found in Day (1999)[3] and Day & Karayiannis (1999)[4]. These suggest that the heavier the building, the less effective depth is used in the calculation of C. Levermore[5] suggests that for single skin homogeneous elements the effective mass is half the actual mass, although Day[3] suggests that this may be an overestimate in cases of heavy buildings. As a rule of thumb it should be assumed that no more than the first 100 mm of the inner skin should be taken for calculating the effective mass.

Table 9.1 Set of results using the building cooling equation (equation 9.5).

θ_i	18	17	16	15	14	13	12	11
t_2-t_1	0	2.1	4.2	6.6	9.0	11.7	14.6	17.7

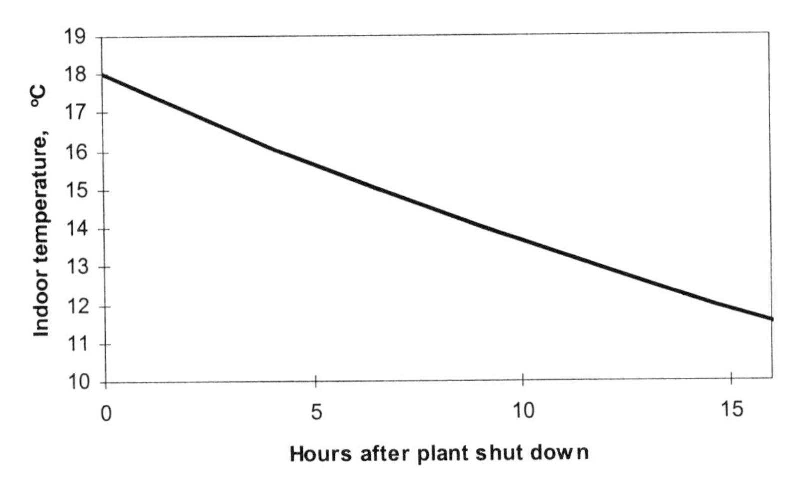

Figure 9.3 Building cooling curve.

Example

A building must be maintained at 18°C from 08.00 to 18.00. Investigate the savings from intermittent operation for an outside temperature of 0°C.

Building heat loss co-efficient (U′) = 50 kW K^{-1}

Building heat capacity (C) = 6.48 GJ K^{-1}

Building time constant (C/U′) = $\dfrac{6.48 \times 10^6}{50 \times 3600} = 36\,\text{hrs}$

The cooling rate of the building is found from equation 9.5:

$$t_2 - t_1 = -36 \ln\left[\frac{\theta_i}{18}\right]$$

which gives a set of results as shown in Table 9.1, and these values can now be plotted on a graph as shown in Figure 9.3.

The steady state heat requirements are found from:

$$50 \times (18 - 0) = 900\ \text{kW}$$

Q_p must be greater than 900 kW if the building is to heat up at a reasonable rate and reach the required set point temperature. A value of Q_p = 1200 kW will be used here, which corresponds to a plant ratio of 1.33. As t_3 and t_1 are known we can use equation 9.8 to determine θ_{so} at which the heating must come on, and hence the preheat time $t_3 - t_2$ from equation 9.7:

$$t_{unocc} = -36 \ln\left[\left(\frac{\theta_{so}}{18}\right) \times \left(\frac{1200 - 50 \times 18}{1200 - 50 \times \theta_{so}}\right)\right]$$

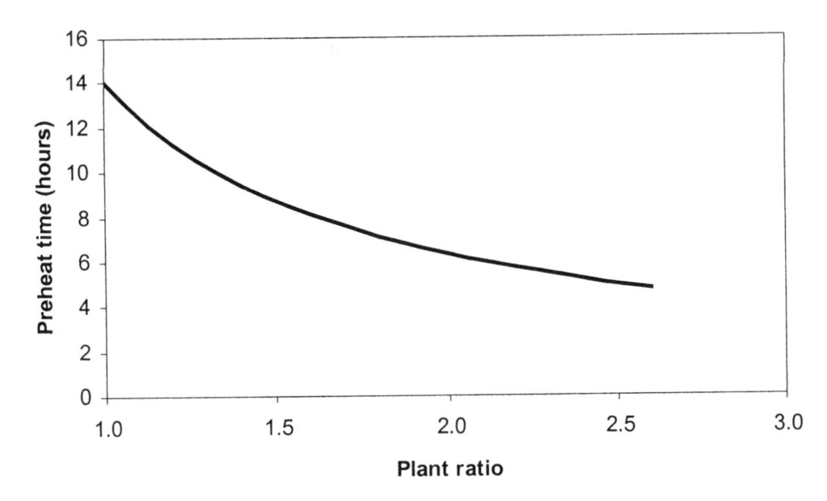

Figure 9.4 Preheat time against plant ratio.

where $t_{unocc} = 14\,hrs$

$$-\frac{14}{36} = \ln\left[\left(\frac{\theta_{so}}{18}\right) \times \left(\frac{300}{1200 - 50\theta_{so}}\right)\right]$$

$$\frac{300\theta_{so}}{1200 - 50\theta_{so}} = 18e^{-\left(\frac{14}{36}\right)}$$

$$\theta_{so} = 16.0°C$$

(Note also that equation 9.9 is a direct solution for θ_{so}.)

This value can either be checked against the cooling curve (Figure 9.3), or the preheat time $t_3 - t_2$ can be calculated from equation 9.7:

$$t_3 - t_2 = -36\ln\left[\left(\frac{1200 - 50 \times 18}{1200 - 50 \times 16}\right)\right] = 9.95 \text{ hours}$$

This process is repeated for all selected values of Q_p and a graph of $(t_3 - t_2)$ v p can be plotted as shown in Figure 9.4. From this curve it can be seen that there are diminishing returns with increased plant capacity.

The percentage energy savings are calculated as follows using equation 9.11:

$$\frac{(50 \times 14 \times 18) - (1200 \times 9.95)}{(50 \times 14 \times 18)} \times 100 = 5.24\%$$

Repeating the above example for an outside temperature of –2°C and plant ratios of 1.2, 1.5 and 2.0 gives the results shown in Table 9.2.

9.2.3 An economic approach to sizing

We can also conduct a more detailed analysis of the economic impacts of plant size using these equations. We have seen that preheat time reduces with

Table 9.2 Typical savings for varying plant sizes.

Plant ratio	Plant on temp, θ_{so} (°C)	Preheat time (hours)	% savings
1.2	16.5	11.3	3.14
1.5	15.3	8.7	6.78
2.0	14.2	6.3	10.0

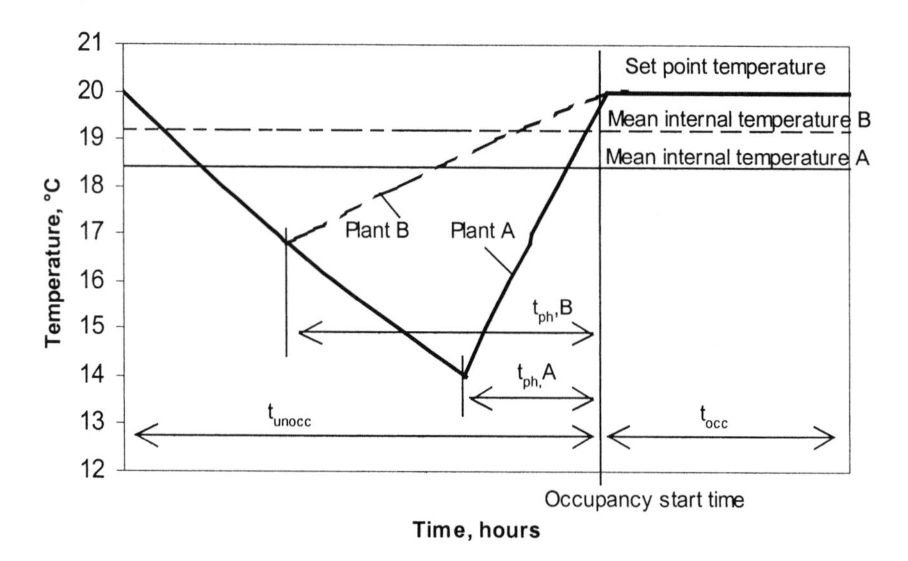

Figure 9.5 The relationship between plant size and mean internal temperature.

increasing plant size. Not only this, but the preheat energy decreases with increasing plant size. To understand why, consider Figure 9.5 which shows the response of two plant sizes. The larger plant A comes on later, by which time the building has cooled further, and the 24-hour mean internal temperature is lower. This means that the 24-hour average rate of heat loss for the building is less with plant A than with plant B.

In the worked example above the preheat energy output from the boiler was:

$$1200 \times 9.95 = 11940 \text{ kWh}$$

But a 1500 kW plant would have a preheat time of 7.7 hours giving:

$$1500 \times 7.7 = 11550 \text{ kWh}$$

This gives a saving of 390 kWh per day (provided the outdoor temperature remained the same). However, larger plant costs more to install, and there must be a point at which increasing the size of plant becomes uneconomic. In order to compare different plant sizes and their economics, it is reasonable to assume that different plant will have similar full load efficiencies, and if multiple boiler sets are used then part load efficiencies should remain high and be the same for each size of installation. Given the latter it is only necessary to compare delivered preheat energy consumption.

We can calculate the annual preheat energy savings, S_A, for a typical outdoor temperature (for example the seasonal mean outdoor temperature) when compared to a reference plant size, $Q_{p,\,ref}$. This reference plant should be the design heat loss of the building, i.e. a plant ratio of 1. The annual savings are thus:

$$S_A = [\theta_{p,ref} t_{ph,ref} - Q_p t_{ph}] \frac{N_d}{\eta} \tag{9.12}$$

where N_d is the number of days in the heating season, η is the system seasonal efficiency and t_{ph} is the preheat time ($t_3 - t_2$). The cost of the reference plant is unavoidable, but the larger plant will have an extra associated capital cost. This extra cost will be defined in terms of the marginal capital cost of the plant, M_C, given in pounds sterling per extra kW installed. This extra cost should be exceeded by the total discounted savings taken over the lifetime of the plant (or other appropriate economic assessment period). The total net discounted savings, S_N, are found by

$$S_N = M_C(Q_p - Q_{p,ref}) - f_c c_f S_A \tag{9.13}$$

where c_f is the cost of fuel in £ kWh^{-1} and f_c is the cumulative discount factor given by:

$$f_c = \frac{1-(1+r)^{-n}}{r} \tag{9.14}$$

where r is the test discount rate expressed as a fraction and n is the lifetime of the plant in years. This can be repeated for a number of plant ratios, and the optimum plant size occurs when S_N is a minimum. (Note that S_N is negative when overall savings are being made; this sign convention has been adopted for the ease of graphical visualisation of the optimum case.)

This procedure requires a fair amount of calculation but can easily be entered into a spreadsheet to give a readily available design tool where the input variables can be changed to assess the effect on optimum plant size. The whole procedure is summarised in Figure 9.6, where each box in the flow diagram represents a cell in a spreadsheet arranged in a row, with a new row for each value of Q_p.

Marginal capital cost

The marginal capital cost is the incremental capital cost of boiler plant over the reference plant size. Unfortunately the installed cost of boilers does not rise uniformly with size; cast-iron sectional boilers can add extra sections for low additional costs, although different burners can provide a step change in overall costs. An analysis of boiler marginal costs has been conducted from *Spon's Mechanical and Electrical Services Price Book 2000*[6], the results of which are shown in Figure 9.7. The cost per extra kW for two sizes of reference boiler is plotted against plant ratio. At low plant ratios there is a wide range in marginal capital cost, but this tends to become more uniform at ratios greater than 1.5. For all reference sizes a marginal cost of £12.5 – £14 kW^{-1} is typical at these higher ratios. The situation can be further

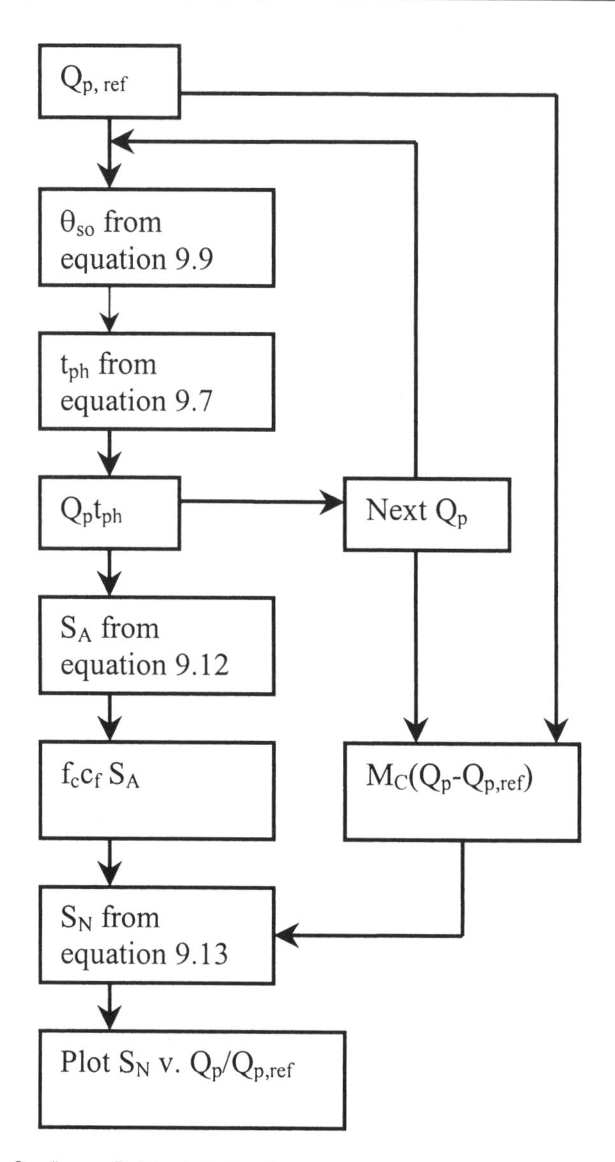

Figure 9.6 Sizing flow diagram. Each box in the flow diagram represents a cell in a spreadsheet arranged in a row, with a new row for each value of Q_p.

compounded when considering multiple boiler installations, but suitable data is not readily available to provide a comprehensive picture of these incremental costs. While it is difficult to provide general guidance on how to determine boiler marginal capital costs, the analysis in Figure 9.7 demonstrates that specific marginal costs can be determined for particular boiler ranges being considered, and that greater variability is evident for smaller boiler plant.

Assessing any capital cost increase arising from larger heat emitters and pipework is more difficult still. With radiators, for example, heat output rate depends on the temperature difference between room and radiator surface; a substantial rise in flow water temperature is required for even a modest

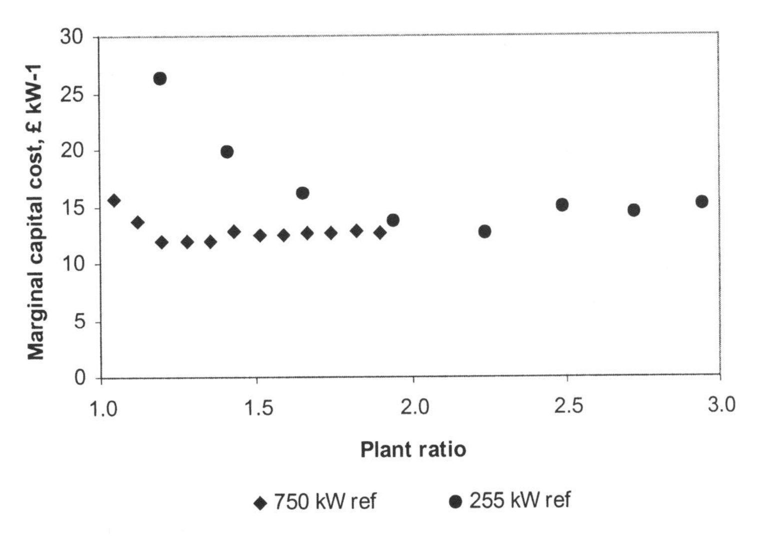

Figure 9.7 Marginal capital cost against plant ratio.

Table 9.3 Example inputs to boiler sizing model.

Internal set point temperature	°C	θ_{sp}	20
Outdoor mean seasonal temperature	°C	θ_o	5
Design outdoor temperature	°C	$\theta_{o,design}$	–1
Building heat loss coefficient	kW K^{-1}	U'	31.5
Active thermal capacity	KJ K^{-1}	C	2500000
Length of unoccupied period	h	t_{unocc}	12
Length of heating season	Days	N	200
Test discount rate		r	10%
Life of plant	years	n	10
Cost of fuel	p kWh^{-1}	c_f	1.0
Marginal capital cost	£ kW^{-1}	M_C	14

increase in output and is accompanied by an increase in return water temperature (as was seen in Chapter 6). It may not be possible to utilise the additional heat output from the boiler(s) without increasing radiator sizes. Higher water flow rates may or may not result in increased pipe sizes in all cases, but if not, there will be an increase in pump energy resulting from the higher water velocities. As with all building services systems, this is a multi-variable design problem, which can often only be assessed by conducting sensitivity analysis. The procedure shown in Figure 9.6 provides a tool for doing this for boiler sizing.

Example

Table 9.3 shows some typical inputs to the model, which represent a building with a design heat loss of 651 kW and a time constant of 22.4 hours. Table 9.4 shows the spreadsheet results for this example, while Figure 9.8 shows the preheat times and energy output for the range of boiler sizes considered. Figure 9.9 shows the net discounted savings against plant ratio;

Table 9.4 Outputs of the building sizing model using the inputs from Table 9.3.

Q_p kW	Plant ratio	θ_{so}	t2-t1 hours	t_{ph} hours	Preheat energy kWh	Saving kWh	Annual kWh	Life time discounted saving £	Extra cost of plant £	Total
651	1.00	17.5	4.1	7.9	6404	0	0	0	0	0
671	1.03	17.3	4.4	7.6	6367	38	7504.4	461.1	280	-181.1
691	1.06	17.2	4.7	7.3	6332	72	14465.0	888.8	560	-328.8
711	1.09	17.0	4.9	7.1	6299	105	20939.0	1286.6	840	-446.6
731	1.12	16.9	5.1	6.9	6269	135	26976.1	1657.6	1120	-537.6
751	1.15	16.8	5.4	6.6	6241	163	32619.1	2004.3	1400	-604.3
771	1.18	16.7	5.6	6.4	6215	190	37905.7	2329.1	1680	-649.1
791	1.22	16.6	5.7	6.3	6190	214	42868.7	2634.1	1960	-674.1
811	1.25	16.5	5.9	6.1	6166	238	47537.0	2920.9	2240	-680.9
831	1.28	16.4	6.1	5.9	6144	260	51936.3	3191.3	2520	-671.3
851	1.31	16.4	6.2	5.8	6124	280	56089.1	3446.4	2800	-646.4
871	1.34	16.3	6.4	5.6	6104	300	60015.8	3687.7	3080	-607.7
891	1.37	16.2	6.5	5.5	6085	319	63734.3	3916.2	3360	-556.2
911	1.40	16.1	6.7	5.3	6068	336	67260.84	4132.9	3640	-492.9
931	1.43	16.1	6.8	5.2	6051	353	70609.9	4338.7	3920	-418.7
951	1.46	16.0	6.9	5.1	6035	369	73794.7	4534.4	4200	-334.4
971	1.49	16.0	7.0	5.0	6020	384	76827.0	4720.7	4480	-240.7
991	1.52	15.9	7.2	4.8	6006	399	79717.43	4898.3	4760	-138.3

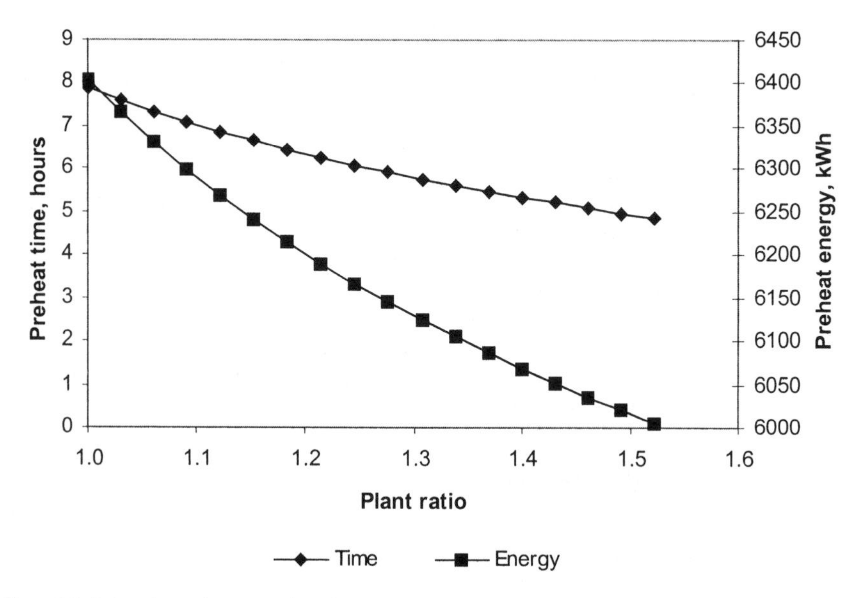

Figure 9.8 Preheat time and energy against plant ratio.

from this it can be seen that the optimum plant ratio for these conditions is 1.25.

The use of such a procedure may at first seem overly complicated for a design decision traditionally made using simple rules of thumb, or based on

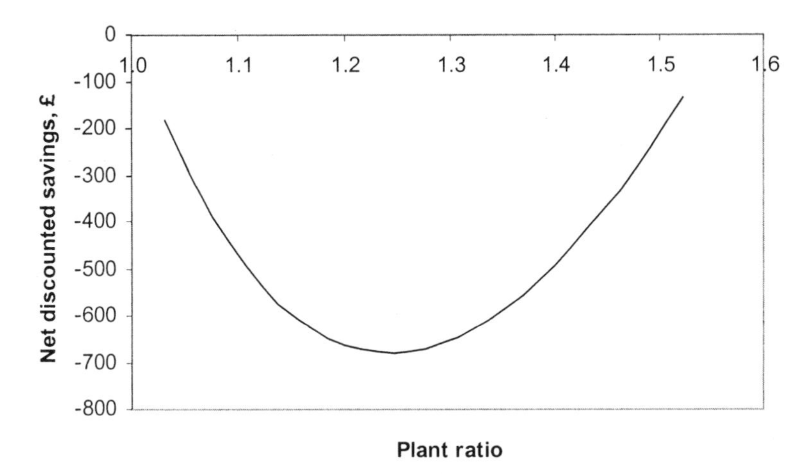

Figure 9.9 Optimum plant ratio.

experience. This may be so, but it can be argued that all decisions should be based on sound principles and should be transparent in their analysis. The only barrier to such an approach is the amount of time and effort required in repetitive and long-winded calculations. Personal computers and (in particular) spreadsheets allow such procedures to be set up once, validated, and used with confidence quickly and effectively. This gives credibility to design decisions. While the use of guides is a demonstration of good practice, so is the demonstration that designs are based on rigorous analysis using first principles. The designer also has the ability to test the robustness of a solution by varying those variables that contain the most uncertainty (e.g. air infiltration, thermal capacity, price of fuel, etc.). An example of the use of sensitivity analysis using this sizing procedure can be found in Day *et al.* (2001)[7].

An important finding of this analysis is that as buildings get heavier, the smaller the optimum plant ratio becomes. For very heavy buildings the best plant ratio approaches 1, implying that it is better to run such buildings continuously with no plant oversize. The analysis shown above helps the designer make this decision about continuous or intermittent operation at an early stage in the design.

9.3 Optimum start control

We have seen that intermittent heating can save considerable amounts of energy for many buildings. There is a theoretical time and temperature at which the plant must start – i.e. the intersection of the cooling and preheat curves. For a given building and installed plant capacity there will be a range of switch-on points for different outdoor temperatures, as shown in Figure 9.10. The locus of these switch-on points may be considered as a response characteristic for a building and system.

It is this characteristic that optimum start controllers attempt to capture. Figure 9.10 has been constructed using the idealised thermal response equations in section 9.2.2. Buildings do not follow such ideal responses, and

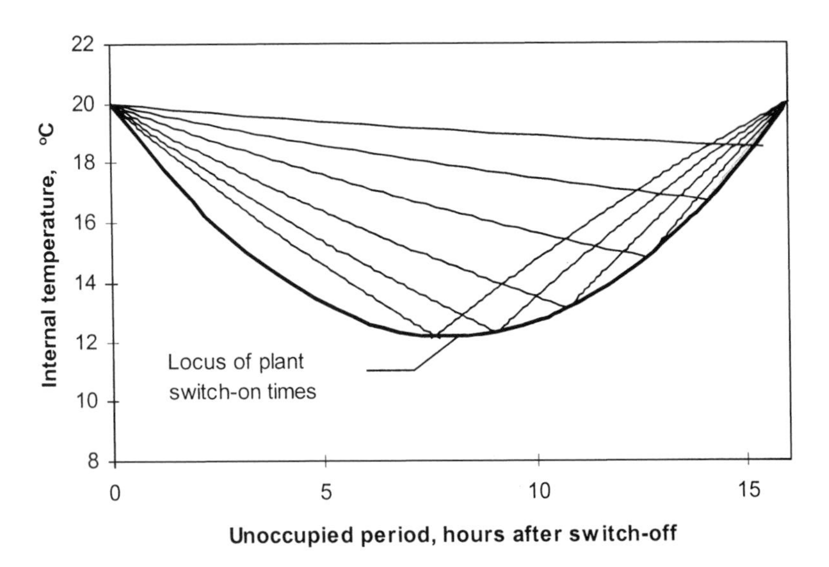

Figure 9.10 Ideal optimum start characteristic.

algorithms based on equations 9.5, 9.7 and 9.8 are unlikely to be successful, as they presume the building properties fully define the response for a given outdoor temperature. Further, if the plant was switched on according to outdoor temperature, the optimiser would effectively be open loop. However, Figure 9.10 shows that a building is likely to have a characteristic relationship between switch-on time and preheat period (the locus of the switch-on times), which allows a closed loop control to be developed – one that is based on the real response of a building. If we take the locus of Figure 9.10 and reverse the axes, we get a curve of preheat time as a function of internal temperature, as shown in Figure 9.11. (This orientation is more useful as preheat time is the dependent variable, whereas the internal temperature is the measured, or independent, variable.)

Compare this to a line of the form:

$$\ln(t_3 - t_2) = A(\theta_{sp} - \theta_i) + B + C\theta_o \qquad (9.15)$$

as shown in Figure 9.12 (note the curve of Figure 9.11 is shown as a broken line for comparison). The two lines are not identical, but it has been shown by Birtles and John[8] that a schedule of this kind produces successful optimum start operation. The inclusion of the outdoor temperature term is necessary to deal with sudden cold snaps, which would otherwise be responded to too slowly. (Equation 9.15 and the ideal locus in Figure 9.11 actually depart radically from each other at long preheat times.) The success of an optimum start controller is measured in terms of the probability that building set point temperature will be reached by the time the occupants arrive. If the temperature is reached too soon there will be wasted energy, and too late gives poor service, which undermines the confidence that the occupants have in the heating system. What allows equation 9.15 to work is that A, B and C are

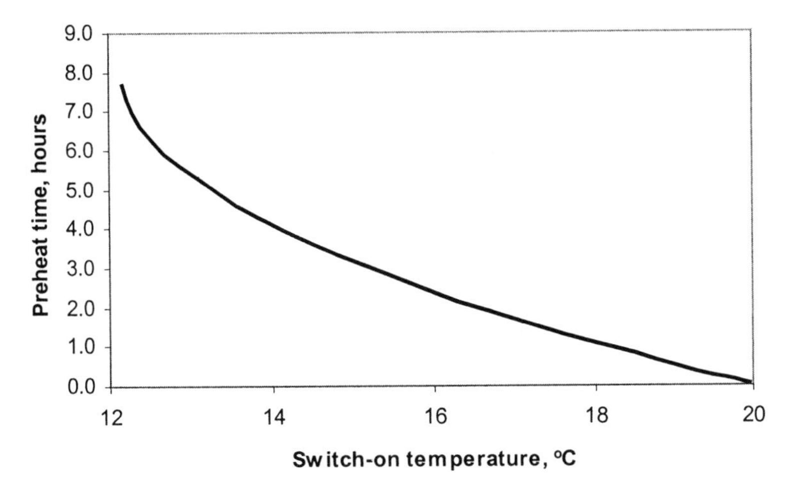

Figure 9.11 Idealised preheat time against switch-on temperature. This is a re-orientation of the locus in Figure 9.10.

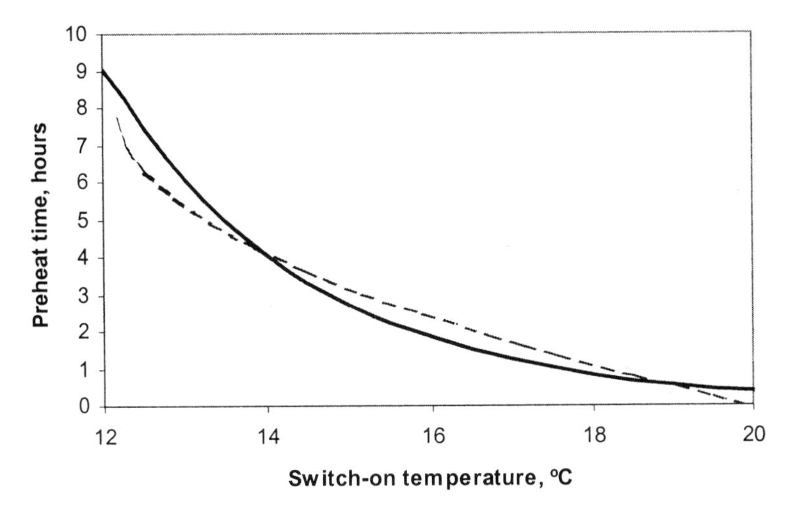

Figure 9.12 Practical optimum start schedule (broken line shows ideal locus of Figure 9.11).

learned by the controller from the actual response of the building, a process known as self-learning or self-adaptation. Self-adapting optimum start algorithms have been used for some years. Originally they used linear schedules of the form:

$$(t_3 - t_2) = A\theta_i + B$$

with A and B recalculated every day and compared against the theoretical ideal values that would have just met target. A typical linear schedule is shown in Figure 9.13. The ideal slope, A, can be obtained by dividing the actual temperature rise the building went through during preheat, by the time it took to reach the set point temperature (in fact A is the inverse of this with a negative slope). B can be found by setting $t_3 - t_2$ to zero when $\theta_i = \theta_{sp}$. This

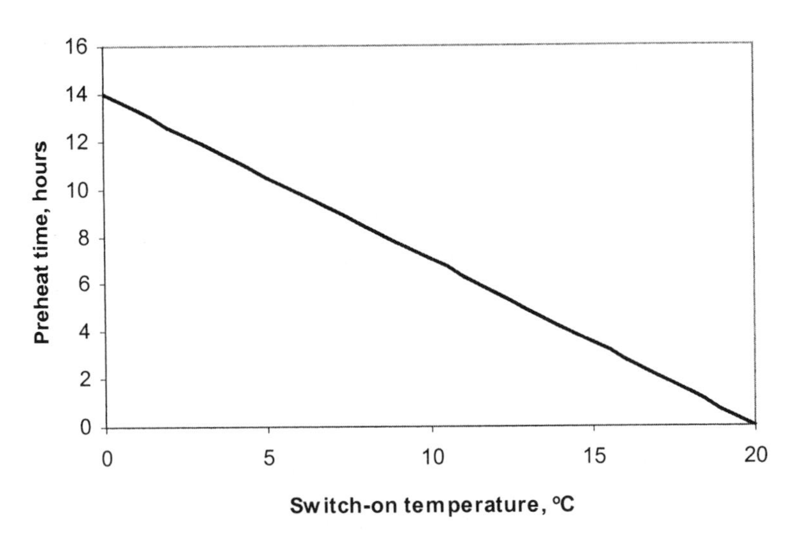

Figure 9.13 Linear optimum start schedule.

process cannot yield the precise desired optimum switch-on time, but with successive learning the algorithm can get close. To prevent too much day-to-day variation a forgetting factor, w, is employed such that some memory of building performance is retained:

$$A_i = wA_e + (1 - w)A_{i-1}$$

where A_i is the value of A on day i, A_e is the ideal value of A that would have achieved the target temperature at the correct time on the previous day, and A_{i-1} is the actual value of A used on the previous day. A similar factor is also applied to the parameter B, and in this way the schedule parameters are adjusted gradually, retaining some 'memory' of how the building responded on previous days. This forgetting factor is typically set to 0.5, and the higher the factor the more sensitive the optimiser will be to changes in weather. This approach takes advantage of the fact that the weather on any given day is similar (temperature-wise at least) to the preceding day and to the day that follows, i.e. that is there is a gradual change in temperatures; the building should therefore respond similarly to yesterday and tomorrow (but not necessarily similar to two days away). When rapid changes occur – for example when a cold front moves in – then the schedule needs to be able to respond (hence the parameter C in equation 9.15).

With linear algorithms it was usual to have one schedule for weekdays and another for Mondays after a weekend shut down. The non-linear schedule does not need an extra Monday schedule where fabric protection is used. Fabric protection brings on the heating system when the temperature in a representative room falls below some preset temperature – usually between 10° and 13°C – to prevent surface and interstitial condensation. This ensures that the building is maintained within the temperature limits of the optimiser schedule. The non-linear nature of the Birtles and John equation makes self-learning of the parameters a little more complex, and they adopted a method of recursive least squares (a discussion of which is beyond the scope of this book). This algorithm is known as the BRESTART algorithm (as it was developed at the Building Research Establishment (BRE)), and is now the basis for many commercial optimum start controllers.

9.3.1 Optimum start and compensator interaction

Where optimum start is employed in conjunction with a weather compensator (see Chapter 10) it is important to disable the compensator during the unoccupied hours. The purpose of the optimiser is to get the building up to correct conditions in the shortest possible time in order to minimise energy consumption. Most of the time the outdoor temperature will be above the design temperature, and if enabled the compensator will adjust the flow temperature according to its programmed schedule. This will reduce the heat output from the emitters and slow the building heat-up. It will thus take longer to reach the desired internal temperature. The optimiser will adjust for this and alter the parameters A, B and C to suit the building/system response. Subsequent days will have different outdoor conditions, with different flow temperatures resulting and different building/system responses. This is not conducive to the optimiser being able to learn the true building response under full output. Disabling the compensator function is a matter of using a time clock to switch the function off, and only allowing it to vary the flow temperature once the indoor set point temperature has been reached.

9.3.2 Optimum stop

It is also possible to shut the boiler plant down ahead of the occupants' departure. Where buildings have reasonable thermal mass the surface temperatures cool slowly such that the occupants do not notice. This is more likely to be acceptable where the heating system has a fairly high time constant – e.g. wet radiators or radiant panels – as the surface temperature of these falls slowly enough for occupants to believe the system is still on (psychology can play a large part in the success of heating systems).

The opportunities for early plant shut-down may be limited. The problem is that the air temperature will fall fairly rapidly, especially in an all-air heated system (including forced convectors). It will only be successful if the mean radiant temperature remains high enough to maintain a reasonable dry resultant temperature (see Chapter 6 for a discussion of these temperatures); it is difficult to set up controls that can account for this, and it is also difficult to design for. Consequently optimum stop is not commonly found to be successful, and becomes even more difficult to implement where the building use patterns are variable.

References

1. Chartered Institution of Building Services Engineers (CIBSE) (1999) Air Infiltration and Natural Ventilation, Section A4 of the CIBSE *Guide*, Table 4.10. CIBSE, London.
2. Chartered Institution of Building Services Engineers (CIBSE) (1986) Estimation of Plant Capacity, Section A9 of the CIBSE *Guide*. CIBSE, London.

3. Day, A.R. (1999) *An investigation into the estimation and weather normalisation of energy consumption in buildings using degree-days*. PhD Thesis. South Bank University, London.
4. Day, A.R. & Karayiannis, T.G. (1999) A new degree-day model for estimating energy demand in buildings. *Building Services Engineering Research and Technology*, **20** (4), 173–8.
5. Levermore, G.J. (1992) *Building Energy Management Systems: an application to heating and control*. E&FN Spon, London.
6. *Spon's Mechanical and Electrical Services Price Book* (2000) 31st edn. Edited by Mott Green & Wall in association with Davis Langdon & Everest. Spon Press, London & New York.
7. Day, A.R., Ratcliffe, M.S. & Shepherd, K.J. (2001) *Sizing central boiler plant using an economic optimisation model*. CIBSE National Conference, London.
8. Birtles, A.B. & John, R.W. (1983) A new optimum start control algorithm. *Building Services Engineering Research and Technology*, **6** (3), 117–22.

10 Matching Output to Demand

Having determined the output capacity of the plant, the next step is to decide the most appropriate arrangement and operational control of the boilers. This chapter looks at the factors that affect selection of boiler type, and examines those aspects that ensure a safe, stable and efficient operation of the plant and system; namely system and central plant configurations, and control strategies.

10.1 Relationship between heating demand, system sizing and boiler capacity

A well-designed heating system has to provide acceptable internal environmental conditions in an economically and environmentally acceptable manner, and in such a way that the operation of the system is barely even noticed by the occupants. The system has to be designed large enough to cope with likely extremes in external weather conditions (the design conditions), and controlled such that the output varies with changes in the load. In temperate climates like the UK's, part-load conditions will occur for most of the time, which means that for most of the heating season the heating system and plant will be considerably oversized. This is compounded in systems that are operated intermittently (for example, shut-down over night), where plant needs to be further oversized to ensure sensible heat up (preheat) times.

Plant oversize may be inevitable and indeed may be beneficial (see sizing procedures in Chapter 9), but this has important consequences for boiler selection and control. When considering the type and number of boilers to install it is important to have some idea of part-load operation – for example, likely base loads and seasonal variations. The configuration of the central plant then needs to be considered in terms of the way it should be controlled to maximise the plant seasonal efficiency. In this chapter control of the central plant will be examined for different system configurations, together with how system controls can be used to improve overall system performance. The interaction of system and central plant operation are also discussed so that appropriate control strategies can be designed and common pitfalls avoided. It is first necessary to examine the concepts of boiler efficiency and system seasonal efficiency.

10.1.1 Boiler efficiency

A more detailed discussion on boiler efficiency can be found in Chapter 3, section 3.8, but a short discussion of its relevance to this chapter follows. The crudest definition of boiler efficiency is simply the ratio of useful heat delivered to the heat input of the primary fuel, i.e.:

$$\text{Efficiency} = \frac{\text{Useful heat output}}{\text{Heat input}}$$

When a boiler is firing, most of the heat passes through the metal wall of the heat exchanger to the water, but a certain amount (about 10–15%) is carried

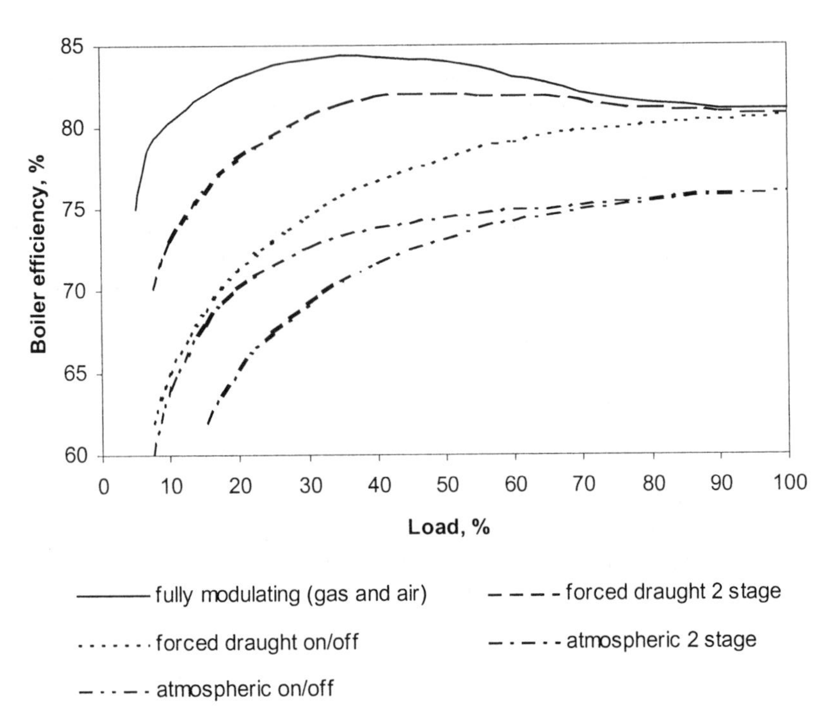

Figure 10.1 Boiler efficiency curves.

away with the exhaust gases, and a further proportion is lost through the boiler casing (about 2–5%). At part-load the boiler is controlled to deliver less heat to the system; however, the losses do not necessarily reduce by the same proportion with the result that the boiler efficiency (i.e. unit of useful energy delivered per unit of fuel energy consumed) will reduce. This is a complex issue and depends on the type of boiler, burner and method of control. For an on/off cast-iron sectional boiler it may be explained as follows: when the boiler is switched off, most of the residual heat retained in the heat exchanger is lost to the surroundings through the flue and casing (some is dispersed to the system for safety reasons); when the boiler fires again this lost heat must be replaced (with associated flue losses) before useful heat is passed to the system. An oversized boiler will have to cycle if system temperatures are to stay within reasonable limits, and the ratio of useful delivered heat to losses decreases with decreasing load (i.e. with decreasing on-time). Figure 10.1 shows a range of boiler efficiency curves (of gross calorific value) for typical modern boiler plant taken from experimental data[1]. Modern boilers have better part load efficiencies than older plant due to improvements in casing insulation, changes to pumping arrangements, and techniques for reducing flue losses when the boiler is off.

In theory, boilers with modulating burners should have constant efficiencies, but this depends on the burner being able to provide optimum air fuel ratios across the operating range of the burner. Modulating burners also keep the boiler heat exchanger at higher temperatures for longer periods of time, leading to higher casing losses (although these are relatively small).

10.1.2 System seasonal efficiency

The concept of *system* efficiency is a little more difficult to define. All the heat from a heating system is lost to the environment, whether or not it is effectively deployed or used, so the definition of efficiency needs to be qualified. The system output provides a useful service, and the overall efficiency can be defined similarly to the boiler efficiency:

$$\text{System efficiency} = \frac{\text{Useful heat delivered}}{\text{Boiler heat input}}$$

The difficulty arises in the definition of useful heat delivered. The heat delivered may result in overheating or underheating, the extent of which may be difficult to assess or quantify for a given building. For any given building the rate of heat loss will depend on how and when it is used (e.g. if windows are left open, or the hours of operation); it is also possible to change the characteristic of a building by changing the fabric performance (adding insulation or improved glazing). This means there is no rigid definition of system efficiency. However, a heating system will be at its most efficient if it uses a minimum amount of energy to just meet the required environmental criteria. It is the job of the system controls to ensure the system delivers just enough heat to meet design conditions; the boiler controls must match the output of the boilers to the system demand; and it is the configuration of the boiler plant and control strategy that will ensure minimum heat input to the building. These distinctions may seem obvious, but it is important to understand the relationship between them for a well-designed system.

10.2 Designing the central plant arrangement

There are two distinct issues in selecting boiler plant:

(1) Sizing the output capacity
(2) The type and arrangement of boilers.

The determination of optimum output capacity has been detailed in Chapter 9; this section will concentrate on selection of boiler arrangements, although the two issues are not entirely independent.

Once the total boiler capacity has been sized, the first decision is the type of boiler. Chapter 2 has described the different types of boiler suitable for different applications, and the reader is referred to that section. The second decision is the number of boilers to install; this is a question of whether to install a small number of large boilers, or a large number of small boilers. The latter option is increasingly popular in the form of modular boiler sets. This allows individual burner/heat exchanger sets to operate at or near their design efficiency for longer periods of time. Modular boilers help to ensure high part-load efficiencies, as each module will operate at the top of its efficiency curve. At the same time this approach allows the purchase of boilers as close to the design output as possible. The selection of a small number of

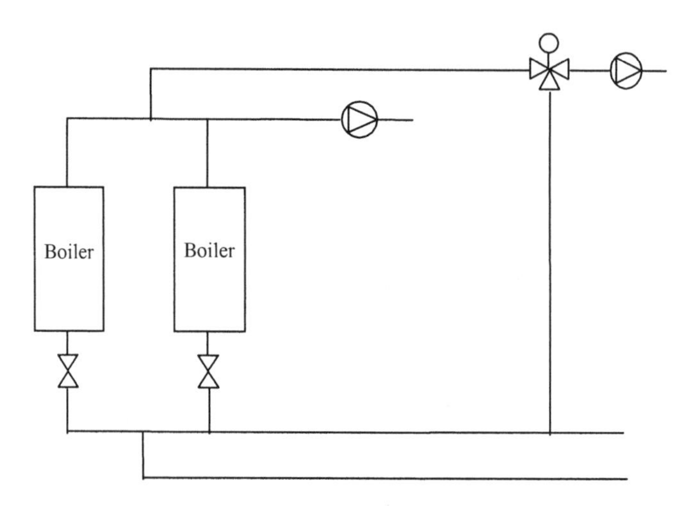

Figure 10.2 Ad hoc connection of system circuits to central boiler plant.

larger boilers may lead to inherent oversizing, as it may not be possible to precisely match output to design capacity and instinct tends to err on the side of caution and select larger plant.

10.2.1 Plant configuration

The configuration of the pipework, and the connection to the rest of the system, may be dictated by the type of boiler selected (modular, modular with common combustion chamber, etc.) or depending on how the system is arranged and controlled. Boilers require stable flow regimes, and there must be a representative way of sensing the system load. Systems may have variable flows due to the operation of two-port valves, or variable temperature controls (which will be examined later in this chapter). There are four broad types of connection arrangement, as follows.

(1) Individual flow and returns from different subsystems connected ad hoc to the boilers (Figure 10.2)

This is not recommended for large systems. There is a danger that varying flow rates from the system sub-circuits, and hydraulic interaction between boilers coming on and off line, can give rise to variable flow rates through the boilers. There is also no common position, either in the flow or return, to sense the total load on the system in order to sequence the boilers.

(2) Split flow and return common headers (Figure 10.3)

This allows for common temperature sensing in flow or return headers, but the boiler circuit is still subject to variations in flow rates from the system, and the return water temperature may not be indicative of the load.

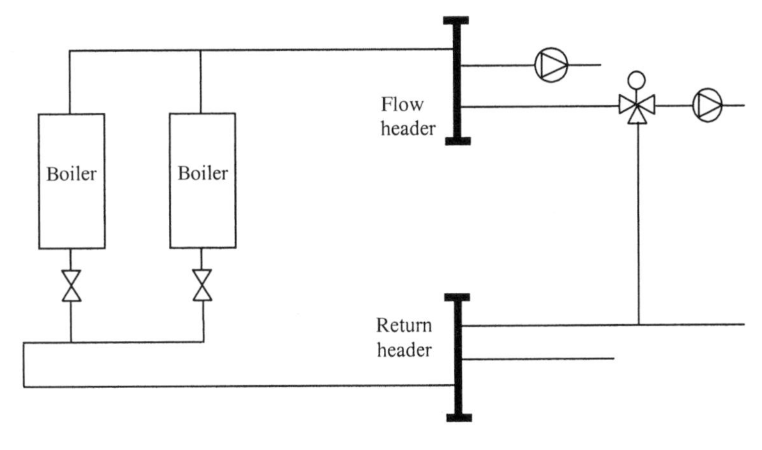

Figure 10.3 Split flow and return headers.

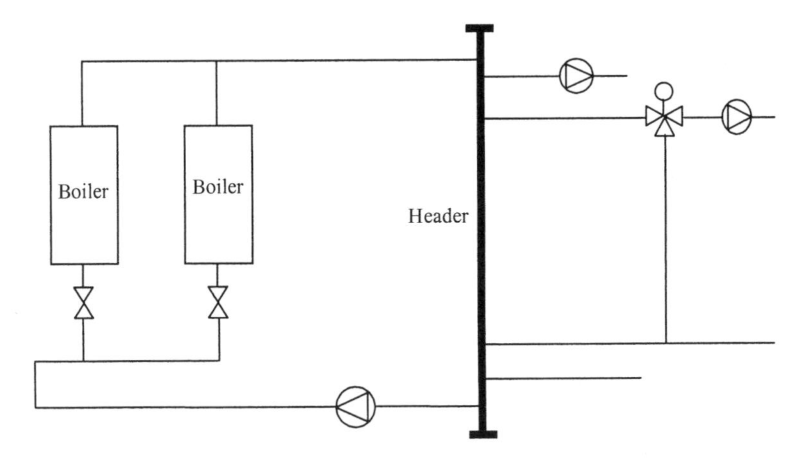

Figure 10.4 Primary ring main with boiler primary pump.

(3) Common header with primary and secondary pumping (Figure 10.4)

This arrangement helps to overcome many of the problems indicated above. The reasons why will be explored in this chapter.

(4) Common header with injection manifold (Figure 10.5)

A common arrangement for modular boiler sets. However, the inlet temperatures to boilers lower down the firing order are increased, leading to increased boiler exit temperatures. Sequence and burner controls must be set up correctly if output is to be utilised fully, and the danger of overheat/lockout is to be avoided.

In order to appreciate these points it is necessary to examine the way in which the system interacts with the central plant.

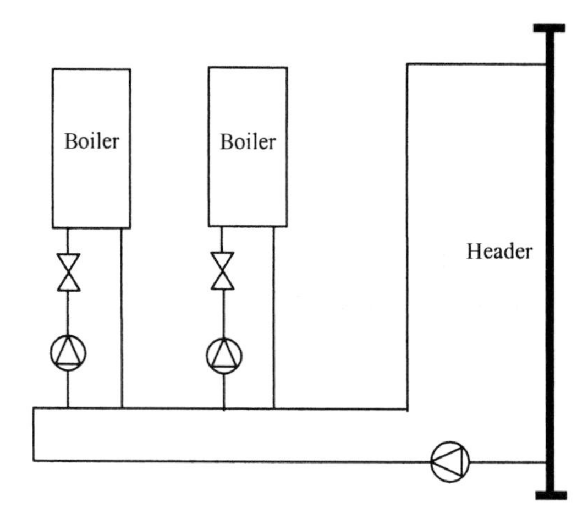

Figure 10.5 Primary ring main with injection.

10.3 System configuration

Systems can be configured and controlled in a number of ways: constant temperature, variable temperature, variable flow (two-port valves), or a combination of all three. This can have profound implications for boiler operation and its controllability. Thermostatic radiator valve (TRV) control has been discussed in Chapter 6; this effects the system flow rate and return temperatures. With a suitable bypass in the system, return temperatures will rise with decreasing load. The use of weather compensated control is a popular alternative for matching system output to the prevailing load, but it can cause difficulties for the central plant operation. This section takes a detailed look at weather compensators.

10.3.1 Weather compensation control

As the outdoor temperature changes, so does the heat loss from the building and there is a need to vary the output of the emitters to match the prevailing load. One way of achieving this is to vary the flow temperature to the emitters, and hence the mean surface temperature. Weather compensation controls, generally referred to as compensators, provide a centralised means of doing this. Figure 10.6 shows a basic compensator arrangement, where the required flow temperature is achieved by mixing boiler flow water with return water from the emitter circuit. The flow temperature is controlled as a function of the outdoor temperature. The principle of the operation can be seen from an energy balance on the heating system and room. Equation 10.1 shows the balance of the heat into and out of the emitters and the heat loss from the space.

$$\dot{m}c_p(\theta_F - \theta_R) = KA_f(\theta_m - \theta_{ai})^n = U'(\theta_{ai} - \theta_{ao}) \qquad (10.1)$$

where $\dot{m}$ is the mass flow rate of water, c_p is the specific heat of water, KA_f is an emitter constant, U' is the heat loss coefficient; θ_F and θ_R are the water flow and return temperatures respectively, and θ_m is the mean water temper-

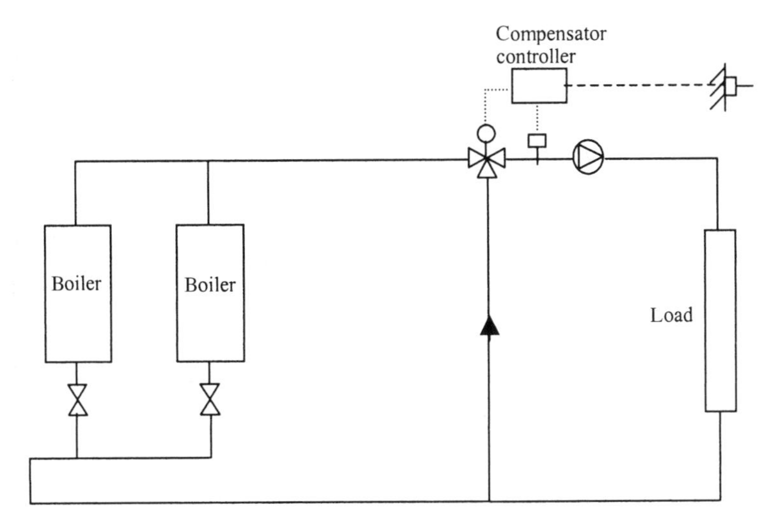

Figure 10.6 Basic compensator arrangement.

ature in the emitter (the definition of emitter output be found in section 6.3.3). Under design conditions, all the values in equation 10.1 are known. If the outdoor temperature is higher than design (which is usual) then we can calculate the theoretical flow temperature required to maintain the desired room temperature. If we divide each expression in equation 10.1 by the design conditions, all the constants cancel out and we are left with a set of temperature ratios:

$$\frac{(\theta_F - \theta_R)}{(\theta_F - \theta_R)_{design}} = \frac{(\theta_m - \theta_{ai})^n}{(\theta_m - \theta_{ai})_{design}} = \frac{(\theta_{ai} - \theta_{ao})}{(\theta_{ai} - \theta_{ao})_{design}} \qquad (10.2)$$

For a given outdoor temperature the theoretical flow temperature can be determined as shown in the following example.

Example

Take the design temperatures to be: θ_F = 82°C; θ_R = 70°C; θ_{ai} = 20°C; θ_{ao} = −2°C. Note that the design mean emitter temperature is θ_m = (82 + 70)/2 = 76°C.

Calculate the flow temperature required to maintain an indoor temperature of 20°C when the outdoor temperature is 10°C. Firstly take the emitter output and room loss and rearrange to find the required mean emitter temperature:

$$\theta_m = \left[(76-20)^{1.3} \times \frac{(20-10)}{(20-(-2))}\right]^{\frac{1}{1.3}} + 20 = 50.53°C$$

The required flow and return temperature difference is found by equating the emitter input to the room loss:

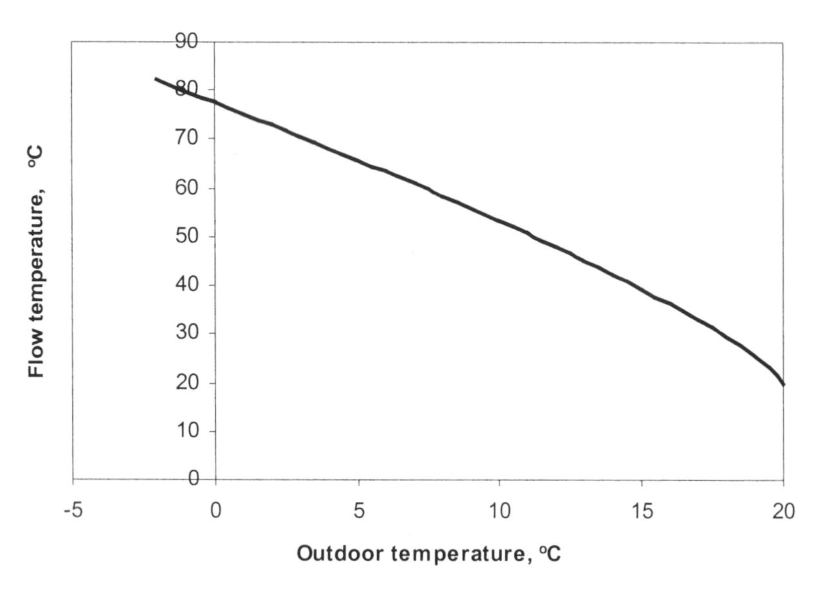

Figure 10.7 Ideal compensator schedule.

$$(\theta_F - \theta_R) = (82 - 70) \times \frac{(20-10)}{(20-(-2))} = 5.5\text{K}$$

and hence the flow temperature is found from:

$$\theta_F = 50.53 + \frac{5.5}{2} = 53.26°\text{C}$$

If this is repeated for all outdoor temperatures between the design outdoor and the design indoor temperatures, an ideal flow temperature schedule can be produced as shown in Figure 10.7. Note this schedule is not linear, due to the output characteristic of the emitter, and the higher the exponent the more marked the curvature of the ideal schedule (this is why radiators and convectors should not be mixed on the same compensator circuit).

However, we rarely deal with ideal schedules. Traditional control methods, particularly pneumatics, meant that linear schedules were easier to use, and the closeness of the ideal compensator schedule to a straight line has meant that this is still common practice. The use of a linear schedule means that the minimum flow temperature must be set somewhat higher than the design room temperature, otherwise the mean emitter temperature will be too low at part-load conditions and the room temperature will fall. Figure 10.8 illustrates how a practical schedule is set up to avoid this. Note that with this practical schedule the room temperature will tend to rise at part-load conditions, because the flow temperature is above the ideal.

10.3.2 Calculating room and emitter temperatures for practical schedules

The use of linear schedules introduces some complexity when analysing system and room temperatures. The problem is that the room temperature

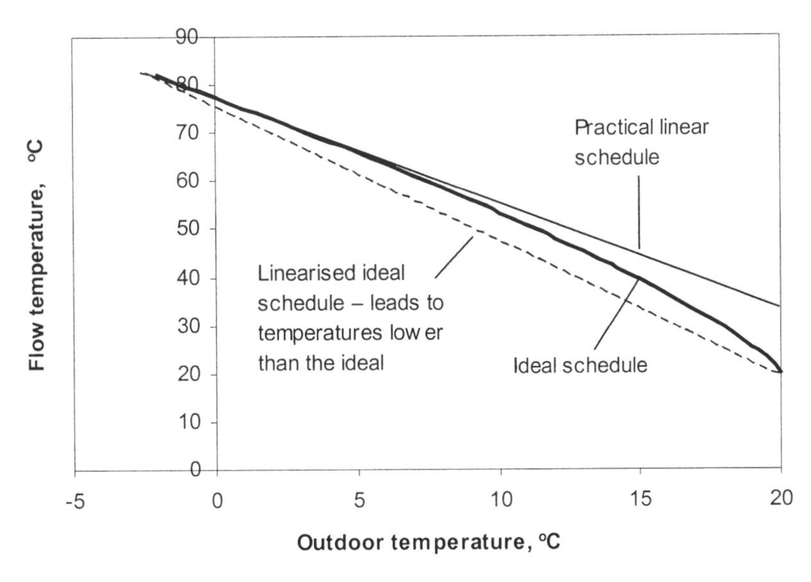

Figure 10.8 Linearised compensator schedules.

Table 10.1 System data for compensator example.

Design inside temp.	20°C
Design outside temp.	−2°C
Heat loss coefficient	10 kW/K
Boiler output temp.	82°C
Boiler design return temp.	70°C
Boiler temp. rise	12 K
Boiler output	220 kW
Mass flow rate	4.39 kg/s

changes and the new system/room energy balance needs to be determined, as the example below illustrates. (A similar analysis can be found in Levermore (1992)[2], but is included here to demonstrate operational pitfalls of heating systems and how to avoid them.)

Example

A heating system has the design characteristics shown in Table 10.1. The compensator schedule has a minimum flow temperature of 40°C, and the emitters have a value of $n = 1.3$ and $B = 1.174\,\text{kW}\,\text{K}^{-1}$. Calculate the room temperature and emitter return temperature when the outside temperature is 10°C.

Note that the compensator schedule can either be shown graphically or given as a formula. In this case the schedule formula is found by calculating the slope of the line as follows:

$$\text{Slope} = \frac{82 - 40}{-2 - 20} = -1.909$$

and the intercept:

$$\text{Intercept} = 82 - 1.909 \times 2 = 78.18$$

Giving the schedule formula:

$$\theta_F = 78.18 - 1.909\theta_o$$

In which case:

$$\theta_F = 78.18 - 1.909 \times 10 = 59.1^\circ C$$

Referring back to equation 10.1, the only temperatures we know are θ_F and θ_{ao}. We now need to find the mean emitter temperature. It is useful to adjust the expression for the emitter input as follows:

$$\dot{m}c_p(\theta_F - \theta_R) = \dot{m}c_p(\theta_F - (2\theta_m - \theta_F)) = 2\dot{m}c_p(\theta_F - \theta_m)$$

If we equate the emitter input to the room heat loss we can get θ_{ai} in terms of θ_m as follows:

$$2\dot{m}c_p(\theta_F - \theta_m) = U'(\theta_{ai} - \theta_{ao}) =$$

$$2 \times 4.39 \times 4.18 \times (59.1 - \theta_m) = 10 \times (\theta_{ai} - 10)$$

Rearranging and simplifying gives:

$$\theta_{ai} = 226.9 - 3.67\theta_m$$

We can now equate the emitter input to the emitter output, and substitute the above expression for θ_{ai}:

$$2\dot{m}c_p(\theta_F - \theta_m) = B(\theta_m - \theta_{ai})^n =$$

$$36.7(59.1 - \theta_m) = 1.174(\theta_m - (226.9 - 3.67\theta_m))^{1.3}$$

$$1847.5 - 31.26\theta_m = (4.67\theta_m - 226.89)^{1.3}$$

This equality cannot be solved analytically and needs a numerical or iterative solution. A Newton-Raphson iteration will be adopted here, where successive approximations of θ_m are found from:

$$\theta_{m(n+1)} = \theta_{m(n)} - \frac{f(\theta_m)}{f'(\theta_m)}$$

where $f(\theta_m)$ is the function of θ_m equal to zero. This is simply a rearrangement of expression for θ_m above with all the terms collected on the right-hand side. The term $f'(\theta_m)$ is the first derivative of this function, and n represents the iteration number. In this case:

$$f(\theta_m) = 0 = 31.26\theta_m + (4.67\theta_m - 226.89)^{1.3} - 1847.5$$

and

$$f'(\theta_m) = 0 = 31.26 + 4.67 \times 1.3 \times (4.67\theta_m - 226.89)^{0.3}$$

A first guess for θ_m is required. This will be less than 59.1°C (the flow temperature), but greater than 53°C (the flow and return temperature difference must be less than the design difference). Putting in a guess of 56°C gives:

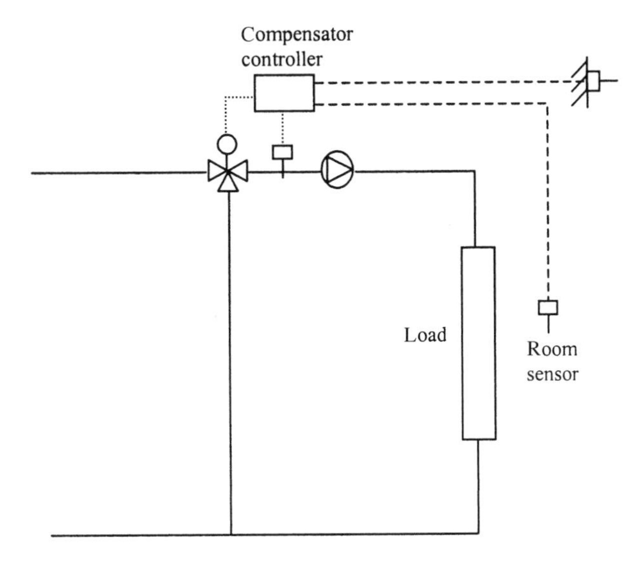

Figure 10.9 Compensator with space temperature reset.

$$\begin{aligned}\theta_{m(n+1)} &= 56 - \frac{1750.6 + (4.67 \times 56 - 226.89)^{1.3} - 1847.5}{31.26 - 6.07 \times (4.67 \times 56 - 226.89)^{0.3}} \\ &= 56 - \frac{3.357}{48.82} \\ &= 55.93^{\circ}\mathrm{C}\end{aligned}$$

Further iteration shows 55.9°C to be a good solution. The return water temperature can be found from:

$$\begin{aligned}\theta_R &= \theta_F - 2(\theta_F - \theta_m) \\ &= 59.1 - 2 \times (59.1 - 55.9) \\ &= 52.7^{\circ}\mathrm{C}\end{aligned}$$

Substituting θ_m back gives:

$$\theta_{ai} = 226.9 - 3.67 \times 55.93 = 21.6^{\circ}\mathrm{C}$$

This example demonstrates that using a linear schedule will give an offset to the room temperature at part-load. This is better than having a loss of service, and room temperatures being too cold. It is possible to improve the situation either by incorporating a room sensor to reset the flow temperature, as shown in Figure 10.9, or by using TRVs. Given that the analysis above does not allow for casual gains, it is highly recommended that one of these extra controls is employed. Note that if a room sensor is used to reset the schedule, considerable care must be taken when choosing a suitable location. A space representative of the majority of the building must be selected; alternatively averaging sensors may be used. The value of the compensator is to provide a means of central control that can be easily monitored for reliability and performance, and to some extent it can reduce losses from pipework at part-load. TRVs can become damaged or removed and are difficult to maintain and monitor, but they allow some user control, which can increase confidence in the heating system.

10.3.3 The implications of compensator control for the central plant

When compensator circuits form a significant proportion of the heating system, then the configuration shown in Figure 10.6 is not satisfactory. The example above shows that return water temperatures reduce as the load decreases, and if further room temperature control is employed to account for internal gains and remove compensator offset, then these temperatures will be lower still. Such temperatures would be detrimental if allowed to return to boilers where the heat exchangers are not designed to cope with condensing from flue gases. (On the other hand considerable advantages can be gained from the use of condensing boilers in conjunction with weather compensators; these are discussed in section 10.5.12.)

Just as significantly, however, the central boiler plant cannot be controlled using return water sensing if the system is configured as Figure 10.6. Boiler controls operate by bringing extra boilers on line as the temperature falls, i.e. falling temperature signifies an increasing load; with compensators just the opposite occurs, and a return sensor cannot be used to sequence boilers without further modification. The problem is compounded by the mass flow rate of water returning to the central plant, which will also reduce. Consider an energy and mass balance on the three-port mixing valve:

$$\dot{m}_F c_p \theta_F = \dot{m}_b c_p \theta_{Fb} + \dot{m}_x c_p \theta_R$$

with

$$\dot{m}_F = \dot{m}_b + \dot{m}_x$$

where $\dot{m}_F$ is the mass flow rate to the emitters, $\dot{m}_b$ and θ_{Fb} are, respectively, the flow rate and temperature from the boiler, and $\dot{m}_x$ is the mass flow rate in the bypass. The specific heats cancel and substituting for $\dot{m}_b$ gives:

$$\dot{m}_F \theta_F = (\dot{m}_b - \dot{m}_x)\theta_{Fb} + \dot{m}_x \theta_R$$

Rearranging gives:

$$x = \frac{\dot{m}_x}{\dot{m}_f} = \frac{(\theta_{Fb} - \theta_F)}{(\theta_{Fb} - \theta_R)}$$

where x is the proportion of system water flowing in the bypass. By definition, therefore, the proportion of water returning to the central plant is (1 – x).

Using values from the example gives:

$$x = \frac{(82 - 59.1)}{(82 - 52.7)}$$
$$= 0.78$$

which indicates that 78% of the system water is flowing in the bypass with only 22% returning to the boiler. Figure 10.10 shows how flow rate and return water temperature vary with load for the schedule used in this example. This has profound implications for safe boiler operation because at

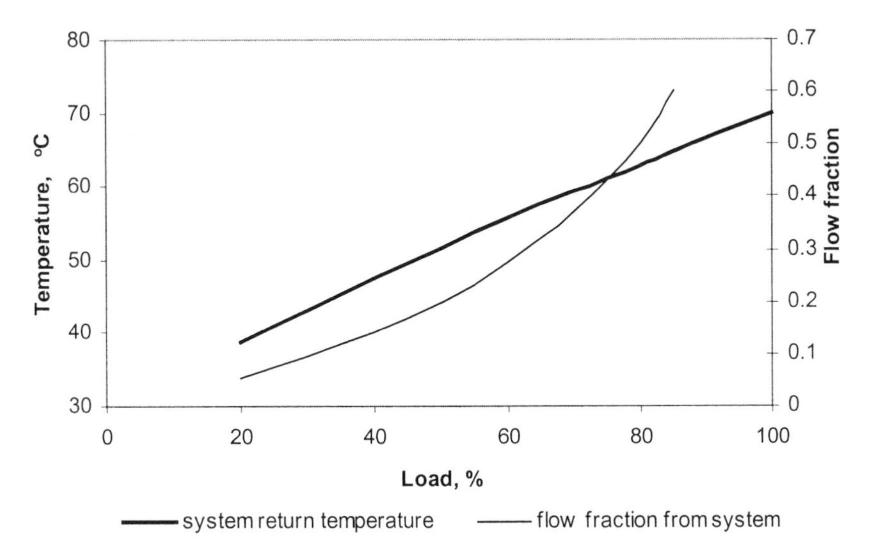

Figure 10.10 System return temperature and flow fraction from a compensator circuit.

reduced flow rates there is a danger of the boiler overheating (a full discussion of the implications of flow rate on boiler operation can be found in section 3.11). The theoretical boiler outlet temperature in our example (assuming on/off firing and no overtemperature control) would be:

$$\theta_{Fb} = 52.7 + \frac{220}{0.22 \times 4.39 \times 4.18}$$
$$= 107.2°C$$

In reality an overtemperature state in the boiler outlet would lock the boiler out. This is clearly not satisfactory.

10.4 Primary ring main

The problems of low return water temperature and reduced flow rates are overcome by the use of a primary ring main, introduced in Figure 10.4. Figure 10.11 shows a single compensator circuit served by a primary boiler ring main. The header must be sized large enough to give a minimal pressure drop across its length to ensure no interaction with the system circuit(s); typically this means twice the size of the rest of the primary pipework.

Note that now the return water flow rate to the boilers is kept constant, as $(1 - x)\dot{m}$ returning from the system mixes with $x\dot{m}$ in the primary header, thus:

$$\dot{m}(1 - x) + \dot{m}x = \dot{m}$$

i.e. the design flow rate is ensured to the boilers at all times. In addition, the mixing at the return end of the header leads to increased boiler return temperatures, θ_{Rb}, at low loads:

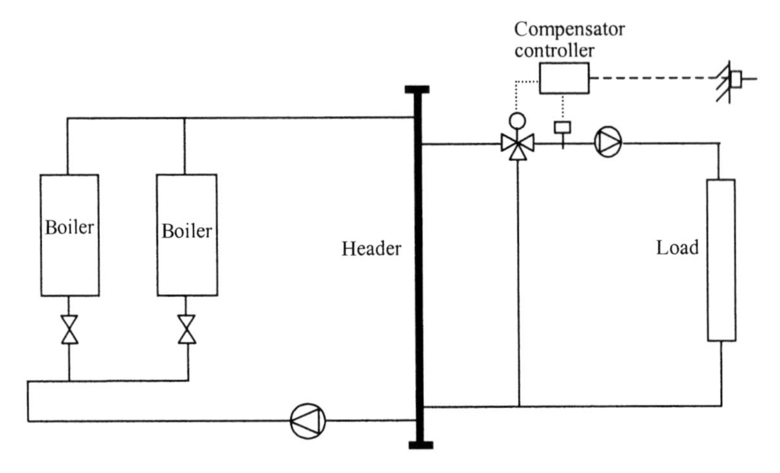

Figure 10.11 Compensator with primary ring main – a solution to variable temperatures and system flow rates.

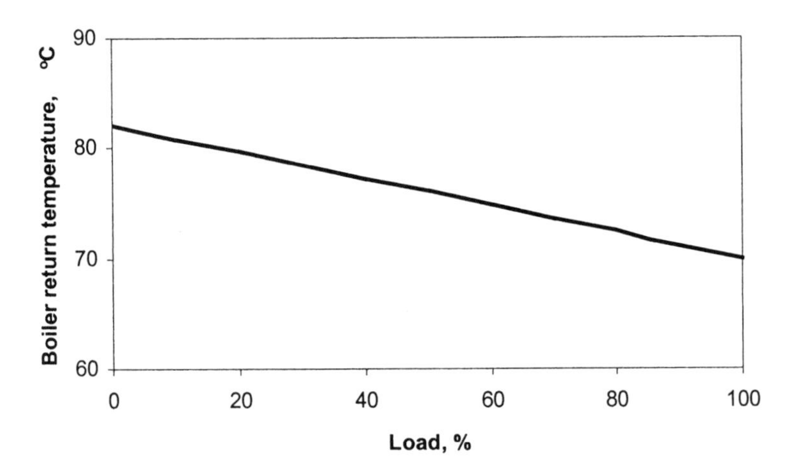

Figure 10.12 Temperature returning to boiler in a primary ring main as a function of load.

$$\theta_{Rb} = x\theta_{Fb} + (1 - x)\theta_R$$

Figure 10.12 shows a plot of boiler return water temperatures and as a function of load. This ensures high back-end temperatures and enables return water sensing for sequencing of the boilers.

10.5 Controlling boiler output

The aim of good central plant control is to ensure that the boilers deliver the amount of energy required by the system. In some cases boiler control can be directly related to system control (e.g. direct flow temperature compensation as discussed in section 10.5.12), but normally these functions are served separately. If inappropriate boiler control strategies are used this can lead to inefficient operation, rapid cycling of boilers or control valves, or frequent boiler lock-out; the latter implies a dangerous system as it relies on safety devices to ensure overheating does not occur. Unstable control, however it is manifest, inevitably leads to poor service (i.e. not enough heat getting to the load) or energy wastage.

The correct choice of control will depend on factors such as size of the system, number and type of heating sub-circuits being served, the type of boilers and burner operation and the configuration of the system. There are two principle methods of matching boiler output to load:

(1) Temperature sensing
(2) Heat load control.

The first of these is most common due to simplicity and cost, while heat load control is generally confined to larger installations. The method of control must ensure safe and stable control, and ensure the boiler output suitably matches the demand.

10.5.1 Temperature sensing

The temperature in some part of the system does not necessarily represent the load at that moment in time, and how true this is depends strongly on the location being sensed. Boilers can be controlled on their outlet (system flow) temperature, or inlet (return) temperature. There are advantages and disadvantages to both flow and return sensing, which can be broadly summed up as follows.

Return sensing

- Can be indicative of changes in the load
- The time lag of the flow around the system can provide greater stability
- Return temperatures must increase with decreasing load (which may not always be the case)
- Requires constant flow rate to the boilers.

Flow sensing

- Ensures flow temperature kept within safe design limits
- Can give close control of flow temperatures where necessary
- Can lead to rapid on/off cycling with low water content boilers
- Affected by cold slugs of water from newly fired boilers
- Control stability compromised by hydraulic interaction.

In practice some combination of the two is often employed. It is not possible to present all the possible permutations, but some typical system configurations will be examined here to illustrate the main influencing factors. A more comprehensive set of system configurations is produced by BRECSU[3].

10.5.2 Single boiler with flow sensing

In order to appreciate how the control arrangement affects the dynamics and operation of a boiler installation, it is helpful to start with a single on/off boiler serving a single heat load (perhaps domestic or small commercial), as

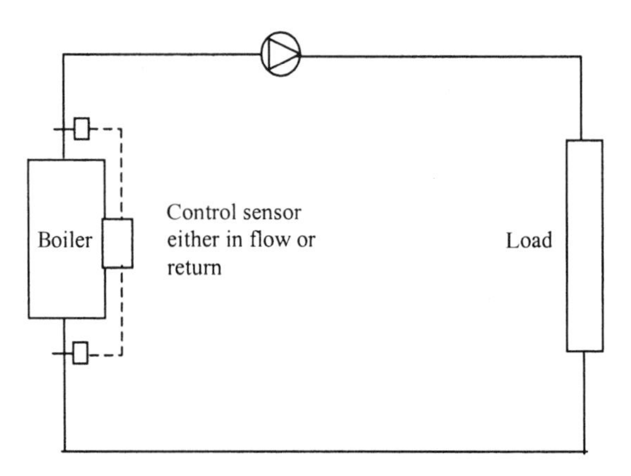

Figure 10.13 Single boiler controlled from either flow or return temperature.

shown in Figure 10.13. It is possible to control the burner operation from a sensor either in the flow or the return.

Consider first flow sensing. On/off control requires a set point at which the burner is fired, and a switching differential set such that cycling is not too rapid but the temperature fluctuations are not too great. Figure 10.14 shows typical flow and return temperatures under 75% full load for (a) a 4 K switching differential (80°C on, 84°C off) and (b) a 2 K switching differential (81°C on, 83°C off). (These temperature profiles have been generated from a mathematical dynamic model of a single boiler operation.) The rate of cycling reduces with wider differential at the expense of close control, and a balance must be drawn between these two requirements. This cycling rate is also strongly dependent on boiler thermal capacity, with lightweight heat exchangers and low water content boilers being more susceptible to rapid cycling; such boilers may need wider switching differentials.

The problem can be somewhat mitigated by using return sensing. Figure 10.15 shows the same boiler controlled by a 1 K switching differential (71°C on, 72°C off). This has a cycling rate similar to 10.14a, but with reduced temperature swings. This occurs because of the time lag and thermal capacitance effects of the system. When the sensor is in the flow the signal from the load is masked by the operation of the boiler, but in the return it detects changes in the load directly. However, return sensing is only possible when the return water temperature rises as the load falls; we have seen that with weather compensation of flow temperature (or even TRV operation) this may not be the case.

For domestic applications some boilers incorporate an internal bypass that operates in the same way as a primary ring main, ensuring a constant flow rate to the heat exchanger, with a mixing of boiler flow and system return temperatures that enable the burner to be controlled from a return temperature sensor. Such an arrangement still requires a flow sensor to ensure the boiler does not overheat due to low flow conditions or control failure.

This principle of operation can be extended to modulating output burners, which vary their output according to flow and return temperature difference via a proportional controller. Figure 10.16 shows the relationship between

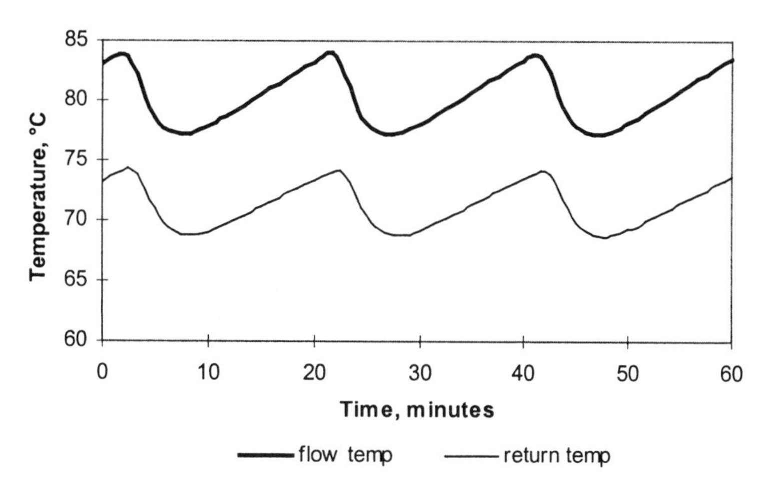

Figure 10.14a Flow sensing with 4 K differential.

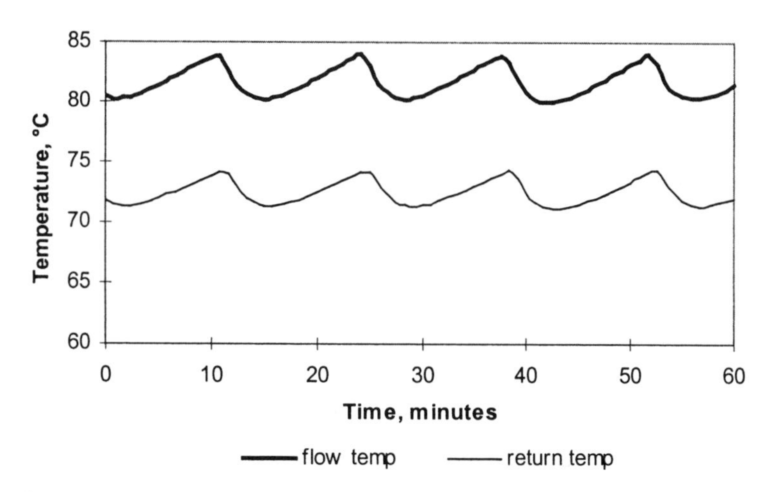

Figure 10.14b Flow sensing with 2 K differential.

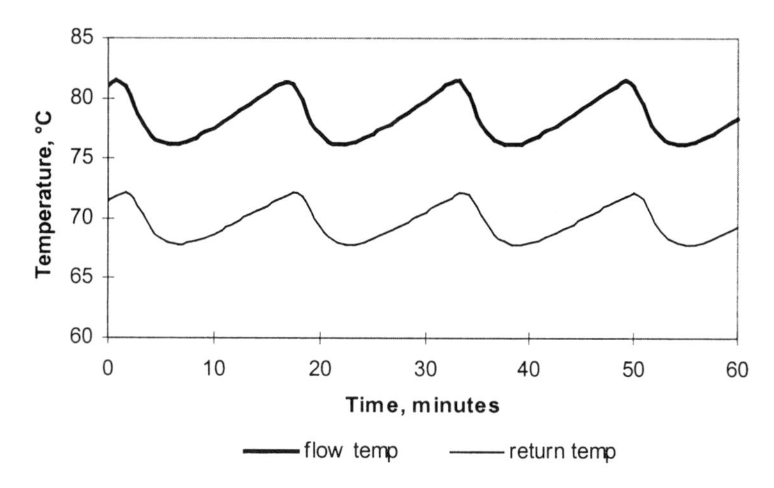

Figure 10.15 Return sensing with 1 K differential.

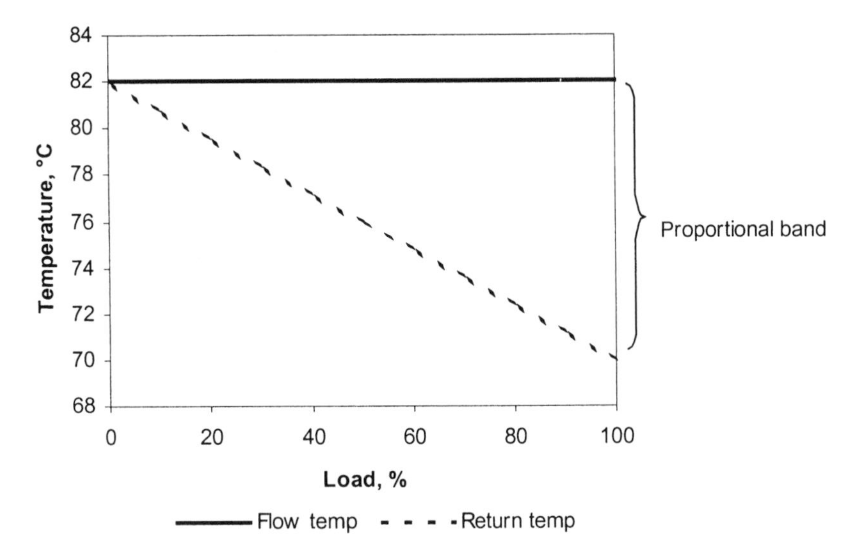

Figure 10.16 Modulating control schedule – return temperature control.

temperature difference and load, although turn down is often limited to 50%, and the flow sensor acts as a temperature limiting sensor that will turn the burner off if the flow temperature starts to rise. Modulating burners reduce the amount of on/off cycling and can help to maintain more consistent flow temperatures out to the system.

10.5.3 Multiple units with common header

For reasons described in section 10.4, larger installations are likely to have a number of boilers connected into a primary ring main. The primary ring will be essential if return sensing is to be employed. Secondary circuits with variable flow and variable temperature controls (e.g. TRVs and weather compensators) will send back ambiguous signals about the change in load (i.e. temperature falling with reducing load).

However, the use of a primary ring will not in itself solve all the problems of excessive cycling, nor ensure stable temperature patterns where on/off boilers are used. Oscillations in temperature caused by burner firing patterns are experienced in all parts of the ring with more or less equal amplitude, which will also subject return temperature sensing to problems of stability and dynamic interaction between boiler modules. The oscillations occur at low loads when the majority of the water leaving the boiler flows around the ring without passing out to the load. At high loads there is more inherent stability due to the larger temperature differences, and longer system time lags.

Figure 10.17 shows a primary ring main with two boilers (the principles can be extended to any number of boilers). A typical control strategy would be to enable each boiler according to some sequenced return temperature, while the individual burners are turned on and off from individual flow temperature sensors. Water flow is maintained through each boiler at all times

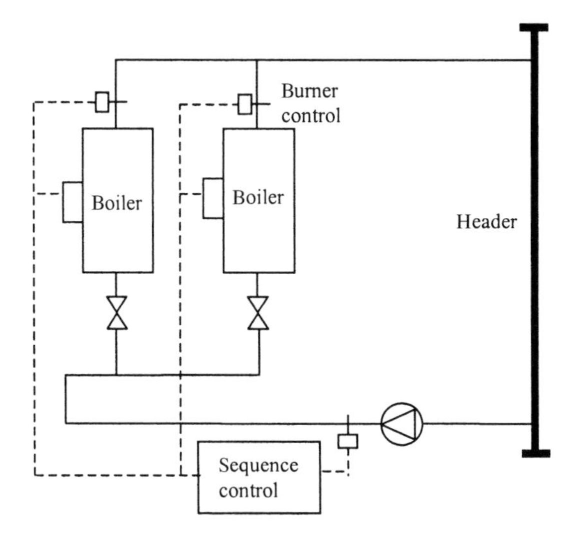

Figure 10.17 Two boilers in a primary ring main, sequenced from return water temperature.

to keep a constant flow rate around the primary circuit (flue dampers can be used to reduce standing losses when a boiler is off). The advantage of this strategy is that the return sensor detects variations in the load (from the temperature drop across the header), while the flow temperature is not allowed to become excessive.

10.5.4 Sequencing

Sequencing refers to the temperature settings that will switch each boiler on or off. As with the individual boiler there may be a switching differential for each boiler, and an additional dead band between the enabling of each boiler. There is no simple rule for determining these settings as it will depend on number of boilers, boiler thermal capacity, plant configuration, and the service required over the operating range. However, it is important to consider that the closer the required control (i.e. smaller switching differentials and dead bands) the greater the cycling rates. Heating systems rarely require very close control, although it is important to limit maximum temperatures from the boilers (which may introduce further cycling). Figure 10.18 shows typical sequencing temperatures for the arrangement in Figure 10.17 (considered for a LPHW application). Return temperature sequence control has the characteristics listed below.

- The most critical periods are when the load is close to the output of the lead boiler (or multiples of boiler output in the case of multiple boiler installations), where control interaction will cycle both boilers.
- This control interaction is reduced by increasing the dead band.
- Wider dead bands are more important for control stability than individual boiler switching differentials. In our example of Figure 10.17 the switching differentials can be reduced to 1 K and the dead band increased to 4 K, giving the same overall control limits but improving firing patterns (at 45% full load only boiler A operates; under the regime in Figure 10.18 both

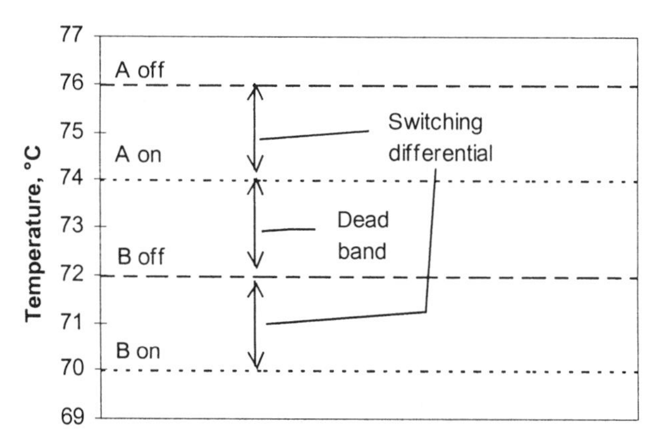

Figure 10.18 Return temperature on/off control sequence for two boilers.

boilers cycle at this load). It is possible to set the switching differential to zero, provided there is suitable thermal capacitance in the boilers to prevent excessive cycling.

- The wider the dead band the higher the flow temperatures from the lead boiler. This becomes exacerbated at low loads where the higher control limits dominate the operation.
- To obviate high temperatures (say over 85°C, or as prescribed by the required anti-flash margin) it is necessary to set the flow temperature sensor from each boiler to shut the burner off at this temperature. This will inevitably increase the cycling rate of the lead boiler in particular. It may be advisable to increase the system pressure and thus allow flow temperatures to elevate slightly (dilution effects and part-load dynamics mean that temperatures rarely exceed 90°C for a correctly set up return water sequence).

Because of the discontinuities in operation (i.e. boiler firing patterns) it is not possible to ascertain the dynamics of such a system using classical control theory. Dynamic modelling has been used here to explain the behaviour of the system. Due to the number of permutations of possible control settings it would not be helpful to give specific examples of control behaviour here. Dynamic boiler models do not have to be hugely sophisticated, but need to incorporate boiler thermal capacitances and system time delays. There are no proprietary models known to the authors, but with a little thought it is possible to develop such models using standard spreadsheet software packages. Such models have been developed at South Bank University. Such models may not be necessary for the design process (although designers would benefit from an appreciation of dynamics). What is essential is that a system is properly commissioned and tested under a wide range of operating conditions to ensure that rapid cycling or module interaction does not occur. Such commissioning procedures may not happen in practice, but the discussion here is presented to demonstrate just how important it can be.

10.5.5 Common flow temperature control

An alternative strategy is to place the sequence control sensor in the common flow from the boilers. In this case the required flow temperature in the header is controlled directly. Boiler A is set to switch off at, say, 83°C and boiler B to switch on at 80°C. If each has a switching differential of 1 K this leaves a dead band of 1 K. The same principles of stability apply as for return sensing, i.e. the wider the dead band the less the cycling. However, there is a limit to how wide the dead band can be; if boiler A off temperature is raised too high then the flow temperature from the boiler (due to dilution effects) can become dangerous. If boiler B temperatures are set too low then at higher loads the overall flow temperature becomes depressed. In general, flow sensing is more likely to lead to rapid cycling if close control is required. Thus return sensing allows greater scope for wider dead bands and smaller switching differentials, leading to reduced cycling rates and temperatures closer to the design values across the operating range.

10.5.6 Temperature dilution effects

In multiple boiler installations where flow is allowed through unfired boilers, cooler water from these units mixes with hot water from the fired boiler. It is the mixed water temperature that is used for sequence control. This means that to achieve the required mixed temperature the lead boiler will have to have a temperature higher than this. For example, in our two boiler system (Figure 10.17) if the lead boiler were to be switched off when the mixed flow temperature was 83°C, and the return water temperature was 77°C (at 50% load when only the lead boiler is required), the temperature leaving boiler A would be found from (assuming equal flow through each boiler):

$$\frac{\dot{m}}{2}\theta_{F,A} + \frac{\dot{m}}{2}\theta_R = \dot{m}\theta_{F,mixed}$$

Thus:

$$\theta_{F,A} = 2 \times 83 - 77 = 89°C$$

The general expression for the worst case flow temperature from the lead boiler for n boilers is:

$$\theta_{F,A} = n\theta_{F,mixed} - (n-1)\theta_R$$

where

$$\theta_R = \theta_{F,mixed} - \frac{\Delta\theta_{design}}{n}$$

Note that this last expression means for a large number of boilers (high n) the return water temperature is high, and the dilution effect becomes somewhat mitigated. Temperature dilution becomes a significant problem with low return water temperatures at part-load, especially if flow rates are also reduced (conditions that the primary ring is designed to eliminate).

10.5.7 Flow prevention through unfired boilers

One way to overcome temperature dilution effects is to prevent water flow through unfired boilers, achieved by installing automatic two-port valves at the inlet to the boiler. This achieves two things:

- Boiler temperatures are the control temperatures, eliminating overheating of lead boilers
- Standing losses are reduced from unfired boilers.

Compelling as these reasons may seem, there are drawbacks to such an approach. When each boiler shuts down, the flow rate in the primary ring main will reduce; this can lead to hydraulic interaction between the primary and secondary circuits, with variable flow rates in the primary circuit degrading the control of the boilers. In addition, a single primary pump will provide greater than design flow through boilers at part-load – correct flow can only be ensured with individual boiler pumps (as for example with an injection system discussed below). For a discussion of the relationship between pump and system interaction see Chapter 7. Another difficulty is that when a new boiler is brought on stream from cold, it will inject a cold slug of water into the main. This will be 'seen' by the control sensor as a rapid increase in the load, with the danger of bringing other modules on line (this is especially true of flow sensing). In extreme cases this can bring on each boiler in turn, and just as quickly switch them off. This can be overcome by introducing some time delay into the controls, or having shunt pumps around each boiler (an arrangement discussed in section 10.5.9, dealing with large boiler circuits), which adds both complexity and expense. Finally, the frequent heating up and cooling down of cycling boilers, together with increased number of active moving components (isolating valves, etc.) can shorten the life and increase maintenance requirements of the plant. There is also an increased risk that parts of the heat exchanger will more often pass through the dew point of the flue gases leading to condensing and the risk of corrosion. For smaller systems it is therefore more cost effective to allow flow through unfired boilers, and possibly use flue dampers to reduce standing losses.

10.5.8 Injection systems

A common configuration for overcoming the problems discussed above is the primary ring main with individual boiler injection (Figure 10.19). This enables good hydraulic stability and ensures that design flow rates pass through each boiler, while allowing isolation of unfired boilers. Back-end protection (avoidance of condensation) can be provided by allowing the boiler pump to run for a period before the burner is enabled.

The control sequence has to be set up differently from the previous example if the full output from each boiler is to be achieved. Consider the two boiler system in Figure 10.19 with a design temperature lift across the boilers of 12 K; when operating at full load, assuming a 70°C return temperature, the outlet temperature from boiler A will be 70 + 12 = 82°C. The inlet temperature to boiler B will be (82 + 70)/2 = 76°C, while its outlet temperature will be 88°C (the resulting flow temperature in the ring will be 82°C). If the limit sensor on boiler B was set at, say, 86°C, then boiler B would cycle on and off and the required flow temperature would not be reached (full output of both boilers would only occur when the flow temperature from boiler B was less than 86°C, and the combined flow temperature less than 80°C). This situation is exacerbated with more boilers – the peak boiler outlet temperature for a set of four boilers would be 91°C – and poses problems at part-load when the return temperatures are higher. In a multiple boiler installation more units than normally necessary will be brought on line to cope with a particular load. For example, in a four boiler system at 50% load, the theoretical required temperature leaving the second boiler would be 91°C; if the outlet temperature was limited to 86°C then the ring flow (and subsequently return) temperatures would fall, bringing the next unit on line. This situation would

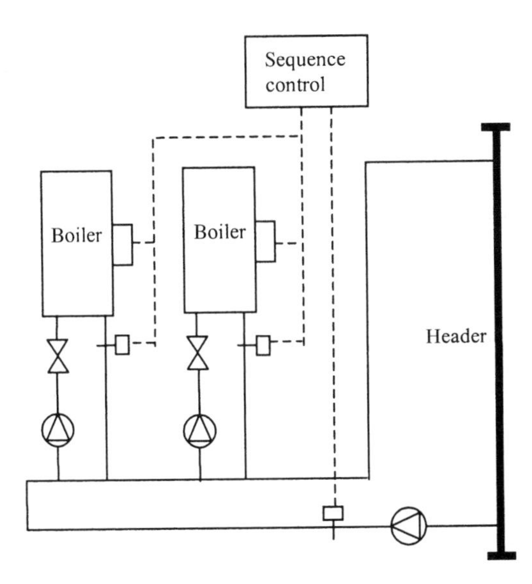

Figure 10.19 Injection circuit with return water temperature sequence control.

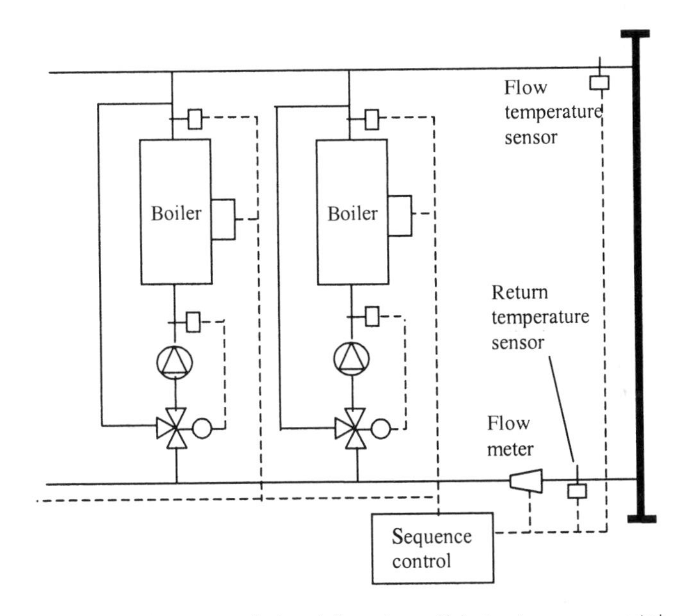

Figure 10.20 Boiler and control arrangement for large boiler systems with heat meter sequence control.

increase cycling rates unless dead bands were set far enough apart. As we have seen, wide dead bands are best achieved with return sensing control. Even so there are practical limits to the width of the dead bands, and it may not be possible to eliminate cycling completely.

10.5.9 Large systems with cascade control or heat metering

Where very large, high thermal capacity boilers are to be installed, plant layout as Figure 10.20 should be used. Each unit has its own pump with a bypass arrangement that allows the boiler to heat up fully before being brought on line; this provides back-end protection and prevents cold slugs of water from entering the system. Only when the back-end temperature is sufficiently high does the three-port valve shut the bypass and open the boiler to the system. The flow in the primary ring is variable, depending on the number of boilers on line, and it is necessary to ensure the header is sufficiently large to prevent primary/secondary flow interaction (i.e. sized with a negligible pressure drop).

The lead boiler needs to be enabled at all times, with the burner being operated by the boiler outlet (flow) sensor. Other boilers are enabled via a heat meter, which measures flow and return temperature difference and flow rate in the primary ring. Heat metering can provide more accurate control than temperature sensing where the flow rate varies, giving more stable flow temperatures to the system. An alternative is to use cascade control where the signal from the lead boiler thermostat is used to enable the second boiler, and the second boiler enables the third, and so on. This, too, gives a control method that is independent of the variable flow rate in the ring.

10.5.10 Heat metering

Heat meter control is a means of sensing the actual energy load on the boilers, and sequencing them accordingly. It must be used where there are variable flow rates in the primary circuit (e.g. with individual boiler pumping or where there is likely to be significant hydraulic interaction, either between boilers or between primary and secondary circuits). The heat load on the boiler at any moment in time is defined by:

$$Q_{load} = \dot{m}c_p(\theta_F - \theta_R)$$

where $\dot{m}$ is the mass flow rate around the system, and c_p is the specific heat of water. As c_p is (for all practical purposes) a constant, the heat load can be determined by measuring flow and return temperature and the flow rate in the primary circuit. Temperatures are measured in the usual way, but there must be a flow measuring device. Flow measurement can be either fixed in line (using venturi or orifice plate meters), or through non-invasive meters. These non-invasive flow meters are extremely useful when retrofitting heat metering, but there have been concerns over their accuracy and reliability, which have largely been eliminated in recent years.

From a control point of view, heat metering relies on a temperature difference, and boiler plant must be up to operating temperature before it can be effective. For this reason at plant start up some other control signal is used to bring the plant up to temperature, for example outdoor air temperature, a representative room temperature or even a fixed plant start time.

Note that if the output from a heat meter is sampled at regular intervals (e.g. every few seconds or every minute) and these values are continuously added together, this measures the energy delivered to the system – a so-called integrating heat meter. This is very useful for monitoring heating system performance. Building (or stand alone boiler) energy management systems can use the same heat metering hardware to carry out both the control and monitoring functions.

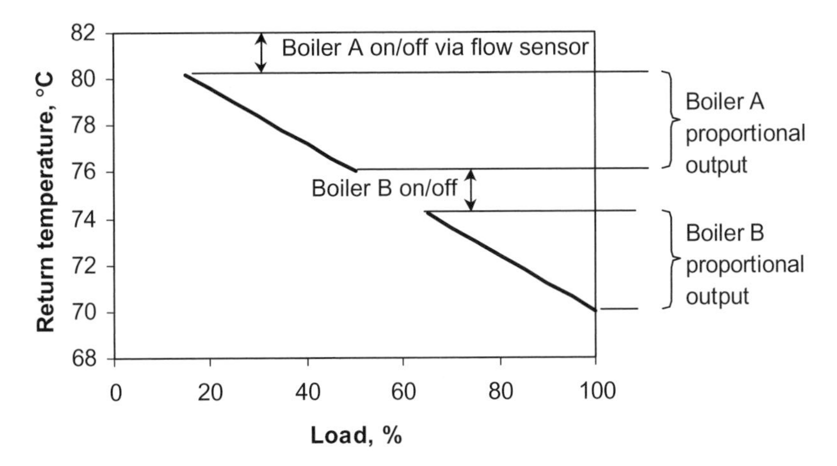

Figure 10.21 Modulating output – two boilers return control.

10.5.11 High/low firing

Control of high/low firing burners can be linked to return sequence or heat load control as described above, but may not be used with cascade control. High/low boilers have different temperature lifts depending on the firing rate (and the different stages may not have equal lifts – see Chapter 2), and the step changes in output make any form of flow temperature control unsuitable. The setting up of temperature or heat load sequence must be done in accordance with the high/low firing ratio.

An alternative to high/low firing is the use of series connected boilers. This can be a very effective solution to staged load matching as constant flow rates can be maintained without temperature dilution effects, leading to reliable and stable control (again using return sensing). The downside can be high pressure drops across the boilers, but this can be mitigated by using increased flow and return temperature differences.

10.5.12 Multiple units with modulating burners

Modulating output burners offer the potential for more stable temperature control and a reduction in cycling as introduced in section 10.5.2 (Figure 10.16). In practice burners do not modulate from 100 down to 0%, but more likely have turn down ratios of 50% or perhaps 30% (see Chapter 2). Figure 10.21 shows a practical return water control schedule for a two boiler installation with burners that modulate down to 30% of their full output. Note that it is essential to have a constant flow rate in the primary ring. With variable flows it is necessary to use heat metering.

An alternative is to use flow temperature sensing with P + I or floating control (only P control is viable for return sensing). Flow sensing can remove

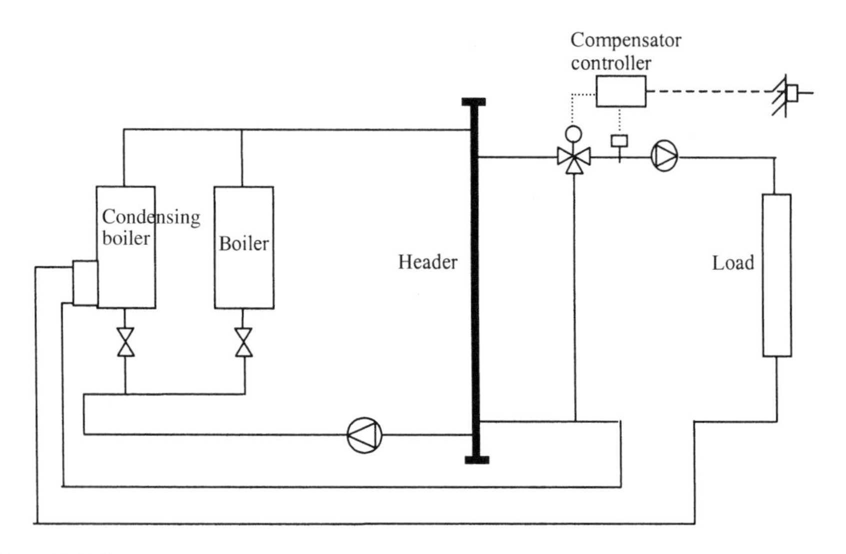

Figure 10.22 Connection arrangement for condensing boiler with a weather compensated circuit.

offset and can tolerate some variations in flow rate (which may occur if any interaction with the system is allowed).

10.5.13 Condensing boilers

The use of condensing boilers introduces other options for boiler control or pipework arrangements. Condensing boilers require low return water temperatures – typically less than 55°C – to recover the latent heat from the flue gases (see Chapter 2). Where a set of modular condensing boilers are used, and if they do not serve the hot water service or other high temperature load, then this can be achieved by directly compensating the boiler flow temperature against outdoor air temperature (using the same sort of schedule as shown in Figure 10.8). This has the advantage that compensator control is achieved with varying the flow rates around the system, giving better controllability, and it reduces the hardware costs of extra pumps and control valves.

Where a single condensing boiler is used in a set of multiple boilers (a more common situation) the condensing heat exchanger should be connected into the system return pipework as shown in Figure 10.22. This takes advantage of the low return temperatures from the emitters (see Figure 10.10), while allowing a common control strategy for the boilers. In such an arrangement the condensing boiler should always be selected as the lead boiler.

10.5.14 Anti-cycling and add-on devices

There have been a number of add-on boiler control devices marketed in recent years claiming to reduce the energy consumption of heating systems. These devices vary in their complexity, but essentially all of them modify the signal from the control thermostat (whether located in flow or return) to prevent the boiler from firing until some predefined criteria are met. The simplest devices employ time delay and simply hold the boiler off from firing for a preset amount of time. Other devices vary the time delay as a function of the load, which is usually determined by the rate of change of the return water temperature, or a measure of the flow and return temperature difference. The newest generation of these devices employs sophisticated sequencing algorithms that attempt to more closely match the boiler output to the measured load.

If used in the absence of any system load control (whether TRVs, room thermostats or compensator circuits) all of these devices have the potential to save fuel. However, this is largely because such uncontrolled systems tend to overheat the space, and reduced boiler cycling will bring the space closer to design requirements. In well-controlled systems the danger of anti-cycling devices is that they will result in loss of service; in other words the savings will be at the expense of reduced indoor temperatures. Some of the more extravagant claims for the early devices (no longer being marketed) suggested 30–40% savings were possible, often accompanied by unsubstantiated explanations. The first law of thermodynamics states that the energy entering a system must equal the energy leaving it if there is to be no increase or decrease in temperature; therefore claims that reducing the energy input from the boilers can maintain the required level of service have to be treated with caution.

One study conducted on three types of device in domestic applications showed that while cycling was significantly reduced, savings were small and that in all cases the savings were as a result of lower room air and water cylinder temperatures[4]. In large capacity systems these devices can show energy savings, but again this must be at the expense of service temperatures. Systems originally fitted with inadequate controls are likely to give rise to overheating in internal space and hot water storage calorifiers, and retrofit of anti-cycling devices can eliminate this overheating. However, this is an open loop control strategy, and the best advice must be to install correct system and boiler circuit controls in the first place.

Development of boiler control devices continues, with more sophisticated algorithms attempting to minimise firing times while maintaining required temperatures within the building. One device has been developed that integrates a number of inputs to provide self-tuning control of the cycle to cycle firing of the boilers to reduce boiler energy output to the building. Every day the controller adjusts the set-points of the controller and its algorithm for the duration of the main occupancy period of the building. Development was based on the following assumptions[5]:

- Energy efficiencies of boilers are generally higher at lower water temperatures
- Unnecessarily high temperature of distributed water increases system standing losses
- Even with local terminal control higher than necessary supply water temperatures will lead to zone overshoot
- Weather compensation can be effectively carried out from within a building if accompanied by an effective process of dynamic measurement, prediction and control of heating load.

The device was tested and developed via simulation and field trials to validate the assumptions and performance. Tests have shown significant savings with no noticeable degradation in space temperatures[6], and that the performance is better than with some state-of-the-art boiler energy management systems. Because of the dynamics of buildings and systems there is no simple theory to explain such savings, although in principle boiler controls that help to make better effective use of thermal storage and casual gains within a building should be able to deliver reductions.

The advent of advanced control sequencing algorithms may offer advantages in a variety of applications, especially in retrofit situations. In theory properly designed boiler and system circuits with correctly installed and commissioned controls should ensure that a building is always maintained within its design limits with minimum energy consumption. Further research is needed to show just what is the most effective way of matching output to demand for a variety of applications, and it is possible that centralised control devices can play a significant role, especially in retrofit applications.

References

1. Stokvis Energy Systems Ltd, West Molesey, Survey.
2. Levermore, G. J. (1992) *Building Energy Management Systems, an application to heating and control*. E&FN Spon, London.
3. Building Research Energy Conservation and Support Unit (BRECSU) (1996) *Heating systems and their control*. General Information Report 40. Best Practice Programme, BRECSU, Garston.
4. Building Research Energy Conservation and Support Unit (BRECSU) (1998) *Domestic boiler anti-cycling controls: an evaluation*. General Information Leaflet 83. Best Practice Programme, BRECSU, Garston.
5. Jennings, P. (2002) Fuelstretcher Ltd. Private communication.
6. Parand, F. & Liao Z. (2001) *Controller efficiency improvements for commercial and industrial gas and oil fired boilers*. EU contract joe3-ct98-7010, final report (the Building Research Establishment) on behalf of the consortium of partners. BRE, Garston.

11 Energy Consumption of Heating Systems

Heating systems consume the single largest amount of energy in the UK – about 30% of all primary energy (transport is a very close second). Most of this is met with fossil fuels – mostly gas or oil, but there is a significant contribution from electricity and in places coal may still be used. This fossil fuel use is a major contributor to greenhouse gas emissions, and designers and building owners alike should be working to minimise the environmental impact of buildings. It is therefore important to have assessments of building energy use at the design stage in order to make judgements about both running costs and environmental impacts of heating systems, and to be able to compare design options. Building owners will not only want to assess the environmental impact but also to set out budgets for operating the building, and to negotiate tariffs with energy suppliers; for new buildings this can only be done from forecasts of a building's performance, which need to be as accurate as possible.

There are a number of ways of estimating energy consumption:

1. Simplified energy estimation methods, e.g. degree-days or frequency of occurrence
2. Steady-state procedural models. e.g. CIBSE Energy Code, SAP (the government's Standard Assessment Procedure), LT (Lighting and Thermal) Method
3. Dynamic simulation.

The last of these is a time consuming and expensive option which would normally only be considered if the building was being modelled for other purposes. The use of steady-state procedures, such as the CIBSE Energy Code (Part 1 for heating), are not, strictly speaking, energy estimation tools. They are, rather, a way of comparing proposed designs against good practice targets in order to assess the effects of individual building and system options (window sizes, day lighting, control options, etc.). They provide a good way of getting sufficient detail for a building into a model in order to compare design decisions, without going to full simulation. Such models have their limitations, but in skilled hands they can provide real insight into a building's energy performance and compare this against realistic benchmark performance. The outputs are normally (as is the case with the CIBSE Energy Code) given in $W m^{-2}$, and it is not valid to convert this to annual energy consumption as the models use seasonal averaging for a diverse range of parameters.

This chapter will focus on the simplified methods, which can be adopted for both energy estimation and monitoring and targeting of on-going performance.

11.1 Degree-day based estimates

The use of degree-days in estimating energy consumption in buildings dates back to the first half of the twentieth century. In the 1960s in the UK techniques were refined[1] which form the basis of guidance given in the 1986 CIBSE Guide[2]; further refinements have since been developed which try to limit the inaccuracies and uncertainties. Degree-days are a measure of the severity of the weather, and when multiplied by the building heat loss factor ($\Sigma UA + 0.33NV$) they can be used to give an estimate of the building energy consumption. This is based on the steady state heat loss equation 6.9, which is fine in principle, but there need to be adjustments in order to give realistic estimates. The sections that follow describe what degree-days are, how they are calculated and how they should be applied. The concept of base temperature is introduced, which accounts for the useful casual gains in a building (solar, people, lights and machines); it is important to remember that gains are only useful if suitable zone temperature controls, for example TRVs, exist to detect them.

Degree-days are a summation of the indoor to outdoor temperature difference that the heating system has to make up. Figure 11.1 shows outdoor temperature over four days, the building set point temperature, and the degree-day 'base' temperature. The degree-day total for these four days would be the sum of the difference between the base temperature and the average daily outdoor temperature. This definition has to be adjusted slightly when the outdoor temperature rises above the base temperature for part of the time (more formal definitions are given later on in the chapter).

The base temperature is the theoretical outdoor temperature at which there would be no need for the heating to come on, with the indoor temperature being raised to the required conditions by casual gains (people, lights, machines and solar) only. If these gains are reasonably constant the base temperature remains constant, and the energy output from the heating system is directly proportional to the summated (base-outdoor) temperature differences – i.e. the degree-days. Therefore multiplying degree-days by the heat loss coefficient of the building ($\Sigma UA + \frac{1}{3}NV$) gives the theoretical building energy demand for the period being investigated. Dividing this value by the average system efficiency therefore gives the theoretical fuel consumption by the boiler plant. There are a number of caveats to this: it is only strictly true for con-

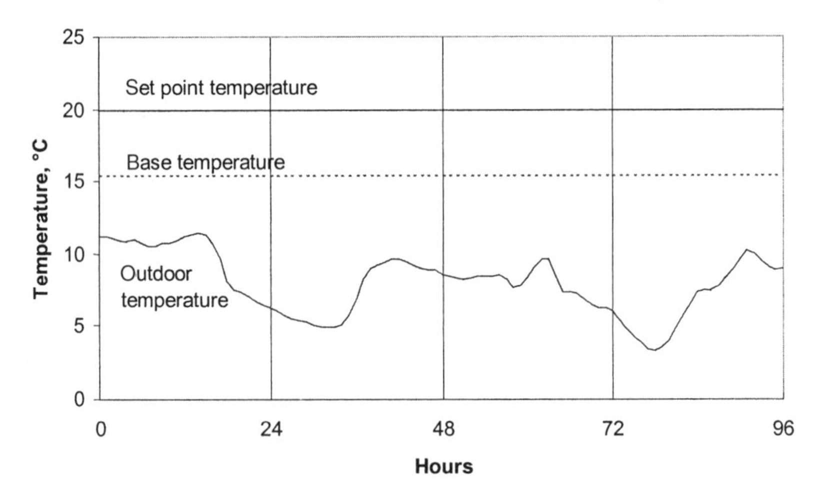

Figure 11.1 Degree-days are the difference between the base temperature and daily outdoor temperature.

tinuously heated buildings with constant heat gains and good space temperature control. Degree-day methods become less certain when dealing with intermittent buildings with variable gains and system efficiencies – the majority of buildings in the UK.

CIBSE Guide 1986 section B.18 gives tables of factors to account for variations in base temperature and intermittent operation. Such an approach is simple to use but the results are uncertain with no way of assessing the accuracy of the results. Simple degree-day estimates may not be precise, but they do indicate the order of magnitude of a building's energy consumption. Where greater accuracy is required a more rigorous method is needed, which will be described below.

A note on base temperature before proceeding. Studies in the USA in the 1920s suggested that casual gains accounted for about a 2.5°C lift in domestic buildings. When introduced on this side of the Atlantic (c.1934) it was assumed the same was true of UK building stock, which being heated to about 18°C gave a base temperature of 15.5°C. Degree-days have been calculated and published to this base temperature ever since. There are good reasons not to change this practice, but we shall see that buildings are unlikely to have this base temperature in reality.

11.1.1 Calculating degree-days

Before considering the application of degree-days to building energy demand we need to see how they are calculated in practice. The most accurate method for calculating degree-days is to sum the hourly temperature differences and divide by 24:

$$D = \frac{\sum_{24} (\theta_b - \theta_{o,h})^+}{24} \quad (11.1)$$

where $\theta_{o,h}$ is the hourly outdoor temperature, and the symbol $^+$ indicates that only positive values are summed – when the outdoor temperature exceeds the base the difference for that hour is taken as zero. This method requires the continuous logging of outdoor temperatures and results in large data storage requirements. Before the advent of electronic storage media it was necessary to calculate degree-days from daily maximum and minimum temperatures. The most common practice (used since 1928) has been to use the Meteorological Office equations as shown in Table 11.1. Each equation approximates a particular relationship between outside temperature profile and the base temperature, as shown in Figure 11.2. These equations still form the basis

Table 11.1 Meteorological Office equations (see Figure 11.2).

Case	Condition	Approximation to $\int(\theta_b - \theta_o)dt$
1	$\theta_{max} \le \theta_b$	$\theta_b - \frac{1}{2}(\theta_{max} + \theta_{min})$
2	$\theta_{min} < \theta_b$, and $(\theta_{max} - \theta_b) < (\theta_b - \theta_{min})$	$\frac{1}{2}(\theta_b - \theta_{min}) - \frac{1}{4}(\theta_{max} - \theta_b)$
3	$\theta_{max} > \theta_b$, and $(\theta_{max} - \theta_b) > (\theta_b - \theta_{min})$	$\frac{1}{4}(\theta_b - \theta_{min})$
4	$\theta_{min} \ge \theta_b$	0

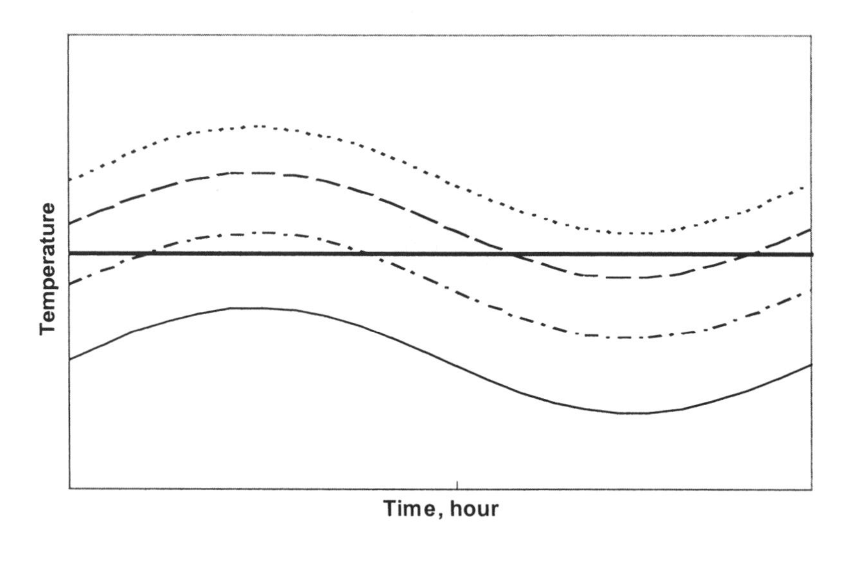

Figure 11.2 Temperature relationships for the Meteorological Office equations in Table 11.1.

for published degree-day values in the UK. (In the USA only equation 1 is used.)

An alternative method is the use of Hitchin's equation[3], which calculates monthly degree-days from the mean monthly outdoor temperature:

$$D_m = \frac{N(\theta_b - \bar{\theta}_{o,m})}{1 - e^{-k(\theta_b - \bar{\theta}_{o,m})}} \qquad (11.2)$$

where N is the number of days in the month and k is a constant that depends on the standard deviation of temperature distribution in a given location. It is calculated from:

$$k = \frac{2.5}{\sigma} \qquad (11.3)$$

The average value of k for the UK is 0.71. The advantage of this method is that it requires easily obtainable temperature data and can readily be used in energy forecasting procedures.

11.1.2 Calculation errors

Since the Met Office equations and Hitchin's equation are approximations they necessarily incur errors. The magnitude will vary with location, but it is possible to quantify the error. Ten years of temperature data from Stansted were used to calculate actual and approximate degree-days to different base temperatures, and the differences evaluated. Figure 11.3 shows the results where:

$$\delta = \frac{D_{actual} - D_{approx}}{D_{actual}} \times 100 \qquad (11.4)$$

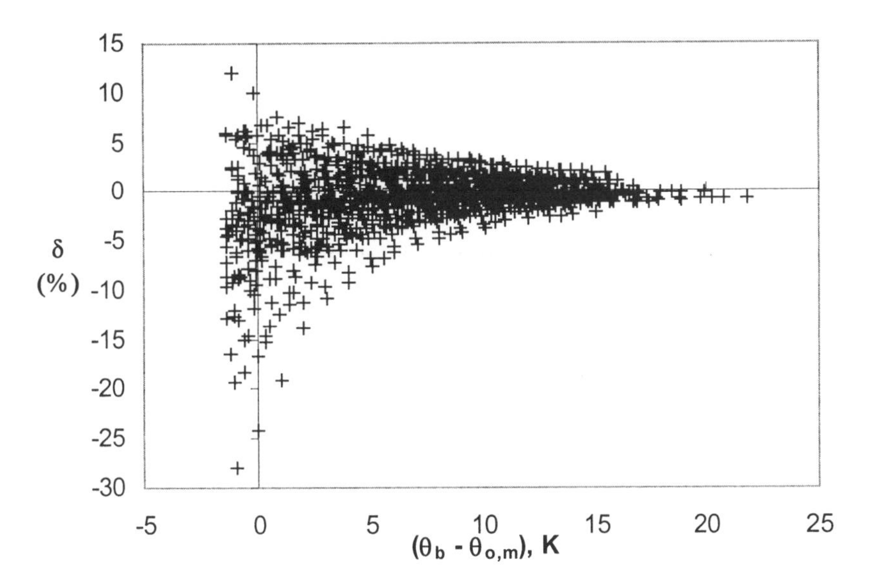

Figure 11.3 δ against (base – mean monthly temperature) for Stansted.

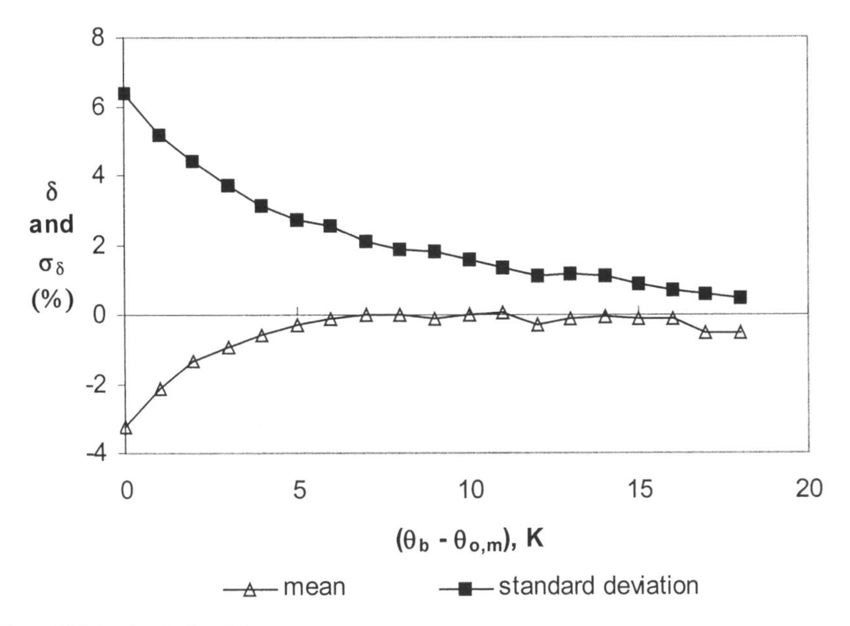

Figure 11.4 δ and σ_δ for Stansted.

where δ is the % difference between actual degree-days (D_{actual}) and approximate degree-days (D_{approx}). These differences can be represented in a simpler format shown in Figure 11.4 where the mean value of δ and the standard deviation, σ_δ, can be used to describe the error for particular base temperature and mean monthly temperature combinations. This gives rise to an uncertainty relationship of:

$$\Psi_{D(95\%)} = \bar{\delta} + 2\sigma_\delta \quad (11.5)$$

11.1.3 Base temperature correction

Traditionally the conversion of degree-day values from one base to another has been given by single correction factors. However, these are only valid for a given time frame and are location specific. Figure 11.5 shows that the ratio between different base temperature degree-days will be different for different time frames; as such curves are different for different locations, it follows that the degree-day ratios will differ from region to region.

More accurate base temperature correction procedures have been developed, for example the BRE Domestic Energy Model (BREDEM)[4] uses linear correction factors. Another alternative is to use Hitchin's equation (11.2) using a numerical iterative technique to find the mean monthly temperature (e.g. Newton-Raphson), and then enter the new required base temperature to calculate corrected monthly degree-days.

Example

Example of using Hitchin's formula to convert monthly degree-days of 115 at 15.5°C base to a base of 13°C, assuming k = 0.71

Newton-Raphson iteration is found from:

$$\bar{\theta}_{0,(n+1)} = \bar{\theta}_{0,(n)} - \frac{f(\bar{\theta}_o)}{f'(\bar{\theta}_o)}$$

where

$$f(\bar{\theta}_o) = \theta_b - \bar{\theta}_o - \frac{D_m}{N} + \frac{D_m}{N} e^{-k(\theta_b - \bar{\theta}_o)}$$

and

$$f'(\bar{\theta}_o) = 1 - \frac{D_m}{N} k e^{-k(\theta_b - \bar{\theta}_o)}$$

Best first guess:

$$D_d = D_m/N = \theta_b - \bar{\theta}_o$$
$$= 115/31 = 3.71$$
$$\theta_{o(1)} = 15.5 - 3.71 = 11.8$$

Use this value to start the iteration process to obtain:

$$\theta_o = 12.1°C$$

New degree-days:

$$D = 31 \times (13 - 12.1)/[1 - \exp(-0.71(13 - 12.1))]$$
$$= 59 \text{ K-day}$$

11.1.4 Application of degree-days

The section that follows derives the degree-day estimation model based on the energy balance of a building; however the important result is equation 11.14 which can be used without a full understanding of this derivation.

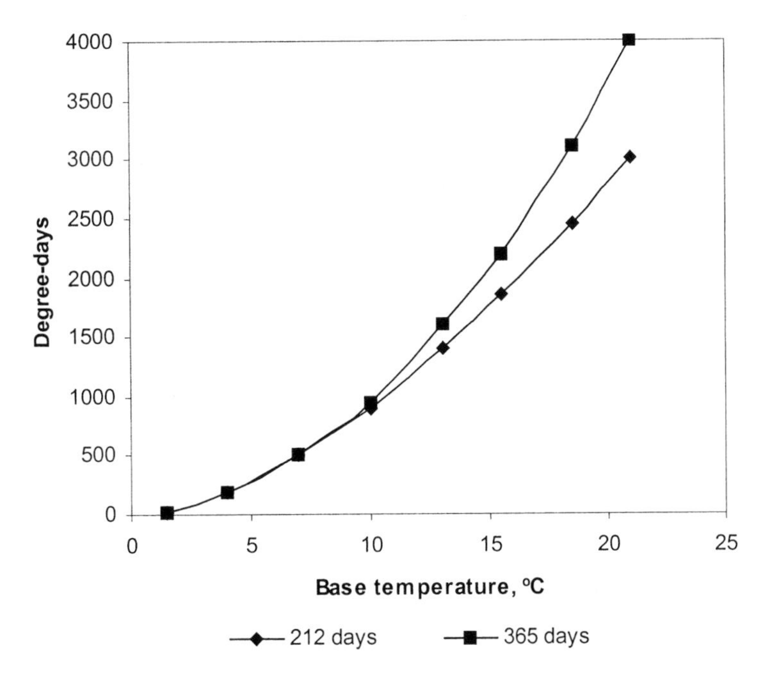

Figure 11.5 Degree-days versus base temperature for different time periods.

Of all the heat lost from a building to its surroundings, some is made up by casual gains (solar, people, lights and machines), and the rest is made up from the heating system; there will also be some heat flowing into and out of the building structure at any moment in time. At a given moment in time this can be expressed as a heat balance on the heating system:

$$Q_E = U'(\theta_i - \theta_o) + Q_C - (Q_S + Q_I) \tag{11.6}$$

where Q_E is the heat output of the system in kW; Q_C is the heat flow into (or out of) fabric storage; Q_S and Q_I are the solar and internal gains respectively; $U' = (\Sigma UA + \frac{1}{3}NV)/1000$ and is the heat loss coefficient (kWK^{-1}); θ_i is the internal temperature (usually taken as the control set point); and θ_o is the outdoor temperature.

These rates of energy flow can be summed over a period of time to give total energy:

$$\int Q_E dt = \int U'(\theta_i - \theta_o)dt + \int Q_C dt - \int (Q_S + Q_I)dt \tag{11.7}$$

Over a suitably long period of time the storage term, Q_C, becomes negligible and can be taken out of the expression (in fact its effects cannot be ignored, but this will be dealt with later). The expression for the heating system energy, now given the symbol E, can be simplified thus:

$$E = U'\int\left(\theta_i - \theta_{ao} - \frac{Q_G}{U'}\right)dt \tag{11.8}$$

where $Q_G = (Q_S + Q_I)$. From this we can derive the concept of a base temperature, θ_b, as:

$$\theta_b = \theta_i - \frac{Q_G}{U'} \tag{11.9}$$

In which case:

$$E = U'\int(\theta_b - \theta_o)dt \tag{11.10}$$

The integral in equation 11.10 can be expressed as a summation and put in terms of degree-days:

$$\int(\theta_b - \theta_o)dt = D_d = \frac{1}{24}\sum_1^n \sum_1^{24}(\theta_b - \theta_o) \tag{11.11}$$

where n is the number of days in the month or heating season.

The heating system fuel consumption, F, is given by:

$$F = \frac{24U'D_d}{\eta} \tag{11.12}$$

where η is the system efficiency

This approach is all very well for a continuously heated building, but for intermittently heated buildings it overestimates energy consumption. The traditional approach has been to apply correction factors (see text box on p. 304) which depend on building mass and occupancy patterns. However, it has been shown[5] that correction factors can be dispensed with if the mean internal temperature, taken over a suitable period of time, is used. Figure 11.6 shows how the temperature fluctuates in an intermittently heated building. The *average* rate of heat loss from the building over, say, a month is related to the average indoor temperature. This mean internal temperature is in turn dependent on the fabric thermal properties, including storage. It has further been shown that the mean internal temperature can be approximated by an expression derived from the ideal cooling and heating equations (9.4 and 9.6). This is given by:

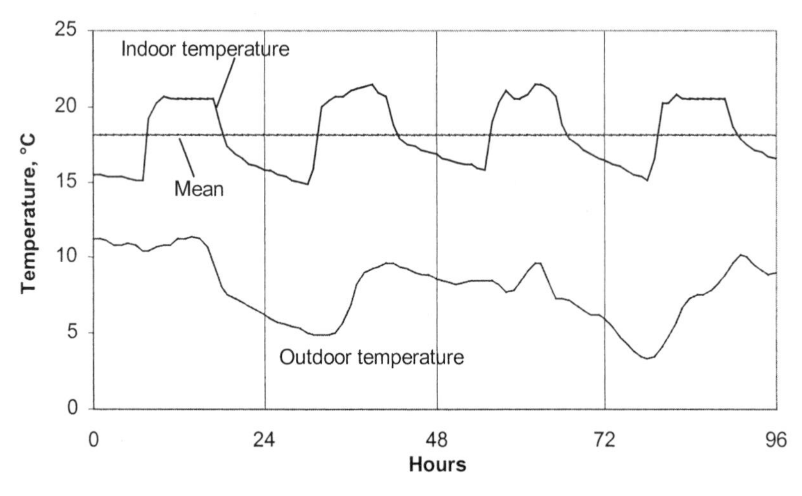

Figure 11.6 Temperature profile for an intermittently operated building.

$$\bar{\theta}_i = \frac{\sum_{t_1}^{t_3} \theta_i + \theta_{sp}(24 + t_1 - t_3)}{24} \tag{11.13}$$

where

$$\sum_{t_1}^{t_3} \theta_i = \theta_o(t_3 - t_1) + \tau(\theta_{i,1} - \theta_o)\left[e^{\left(\frac{t_3 - t_2}{\tau}\right)} - e^{-\left(\frac{t_2 - t_1}{\tau}\right)}\right]$$
$$+ \frac{\tau Q_p}{U'}\left[1 + \left(\frac{t_3 - t_2}{\tau}\right) - e^{\left(\frac{t_3 - t_2}{\tau}\right)}\right] \tag{11.14}$$

With t_2, the optimum switch on time, calculated from:

$$t_2 - t_1 = -\frac{C}{(U' \times 3600)} \ln\left[\frac{(\theta_{i,2} - \theta_o)}{(\theta_{i,1} - \theta_o)}\right] \tag{9.5}$$

with:

$$\theta_{i,2} = \theta_{so} = \frac{e^{-\left(\frac{t_{unocc}}{\tau}\right)} Q_p(\theta_{sp} - \theta_o)}{Q_p - U'(\theta_{sp} - \theta_o)\left(1 - e^{-\left(\frac{t_{unocc}}{\tau}\right)}\right)} + \theta_o \tag{9.9}$$

For an explanation of these two equations see Chapter 9. These expressions may look difficult to use, but when entered into a spreadsheet they never need to be solved manually. It is possible to produce curves of (θ_{sp} – $\bar{\theta}_i$) against τ as shown in Figure 11.7, however the mean internal temperature is significantly dependent on plant size and occupied hours so it is difficult to provide definitive sets of curves; the most reliable way is to use the equations as given.

So the base temperature is now defined as:

$$\theta_b = \bar{\theta}_i - \frac{Q_G}{U'} \tag{11.15}$$

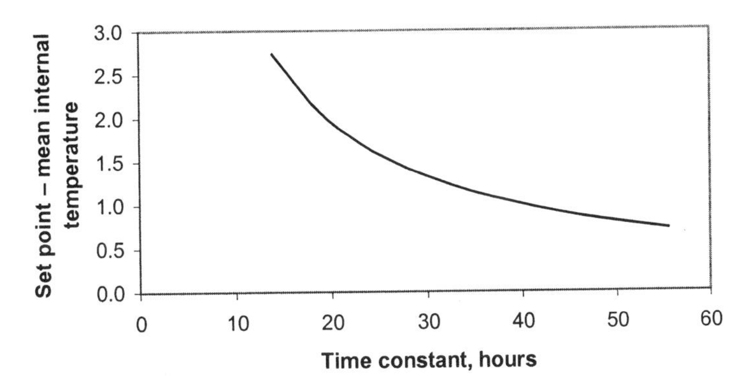

Figure 11.7 Relationship between mean internal temperature and building time constant.

This can then be used to calculate degree-days, either from historical temperature data or by converting published degree-day values (usually to base 15.5°C) using Hitchin's formula (equation 11.2). Where Q_G changes from month to month this can result in a variation in base temperature, and as Hitchin's formula also works for monthly time-scales it is valid to conduct the calculation for each month of the heating season. These values can be summed to give the expected heating season energy consumption. This energy figure can be translated either into money or tonnes of CO_2 (1 kWh of gas gives rise to approximately 0.21 kg CO_2).

11.1.5 Uncertainty

Degree-day estimates cannot give accurate predictions about a building's energy consumption; future weather patterns or building use patterns cannot be predicted. However, such an estimate gives a good indication of what can be expected for a given set of conditions, and it makes sense to make this estimate as accurate as possible and also define the expected accuracy – or the expected uncertainty – of the result. This uncertainty increases as the time-scale gets shorter, for example daily energy estimates are likely to be much less accurate than monthly estimates, and monthly are less accurate than seasonal. Hourly estimates are the most inaccurate of all as they cannot account for the thermal storage and time lag effects of the building. For the estimation method described here this uncertainty can be quantified[6] as follows; monthly energy estimates have an uncertainty, ψ_m, of:

$$\Psi_m = 130D_m^{-1.3} \tag{11.16}$$

and a seasonal uncertainty, Ψ_s, of:

$$\Psi_s = 1600(\Sigma D_m)^{-1.35} \tag{11.17}$$

This information is useful as it tells the user how much confidence one can have in the final result.

Using degree-days

The accepted method for using degree-days is that set out in the 1986 CIBSE Guide Section B.18[2]. This uses the concept of equivalent hours of operation of a plant size equal to the design day heat loss. It is effectively the same as equation 11.12, but the use of equivalent hours of operation – to which correction factors are applied – has a sense of practicality and is intuitively easier to visualise. However, such an approach only allows the use of correction factors to account for intermittency and thermal characteristics. It has been shown that the mean internal temperature approach is more accurate and reliable than the use of published correction factors in B.18[6].

The basis of equivalent hours of full load operation is explained as follows. The fuel consumption can be expressed as the full load output of the plant multiplied by the hours which it operates over the heating season, and divided by the seasonal efficiency:

$$F = \frac{h_{eq} Q_d}{\eta} \tag{11.18}$$

with h_{eq} being the equivalent hours of operation, defined as the energy delivered (in kWh) divided by the design plant output, Q_d, in kW:

$$h_{eq} = \frac{24U'D_d}{Q_d} = \frac{24D_d}{(\theta_i - \theta_{o,d})} \quad (11.19)$$

Substituting equation 11.19 into 11.18 will yield the same answer as equation 11.12, and it can be argued that the equivalent hours concept can confuse rather than clarify. The published correction values are also a source of concern, particularly in that the user is unaware of their origin and is constrained in their selection. This book will not consider this approach further, and the reader is encouraged to go to the original text to assess the difference in the methods.

Figure 11.8 summarises the method for calculating energy estimates together with the associated uncertainty. An example is given below. Such a calculation makes a number of assumptions:

- The building operates under optimum start conditions
- The system is controlled such that the set point temperature is maintained during occupied hours, and the building can make full use of the casual gains without overheating
- Occupancy patterns are relatively stable
- Overall system efficiency remains constant
- The building heat loss coefficient has been accurately determined and is constant.

When assessing how differences to the building or system design will affect energy consumption it is important to note that changes in the fabric U value will affect both the heat loss coefficient and the building base temperature (see equation 11.8). Changes in occupancy patterns will affect the mean internal temperature, and hence the base temperature; this in turn affects the degree-day total and the associated uncertainty. Changes to the way the system operates may either change the seasonal efficiency or the degree-day totals. An example of the former would be setting up the boiler controls to reduce standing losses, whereas boiler anti-cycling devices will reduce the building internal temperature and hence degree-days. Thus equation 11.12 is a simplification of all the complexities of a building's operation, but used correctly can give good estimates of energy consumption.

Example

A building has a heat loss coefficient of $30\,kW\,K^{-1}$ and an effective thermal capacity of $4\,GJ\,K^{-1}$ (see Chapter 9 for the determination of thermal capacity). The internal set point temperature is 20°C and the building is occupied for 8 hours per day (7 days per week) and has an installed plant capacity of 900 kW and mean internal gains of 150 kW. Find the energy demand and carbon dioxide emissions for January if the mean outdoor temperature is 5°C. Take the system efficiency as 75%.

First we need the time constant, found from:

$$\tau = 4 \times 10^6/(30 \times 3600) = 37.04 \text{ hours}$$

The switch-on time is calculated from equations 9.5 and 9.9 by first calculating the switch-on temperature and substituting back into equation 9.5:

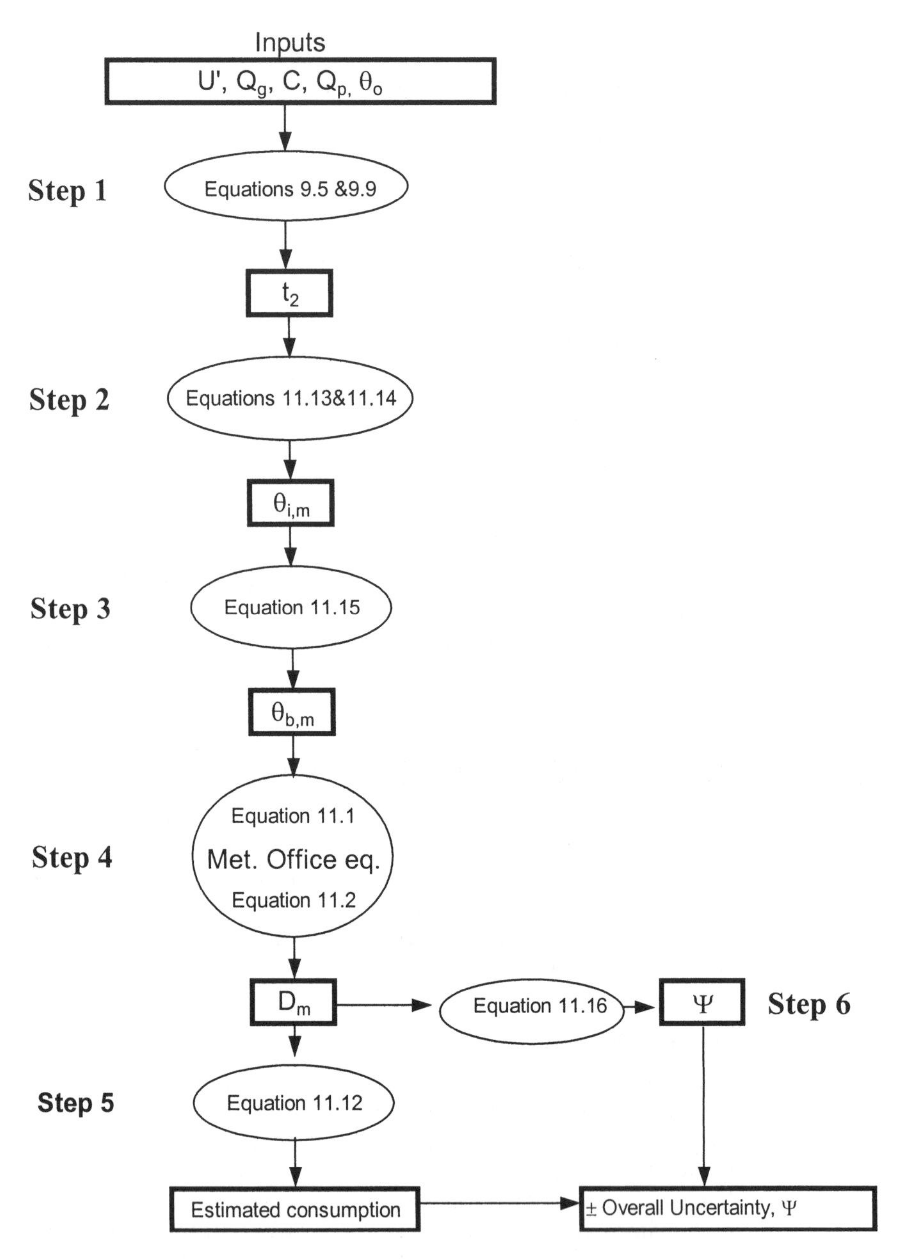

Figure 11.8 Space heating energy estimation algorithm.

$\theta_{so} = 16.8°C$

$t_2 - t_1 = 8.86$ hours

$t_3 - t_2 = 7.14$ hours

These values are then used in equations 11.13 and 11.14 to find the mean internal temperature:

$\bar{\theta}_i = 18.93°C$

The temperature rise due to gains is:

$150/30 = 5 K$

So the base temperature is:

$\theta_b = 18.93 - 5 = 13.93°C$

Using Hitchin's formula, and assuming k = 0.71, the monthly degree-days are:

$D_m = 277$ K-day

and the fuel consumption from equation 11.12 is:

F = 266185 kWh.

The uncertainty in this estimate can be found from equation 11.16, which in this case is:

$\Psi = 130 \times 277^{-1.3} \times 100 = 8.7\%$

This uncertainty ignores errors in the numerical calculation of degree-days.
The energy consumption can also be translated into CO_2 emissions via a suitable factor. For gas this is 0.21 kg CO_2 kWh^{-1}. In this case:

$266185 \times 0.21/1000 = 55.9$ tonnes of CO_2.

11.2 Monitoring and targeting of existing systems

Degree-days are also used in the monitoring and targeting of building energy use. Because the outdoor temperature is constantly changing and no two months from one year to the next are identical, if energy consumption is to be compared year on year then allowances must be made for the weather. Degree-days allow this comparison to be made, which is a process called weather normalisation.

11.2.1 Performance lines

If monthly energy consumption is plotted against degree-days (usually to base 15.5°C) in a scatter diagram, this should show a reasonably linear relationship where energy consumption is largely dictated by space heating requirements. Figure 11.9 shows just such a relationship for a building, with a line of best fit drawn through the points (the equation of the line is also shown). This best fit line can be readily added to graphs in most spreadsheet packages. (For manual calculation of best fit lines using least squares analysis, see for example Draper and Smith[7]). This line is the so-called performance line of the building, and the slope of the line is the rate of increase of monthly energy use per degree-day; the intercept is non-weather related energy consumption, although this needs to be qualified. Therefore monthly energy consumption for the building, if everything remains as it is, can be expected to be:

Fuel consumption = slope × degree-days + intercept

In theory, if degree-days were calculated to the actual base temperature of the building, the slope of the line would be directly related to the heat loss

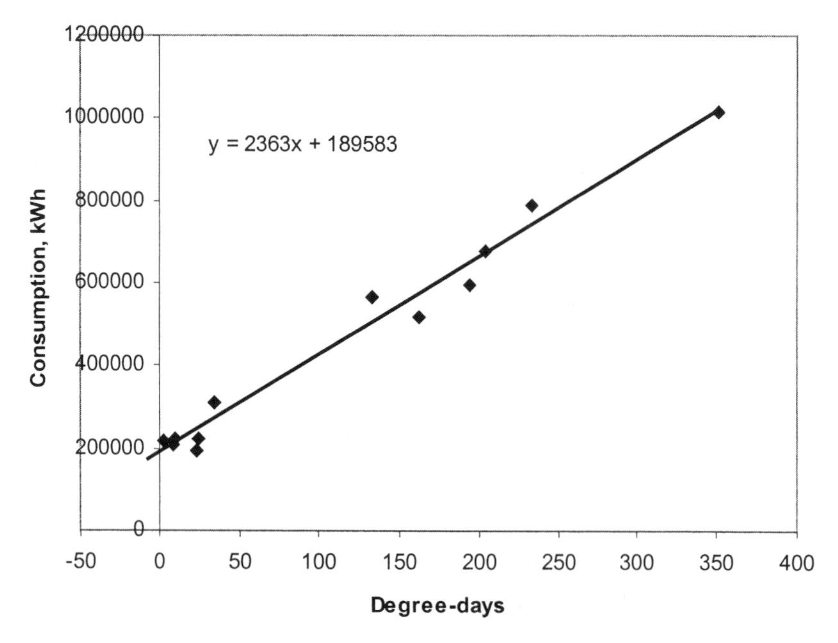

Figure 11.9 Building energy performance line.

coefficient (for energy given in kWh this would be $(\Sigma UA + \frac{1}{3}NV) \times 24/1000$), and the intercept would be the monthly base load (HWS, standing losses, and any other processes).

11.2.2 Scatter in the performance line

It would be very unusual to see all the data lie on a perfect straight line, such building energy data always exhibits scatter. This scatter can be attributed to a number of factors:

- Erratic energy data collection. Data needs to be collected for the same period as degree-days, i.e. every calendar month. Two to three days either side can result in up to 10% discrepancies.
- Poor system control. If the system is controlled within wide tolerances, or not controlled at all, then energy consumption will not correlate well with weather changes.
- Variations in heat gains or heat loss coefficient (particularly infiltration). Ironically in well-controlled buildings, where large and sustained variations are seen in these factors, the base temperature will vary from month to month (e.g. in buildings with high solar gains). When using a single base temperature to calculate degree-days this can result in an increase in the observed scatter. Note that this scatter occurs in the x-axis data.

The degree of scatter in the data is described by what is known as the coefficient of determination (also called the R^2 value) which can be used to say how reliable the performance line is as a monitoring tool. R^2 is a number between 0 and 1, with a value of 1 indicating all the data lies on a perfect straight line (no scatter), and 0 suggests there is no correlation in the data.

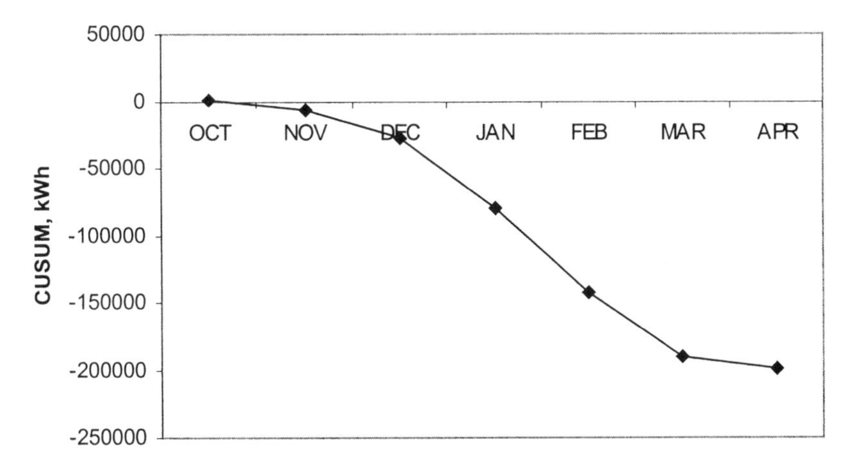

Figure 11.10 CUSUM graph.

We would normally want to see R^2 values greater than 0.75 for use in performance lines – but preferably higher. R^2 can be calculated using standard facilities in spreadsheet charts.

11.2.3 Applying the performance line

Whatever method is used to establish the performance line (and simply using 15.5°C degree-days is often good enough), the next step is to use the equation of the trend line to compare subsequent energy consumption. This is done by entering degree-days for each month into the performance line equation to obtain a prediction of how much energy the building would use according to its historical performance. This value can be subtracted from the actual monthly energy consumption to find the difference (if savings are being made in a given month this difference will be a negative number). Finally these differences are summed (i.e. the cumulative sum of the differences (CUSUM)) to give a running total of savings to date. Table 11.2 shows this process using the performance line in Figure 11.9 as the basis. Figure 11.10 shows a plot of CUSUM against month, which in this case shows an encouraging downward trend. Should this flatten out, or indeed reverse direction, it indicates that the building is using more energy than would be expected and investigation is necessary to determine the cause of the change. (Note that the magnitude of savings will normally be greater in the winter than in the summer, which will lead to steeper CUSUM gradients in the colder months and shallower ones at times of low consumption; such seasonal variations are to be expected.) CUSUM can be a very powerful energy management tool in this respect. It is also a valuable way to present and report savings made as a result of implementing savings measures; without such a technique of weather normalisation it is impossible to prove the magnitude of such savings.

11.2.4 Base temperature and the performance line

For most on-going monitoring applications it is usually adequate to use published (15.5°C base) degree-days, but if the performance line is to be used to

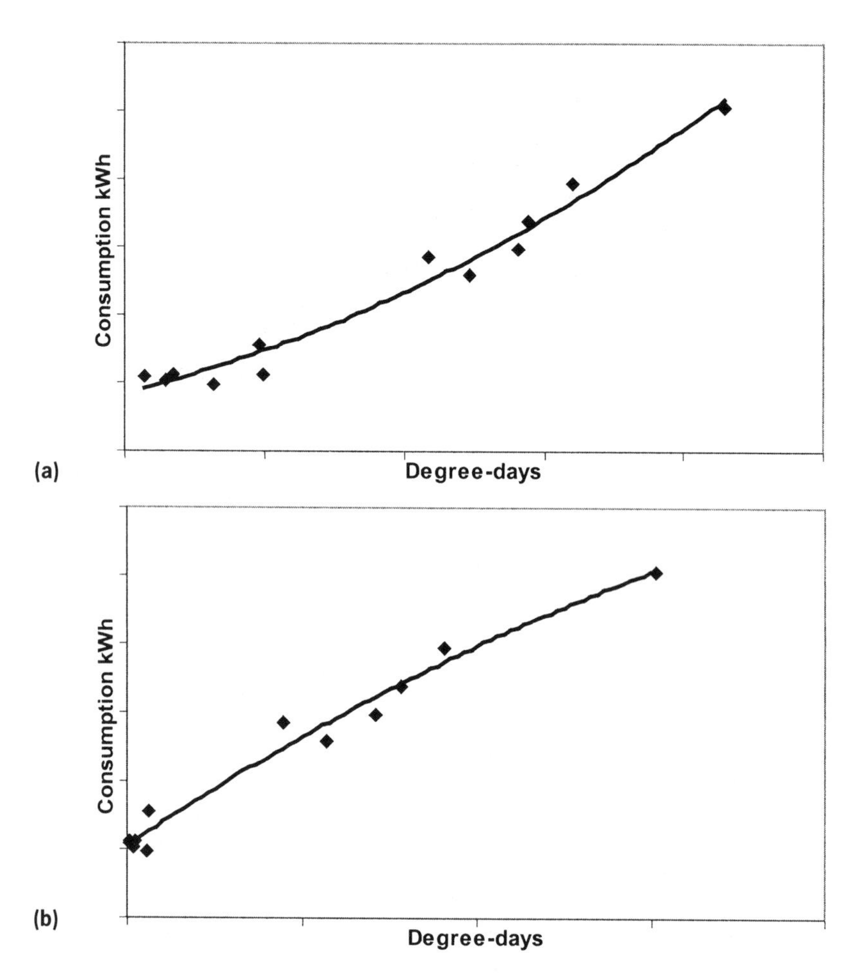

Figure 11.11a Building base temperature less than degree-day base temperature. **b** Building base temperature exceeds degree-day base temperature.

assess base load (which may also be found from summer energy bills) then it is necessary to use degree-days to the correct base temperature. X–Y scatter plots can show a variety of characteristics, and there have been claims that it is possible to infer operational faults of heating systems from such data. Examples of this include negative intercepts or apparent curvature in the data; however, both of these can be attributed to the use of incorrect base temperature, and any attempts at diagnostics should be done with a good knowledge of the building under investigation. When the actual base temperature of a building is lower than 15.5°C, then published degree-days will be too high and the performance line will be shifted too far to the right; in these cases the data may show some curvature as demonstrated in Figure 11.11a, and a best fit straight line may give a negative intercept in extreme cases. When the base temperature of a building is higher than 15.5°C, the observed curvature is in the opposite direction, as shown in Figure 11.11b, as all the data points are shifted towards the y-axis.

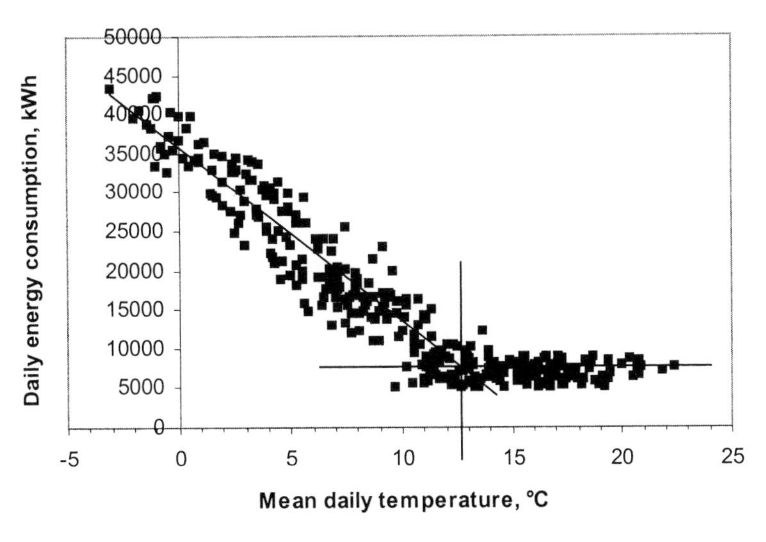

Figure 11.12 Jacobson building energy signature.

There are two ways in which the actual base temperature of a building may be found, the first using daily energy data, and the second exploiting the curvature discussed above. If daily energy and outdoor temperature data exists (as may be the case with a building energy management system), then it is possible to plot the daily energy against mean daily outdoor temperature as shown in Figure 11.12. This shows two distinct regions: a slope where weather influences consumption, and a horizontal portion that is daily base load. The intersection of these two regions occurs at the balance point of the building – the base temperature. This can be used to calculate building specific degree-days. This method was first demonstrated by Jacobson[8], and is known as the building energy signature.

Where daily data is not available then monthly energy consumption can be plotted against degree-days in the usual way, only this time fitting a second order polynomial trend line. If the x^2 component is negative the curvature will be convex; this is due to the base temperature being too low and the data points being pushed towards the y (i.e. energy) axis. If the base temperature is changed, changing the degree-day values, the x^2 component changes also and becomes zero when the correct base temperature is entered (the trend line is now effectively a straight line). It can be shown that this method yields the same answer as a Jacobson energy signature. The degree-day values may be obtained from monthly published values, and applying Hitchin's formula to change the base temperature was discussed in section 11.1.3.

11.3 Benchmarking

Energy benchmarks are a way of comparing the overall energy performance of a building against typical and good practice. These are usually given in $kWh\,m^{-2}$ of treated floor area per annum. A good example is Energy Consumption Guide 19 (Econ 19) of the Energy Efficiency Best Practice Programme[9]. Econ 19 gives normalised (degree-day corrected) benchmarks for all end uses in four different types of offices, for both typical and best practice found in the UK. The best practice programme also gives benchmarks for other types of buildings. One of the problems of using benchmarks is deciding the classification of a particular building, i.e. matching your building to the published benchmark. This is not always as straightforward as it may seem as there does not appear to be such a thing as a typical building. However, such benchmarks are a useful way of quickly assessing a building's overall performance.

It is possible to generate theoretical benchmarks for buildings. The CIBSE Energy Code Part 1[10] is a tool for doing just this. By entering the physical and operational details of a building into a spreadsheet, the Energy Code will return typical expected average heating and electrical power consumption (in $W m^{-2}$) for a heated and naturally ventilated building. It also provides targets for energy consumption based on the specific building geometry. The demands and targets are based on theoretical minimum energy consumption, and the Code was originally designed to assess the energy implications of various design decisions; however, they can also form a good basis for comparison with actual building energy consumption. There are similar assessment tools for air conditioned buildings such as the CIBSE Energy Code Part 2 and the LT Method[11].

11.4 Normalisation

The process of normalisation allows comparison of buildings with similar uses, but in different locations and with different floor areas. The annual heating fuel consumption first needs to be separated into weather and non-weather related components. The non-weather related component can be identified either through the use of performance lines (see section 11.2.4 above) or by inspection of summer energy bills; failing this the CIBSE Applications manual AM5[12] gives suggested HWS loads for typical building types. Once the weather related component is known this can be divided by the degree-days for that year (using base 15.5°C is adequate, and taking the nearest measuring station to the site) and then the result multiplied by the 'standard' degree-day value of 2462 K-day. This standard value is the UK wide 20 year average seasonal degree-days, and this process normalises the building's weather related consumption to a notional standard set of conditions. The use of 15.5°C base temperature degree-days is adequate for this purpose, and under no circumstances should the normalisation be carried out using building specific base temperature degree-days and then multiplied by 2642; this would lead to large errors in the normalisation process. In the case of a building with a base temperature of 12°C, this would result in a normalised consumption far higher than it should be, and lead to the impression that the building was performing poorly.

Floor area can be a significant source of error in normalisation. Building owners and operators seldom have accurate information about treated floor area. Surveys carried out at the BRE have shown that considerable variation exists between declared floor areas (e.g. in cleaning contracts), and floor areas taken from building drawings. In some cases this can be in excess of 10% (and sometimes vastly different), which directly impacts on the accuracy of an area normalised energy consumption. It is important to obtain accurate and up-to-date floor area data; changes in building form or use should be reviewed regularly for an organisation's building stock.

11.5 Minimising energy use in heating systems

Heating systems described in this book will consume fuel to provide the service required in a building. For reasons of environmental impact, economy and plain good management this fuel consumption should be kept to a minimum. Good design and installation, followed by regular and thorough monitoring and maintenance, will achieve this. Much is down to design, with many of the early decisions having a lasting impact on the energy performance of the plant.

11.5.1 Design

The main influences on building energy consumption are heat loss coefficient, system efficiency and effective use of casual gains, all of which are succinctly captured by equation 11.12. The efficiency will be largely dictated by the choice of boilers. The best choice is to employ at least one condensing boiler, although the caveat to this is that the system must be designed and operated to give suitably low return water temperatures for a sufficiently long period. In addition, the extra cost of condensing boilers needs to be weighed up; such a cost benefit analysis can be conducted by changing the expected seasonal efficiency in equation 11.12 to see the effects (for domestic applications, figures for efficiency can be found in the SEDBUK ratings[13], and commercially from manufacturers' literature).

Another influence on seasonal efficiency will be distribution losses. If all the pipework is within a building these will be very small, but on large multi-building sites these can be significant. Where space permits many such sites will employ distributed heat generation (i.e. boilers within each building) as opposed to centralised plant. However, centralised heat generation is essential where combined heat and power can be used; any increase in distribution losses will be greatly outweighed by the efficiency of overall energy delivery.

The selection of the number of boilers in the central plant will also have an impact, as discussed in Chapter 10, with a larger number of smaller boilers increasing the individual performance of each unit. Again the caveat here is that well-designed controls must be installed to ensure correct running of the plant. In theory it is desirable to use modulating burners to reduce on/off cycling, but alternatives of high/low firing or even series connected boilers can be just as appropriate.

The design of space temperature control is also essential (e.g. TRVs or compensators, or both). These ensure no overheating and allow casual gains to be fully utilised. In terms of the degree-day equation, space temperature controls reduce the base temperature (and hence the value of D).

It should go without saying that design of the building envelope can reduce energy consumption. This is a very significant issue for both design and operation (it is necessary to keep infiltration down as well), as a reduced heat loss coefficient helps to utilise more gains, as well as reduce heat loss. In terms of equation 11.12 this gives smaller U' and D. Building Regulations do dictate the minimum requirements for insulation and air tightness, and most designers will work strictly to these requirements; however, arguments can be put forward, on a case by case basis, that show it to be economically beneficial to have better insulation standards than dictated by the Building Regulations over the lifetime of the building. The degree-day estimation procedure outlined in this chapter can be used for just this kind of analysis. The benefits can be improved if carbon savings can be included in the analysis (for example through carbon trading schemes).

11.5.2 Hot water services (HWS)

HWS poses a particular problem for the overall efficiency of a heating system. Where this is met from a central calorifier, served by the central boiler plant, this can add an energy penalty to the system. In summer this means that the central boiler plant will need to be kept on to service this load; the primary circuit will have to be kept up to temperature, irrespective of the demand on the hot water calorifier. Hot water demand can be highly intermittent, and will invariably have periods of very low load, and serving this from the central primary circuit leads to greater circuit and boiler losses. In many circumstances (especially in larger systems) it will be advantageous to serve the HWS calorifier from a dedicated boiler or use a stand alone hot water generator (see section 8.4.5). With the former it is possible to include standby facility from the space heating boilers, but which could only be brought on manually.

The separation of space heating and HWS also allows for an extended period each year when maintenance can be conducted on the space heating plant without loss of service.

11.5.3 Management

Heating systems require thorough and proactive maintenance to stay at their peak efficiencies. This includes regular cleaning of the burner heads and fireside heat exchanger surfaces; any evidence of sooting in the boiler indicates a severe problem at the burner, with incomplete combustion leading to greater fuel consumption to meet demand; it also indicates a danger of carbon monoxide generation. Advice on maintenance regimes can be obtained from boiler manufacturers, and if in-house staff are not available to carry out such work it is essential to set up a service contract.

Controls also need to be checked periodically to ensure they are operating correctly. Boiler firing patterns should be monitored to see whether the sequence controls have drifted out of calibration. Where multiple boiler sets are not being controlled correctly there is a likelihood of excessive cycling and component wear, and most likely increased energy consumption. Temperatures within the circuit should be monitored to ensure these are not drifting too high (an indication of poor control) and return temperatures are within operating limits. In the case of the latter this is important in condensing boiler applications to ensure maximum efficiencies.

The performance of ancillary equipment is of equal importance. Leaks around pump seals, valve glands or through automatic bleed valves should be rectified early, and can have serious consequences for pressurised systems. Pipework should be checked for leaks, and regular inspections carried out to ensure the integrity of the pipework insulation. Chemical dosing pots are a typical area of neglect; these are typically not lagged, and often the valves are left open so that hot water circulates through them – which is a significant source of wastage. Simple vigilance can prevent such problems.

Management of heating systems (an activity combined with all of the building services) is therefore one that requires good knowledge of the systems in place, how they are (or should be) operated, and what needs to be done to maintain their performance. At one level this requires an appropriate regime of maintenance checks, dictated by the technical requirements of the plant and equipment, but at a higher level it requires the appropriate strategy within which these checks are effectively carried out. This may be a choice between in-house or contracted services, with the latter being almost invariably less expensive. However, out-of house maintenance provision needs to be accompanied by sound quality control mechanisms to ensure the work is being carried out effectively. Whichever is used (and this is normally dictated by the complexity and required reliability of the plant) it is therefore essential that there is good engineering knowledge of the plant and systems installed in a building. The design and operational issues set out in this book have been presented in order to provide this required level of knowledge about heating systems. Heating should not be taken for granted simply because every building has a system. On the contrary it is a subject that should be understood in depth because of this; this book has set out to demonstrate that a LPHW heating system is more than just a boiler, a few pipes and a set of radiators, but rather a mix of modern precision technologies that operate together with the building in a complex dynamic way that is often not simple to define.

References

1. Billington, N.S. (1966) Estimation of annual fuel consumption. *Journal of Institution of Heating and Ventilating Engineers*, **34**, 253–6.
2. Chartered Institution of Building Services Engineers (CIBSE) Guide (1986) Owning and Operating Costs, Section B18 of the CIBSE *Guide*. CIBSE, London.
3. Hitchin, E.R. (1983) Estimating monthly degree-days. *Building Services Engineering Research and Technology*, **4** (4), 159.
4. Anderson, B.R., Clark, A.J., Baldwin, R. & Milbank, N.O. (1985) *BREDEM – BRE Domestic Energy Model: background, philosophy and description*. Building Research Establishment, Garston.
5. Day, A.R. (1999) *An investigation into the estimation and weather normalisation of energy consumption in buildings using degree-days*. PhD thesis. South Bank University, London.
6. Day, A.R. & Karayiannis, T.G. (1999) Identification of the uncertainties in degree-day based energy estimates. *Building Services Engineering Research and Technology*, **20** (4), 165–72.
7. Draper, N. & Smith, H. (1981) *Applied Regression Analysis*, 2nd edn. Wiley, New York.
8. Jacobson, F.R. (1985) *Energy signatures and energy management in buildings*. Proc. CLIMA 2000, Copenhagen.
9. Building Research Energy Conservation Support Unit (BRECSU) (2000) Energy Consumption Guide 19, Energy use in offices, Energy Efficiency Best Practice Programme. BRECSU, Garston.
10. Chartered Institution of Building Services Engineers (CIBSE) (1999) *Energy demands and targets for heated and ventilated buildings*. Energy Code Part 1. CIBSE, London.
11. Baker, N. & Steemers, K. (1994) *The LT Method Version 2, and energy design tool for non-domestic buildings*. Cambridge Architectural Research Ltd.
12. Chartered Institution of Building Services Engineers (CIBSE) (1991) *Energy Audits and Surveys*. Applications manual AM5. CIBSE, London.
13. Seasonal Efficiency of Domestic Boilers in the UK (SEDBUK) (2002) Boiler efficiency database: www.sedbuk.com.

Index

NYIT LIBRARIES
0 1955 0091282 7